186
Topics in Current Chemistry

Springer

Berlin
Heidelberg
New York
Barcelona
Budapest
Hong Kong
London
Milan
Paris
Santa Clara
Singapore
Tokyo

Glycoscience
Synthesis of Oligosaccharides and Glycoconjugates

Volume Editors: H. Driguez, J. Thiem

With contributions by
A. Fernández-Mayoralas, W.-D. Fessner,
U. Gambert, M. von Itzstein, T. Kappes,
V. Křen, S. Oscarson, M. Petersen,
B. Sauerbrei, J. Thiem, R. J. Thomson,
H. Waldmann, M. T. Zannetti, T. Ziegler

Springer

This series presents critical reviews of the present position and future trends in modern chemical research. It is addressed to all research and industrial chemists who wish to keep abreast of advances in the topics covered.

As a rule, contributions are specially commissioned. The editors and publishers will, however, always be pleased to receive suggestions and supplementary information. Papers are accepted for "Topics in Current Chemistry" in English.

In references Topics in Current Chemistry is abbreviated Top. Curr. Chem. and is cited as a journal.

Springer WWW home page: http://www.springer.de
Visit the TCC home page at http://www.springer.de/

ISSN 0340-1022
ISBN 3-540-62033-8
Springer-Verlag Berlin Heidelberg New York

Library of Congress Catalog Card Number 74-644622

Printed in Germany

Typesetting: Fotosatz-Service Köhler OHG, 97084 Würzburg
SPIN: 10542981 66/3020 – 5 4 3 2 1 0 – Printed on acid-free paper

Volume Editors

Dr. Hugues Driguez
Centre de Recherches sur les Macromolécules Végétales (CERMAV)
601, Rue de la Chimie, B. P. 53
F-38041 Grenoble Cedex 9, France
E-mail: HDriguez@cermav.grenet.fr

Prof. Dr. Joachim Thiem
Universität Hamburg
Institut für Organische Chemie
Martin-Luther-King-Platz 6
D-20146 Hamburg, FRG
E-mail: thiem@chemie.uni-hamburg.de

Editorial Board

Preface

Within recent years and after a first coverage of carbohydrate chemistry in the series Topics in Current Chemistry (Vol. 154 in 1990) this field has undergone something like a "quantum jump" with regard to interest for a wider community of scientists. Apparently, for most areas of the natural sciences in general and in particular for those bordering natural products chemistry, progress in the saccharide field has attracted considerable attention and, in fact, many bridging collaborations have resulted. Glycoscience is becoming a term to cover all sorts of activities within or at the edge of carbohydrate research.

It was within the course of a sabbatical collaboration that the editors – both "born" carbohydrate chemists – felt it highly appropriate and desirable to compile a number of contemporary reviews focussing on developments in this field. A number of colleagues actively pursueing outstanding research in the saccharide field agreed to discuss topical issues to which they with their groups have contributed significantly. The result is a fresh and contemporary coverage of selected topics including future outlooks in glycoscience.

In this current volume special emphasis has been placed on the formation of complex saccharide structures employing various enzymes found in the carbohydrate metabolism. *A. Fernández-Mayoralas* discusses the use of glycosidases and *U. Gambert* and *J. Thiem* that of glucosyltransferases for glycosylation. *V. Křen* describes the formation of special ergot alkoloid glycosides and *B. Sauerbrei, T. Kappes* and *H. Waldmann* report on useful enzymatic deprotections as well as the synthesis of glycopeptides. *M. Petersen, M.T. Zannetti* and *W.-D. Fessner* present approaches to novel saccharide mimetics employing aldolases. Moreover, there have been novel exciting findings that have contributed nicely to and have even sometimes improved on classical synthetic approaches. Thus, synthesis, modification and glycoside formation of neuraminic acid, the most complex natural saccharide, is the subject discussed by *M. von Itzstein* and *R.J. Thomson*. *S. Oscarson* highlights novel synthetic paths to unusual oligosaccharides with heptose, uronic acid and fructofuranose residues, and finally *T. Ziegler* elucidates the advantages of the pyruvate saccharide approach for the preparation of special oligosaccharides.

The authors, the editors and the publisher hope that by reading this volume many scientists in the natural sciences will acquire a taste for the subject, and joining the growing glycoscience community, will actively contribute to this most fascinating research.

Grenoble and Hamburg
November 1996

Hugues Driguez
Joachim Thiem

Table of Contents

Table of Contents of Volume 187

Synthesis and Modification of Carbohydrates Using Glycosidases and Lipases

Alfonso Fernández-Mayoralas

Grupo de Carbohidratos, Departamento de Química Orgánica Biológica, Instituto de Química Orgánica, CSIC, Juan de la Cierva 3, 28006 Madrid, Spain

This article reviews the main applications of glycosidases in the synthesis of glycosidic bonds and lipases in acylation/deacylation reactions of carbohydrates. The use of these readily available enzymes with polyhydroxylated carbohydrate substrates is sometimes hampered by the low regioselectivity of the reactions. Several features of the reactions are shown, focusing the discussion on the factors that can affect the selectivity. Thus, the enzyme origin, the structure and the substitution of the sugar substrate, and the reaction conditions are important factors of the reaction to be considered when looking for a given selectivity. Some selected examples of di- and oligosaccharide synthesis by glycosidase-catalyzed reactions are included to illustrate the scope of this methodology. The use of lipases in organic solvents for acylation and deacylation reactions of carbohydrates is also described in some detail.

Table of Contents

Topics in Current Chemistry, Vol. 186

1 Introduction

An important number of synthetically useful enzymatic reactions have been carried out in the field of carbohydrate chemistry. Many carbohydrates, such as oligosaccharides, are complex structures with multiple hydroxyl groups of similar reactivities, and their synthesis and modification require the use of highly selective reactions. Four classes of enzymes have been intensively used over the last few years for this purpose [1, 2]: aldolases, glycosyltransferases, glycosidases, and lipases. Aldolases catalyze the stereoselective formation of carbon-carbon bonds through an aldol reaction, allowing the preparation of monosaccharides and polyhydroxy compounds. Glycosyltransferases are the enzymes responsible for the synthesis of oligosaccharides in living organisms, they catalyze the synthesis of glycosidic bonds in a regio- and stereoselective manner. Glycosidases and lipases are two different classes of hydrolytic enzymes that, under physiological conditions, catalyze the hydrolysis of glycosidic and ester bonds, respectively. Under certain conditions, they can also be effective catalysts for the formation of the aforementioned bonds. Glycosidases and lipases have some features that make them attractive to synthetic chemists: they are readily available and inexpensive, easy to handle, and do not need the use of expensive cofactors and substrates. However, they also present important disadvantages. Glycosidases, when used for the synthesis of disaccharides, give low yields of products, and the regioselectivity is often difficult to control due to their low specificity for the acceptor structure. Lipases, although highly selective to acylate or deacylate at the primary position of a sugar, sometimes show a selectivity among secondary hydroxyls that is low and unpredictable.

This article reviews the main applications of glycosidases and lipases in the synthesis of glycosidic bonds and in acylation/deacylation reactions of carbohydrates, respectively. Special attention is given to the factors that can affect the selectivity of the reactions, such as the enzyme origin, the structure of the substrates, and the reaction medium. A number of reviews have appeared in the literature on enzymatic synthesis of carbohydrates that include reactions with glycosidases and/or lipases [1–6]. In this article only a selection of examples are given to illustrate the discussion, rather than an extensive compilation of the work published in the area.

2 Glycosidases in Synthesis of Glycosidic Bonds

As opposed to glycosyltransferases, glycosidases catalyze the cleavage of glycosidic bonds in nature. These enzymes are involved in both anabolic and catabolic metabolism of oligosaccharides. In mammalian cells, glycosidases that occur in the rough endoplasmic reticulum and Golgi complex are involved in glycoprotein biosynthesis [7]. However, most of glycosidases are extracellular, and are involved in catabolic pathways. For example, glycosidases are abundant in saliva and digestive tract and hydrolyze the glycosidic bonds in food oligosaccharides to form monosaccharides, which can be metabolized. Glycosidases

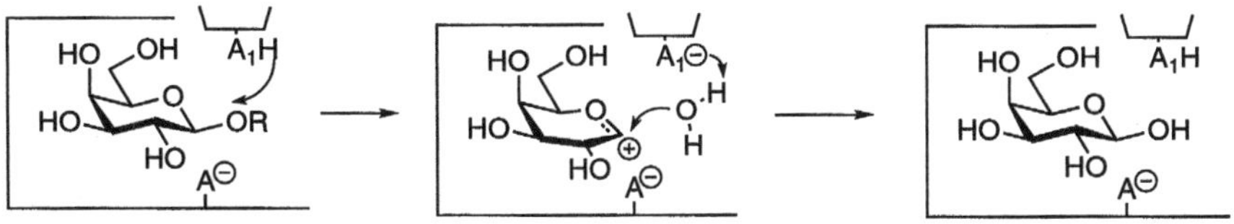

Fig. 1. Mechanism of a retaining glycosidase

can be classified into retaining and inverting glycosidases [8], depending on the stereochemistry of the catalyzed reaction, i.e. either proceeding with retention or inversion of the anomeric configuration. Retaining glycosidases are the enzymes used for the synthesis of glycosides. The mechanism of the glycosidic bond cleavage is presumed to proceed in the same fashion as the acid catalyzed cleavage, via a carbocation intermediate (Fig. 1). The oxocarbenium ion is trapped by a nucleophile (A^-) of the enzyme to form a glycosyl-enzyme intermediate. A water molecule attacks the glycosol moiety of the intermediate on the same face of the pyranoid ring where the aglycon was located, to form the hydrolysis product.

Following the depicted mechanism, glycosidases can be used to catalyze the synthesis of oligosaccharides by two different approaches (Scheme 1): by reversing the normal hydrolytic reaction of the enzyme, or by transglycosylation from a glycoside adding a sugar nucleophile (R_1OH) that competes with water for the glycosyl-enzyme intermediate.

2.1 Reverse Hydrolysis

The synthesis of di- or trisaccharides by condensation of mono- or disaccharides by reversal of the hydrolytic reaction is difficult due to unfavorable equilibrium, and yields of products are usually low. Reaction conditions must be tuned to shift the equilibrium, normally by increasing substrate concentrations. This is feasible, given the high solubility of sugars in buffer solutions. For example, mannose oligosaccharides were synthesised via reversal of the

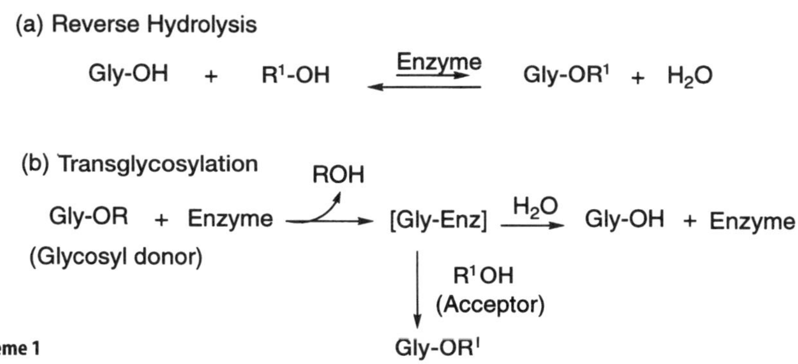

Scheme 1

α-mannosidase reaction using a very high concentration of mannose [9]. The maximum total yield of disaccharides was 37% (w/w) based on the total amount of saccharides.

An original way to increase the yield of disaccharides by condensation of monosaccharides has been introduced [10]. The solution of substrates is circulated through columns of immobilized enzyme and activated carbon connected in series. The procedure is based on the preferred adsorption, by activated carbon, of di- over monosaccharides. By eliminating the products from the system, the equilibrium is continually shifted toward disaccharide formation. This method has also been applied to trisaccharide syntheses [11].

Glycosides of simple alcohols have been synthesized in water-organic two-phase systems in which the organic phase is the reacting alcohol [12]. Yields are variable and depend on the ratio of water to alcohol, and on sugar concentration [12c]. The system has the advantage of providing a simple, easy, and inexpensive procedure to prepare glycosides.

2.2 Transglycosylation

In oligosaccharide synthesis via transglycosylation the reaction is kinetically controlled, and the extent of oligosaccharide formation depends on the partition ratio of the glycosyl-enzyme intermediate between the transfer and hydrolytic reaction (Scheme 1).

Several species have been used as glycosyl donors, mainly substituted aryl glycosides and disaccharides. The only requirement for using a sugar as a glycosyl donor is that it must be cleaved more rapidly than the products formed. Disaccharides are less powerful donors than aryl glycosides, and, therefore, give generally lower yields of transglycosidic products.

In general, synthesis by transglycosylation gives higher yields and the reactions are more rapid than in the reverse hydrolysis. Even though, yields are still low, rarely exceeding 30%. In addition, complex mixtures are usually obtained, making the purification difficult. For example, during the galactosylation of benzyl β-D-xylopyranoside (XylOBn, Scheme 2) using *o*-nitrophenyl β-D-galac-

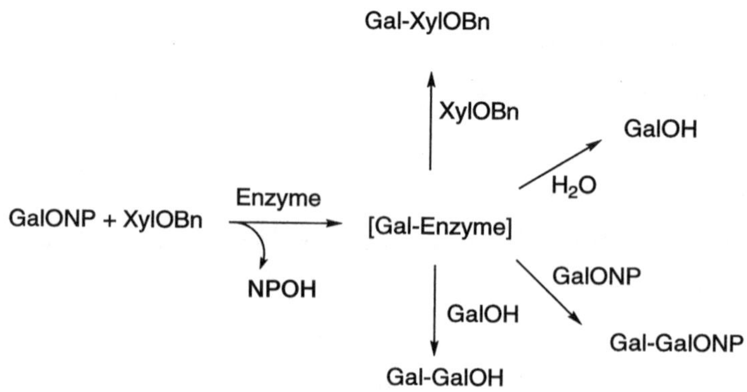

Scheme 2

topyranoside (GalONP) as donor in the presence of a β-galactosidase from *Aspergillus oryzae* [13], apart from Gal-XylOBn disaccharides and D-galactose arising from the hydrolysis of GalONP, the formation of Gal-GalONP disaccharides, for which the donor GalONP behaved as acceptor, was also observed. Even D-galactose, after reaching a certain concentration in the medium, may act as acceptor to give Gal-GalOH disaccharides. The mixture is further complicated since several regioisomers can be formed for each disaccharide family. Given the high number of products that can be formed, one could think that glycosidases would be more useful for combinatorial chemistry than for the synthesis of oligosaccharides of defined structure. In practice, several tricks are employed to drive the reaction toward the direction of the desired products. For example, the desired acceptor is usually employed in higher concentration than the glycosyl donor to minimize secondary transglycosylations. Also, the glycosyl donor can be added over a long period of time to avoid its presence in high concentrations at any time during the reaction [14].

In order to overcome the inconvenience associated with the glycosidase-catalyzed reactions, a great deal of work has been devoted to evaluating the influence of different factors – e.g. the enzyme origin, the reaction medium and conditions, and the acceptor structure – on the selectivity and yield of the glycosylation.

2.2.1
Importance of the Enzyme Origin

Although glycosidases are not specific for the acceptor, they show some selectivity and this may vary depending of the enzyme origin. In the example described above, the β-galactosidase from *A. oryzae* gave, at the time corresponding to the maximum formation of disaccharides, a yield of 36% and 7% of Gal-GalONP and Gal-XylOBn, respectively. However, the enzyme from *E. coli* afforded these disaccharides in 9% and 30%, respectively [13]. Thus, by changing the enzyme, the selectivity for the acceptor was reversed.

In addition to showing different selectivities, the availability of glycosidases from different origins may help to increase the purity of a given regioisomer in a reaction mixture. For instance, the synthesis of the T-antigenic determinant β-D-Gal(1,3)-D-GalNAc was achieved by the sequential use of β-galactosidases from bovine testes and from *E. coli* under synthetic and hydrolytic conditions successively [15]. Using lactose as donor and D-GalNAc as acceptor in the presence of the enzyme from bovine testes, the disaccharide β-D-Gal(1,3)-D-GalNAc was the main product, together with its 1,6 regioisomer. Subsequent treatment of the reaction mixture with the β-galactosidase from *E. coli* led to the preferential hydrolysis of the non-desired regioisomer and of the excess of lactose, and left β-D-Gal(1,3)-D-GalNAc free from contaminant disaccharides.

2.2.2
Influence of the Reaction Medium and Conditions

In a typical reaction, substrates and enzyme are dissolved in a buffer solution at temperatures ranging from 25 to 37 °C. Several attempts to increase the yield of transglycosylations by modifying these conditions did not result in any noticeable improvement. For instance, the addition of cosolvents to reduce the total amount of water did not increase the ratio of disaccharide formation over hydrolysis [13, 16]; on the contrary, a decrease of this ratio is normally observed.

The effect of temperature on the yield and regioselectivity of α-galactosidase-catalyzed formation of disaccharides has also been studied [17]. It was found that total yield and regioselectivity decreased when the temperature was increased from 4 to 50 °C.

In order to lower water activity and therefore favor synthesis over hydrolysis, the effect of the addition of a high concentration of lithium chloride in the β-galactosidase-catalyzed reaction between phenylethanol and lactose was studied [18]. Concentrations of LiCl in the range 0.4 and 0.7 M did not improve the yield of phenylethyl β-D-galactopyranoside; however, it prevented the disappearance of the glycoside product, which may facilitate the control of product formation.

An unexpected result was obtained when the β-galactosidase-catalyzed reaction of GalONP was carried out in the presence of aminoacids [19]. The yield of Gal-GalONP disaccharides increased nearly two-fold when the incubation was performed in the presence of an amphipathic aminoacid, such as Z-L-serine, in comparison with the reaction in absence of aminoacid. The origin of this effect is unknown, and further work should be done to explain this interesting phenomenon.

2.2.3 *Influence of the Acceptor Structure*

In order to characterize the extent of transgalactosylation in a β-galactosidase-catalyzed reaction, Wallenfels et al. [20] defined the transfer number (Y) as follows:

$$Y = [\text{free aglycon}]/[\text{free galactose}][\text{acceptor}],$$

where free aglycon is ROH, and galactose is GlyOH in Scheme 1.

This equation implies the total reaction rate relative to the rate of hydrolysis in the presence of an acceptor. This ratio was found to depend only on the structure of the acceptor but to be independent of both the structure and the concentration of the donor galactoside. Thus, good binding and reactivity for a given acceptor will lead to a large degree of transglycosylation.

A detailed study about the binding and reactivity of the acceptor using the β-galactosidase from *E. coli* has been reported by Huber et al. [21]. They obtained kinetic data for a large number of sugars and alcohols. Two classes of compounds were found to bind well to the galactosyl form of the enzyme (Gly-Enz intermediate in Scheme 1). One class contained sugars and alcohols similar in structure to D-glucose in its pyranose form, and the other class was composed of relatively hydrophobic sugars and alcohols. Large reactivity values were found for straight-chain alcohols as compared to the values obtained for the corresponding ring sugars. Acceptors bearing hydroxyl groups at the end of

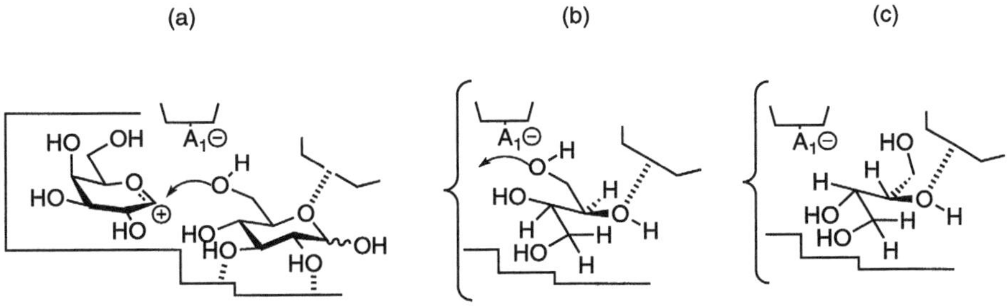

Fig. 2. Postulated interactions of D-glucose (**a**), D-treitol (**b**), and L-treitol (**c**), with the glucosyl site of β-galactosidase from *E. coli* [21]

the molecule showed greater reactivity than molecules with their hydroxyl groups placed in the middle of the chain. The results were interpreted in terms of a hypothetical glucopyranose binding site (Fig. 2a). According to this model, attack by the 6-hydroxyl group of glucose is favored, leading to allolactose, the major product of transgalactosylation with glucose as acceptor. Huber et al. also found that binding at the 5-oxygen position of D-glucose is important for reactivity. This interaction would explain why in the cases in which there was a hydroxyl group at an asymmetric carbon next to the reacting hydroxyl, it must be in the same configuration as D-glucose. Otherwise, the reactivity was low as, for instance, in the case of D- and L-threitols (Figs. 2b and 2c, respectively). Crout et al. made use of this model to interpret the selectivity in the galactosylation of the (*RS*)-butane-1,3-diol [22].

Using the enzyme from *E. coli*, a systematic study of the galactosylation of β-D-xylopyranosides differently substituted at the anomeric position was carried out [23, 24]. The regioselectivity of the reaction depended on small variations in the xylose acceptor, such as the type of atom linked to the anomeric carbon (X atom in compound **1**, Scheme 3). The initial rate of formation of β(1,3) disaccharides decreased with the electronegativity of this atom, while the rate for β(1,4) disaccharides remained constant [24]. To interpret this result it was postulated that two different complexes between the xyloside and the

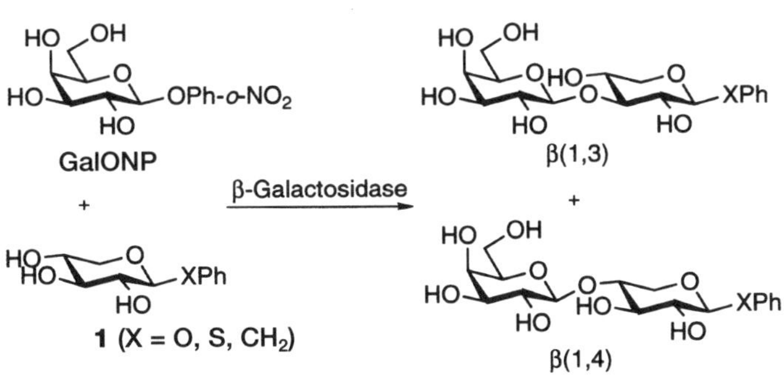

X in **1**	V (μM min^{-1}) β(1,3)	β(1,4)
O	8.7	7.3
S	2.7	7.3
CH_2	1.7	7.3

Scheme 3

(a)

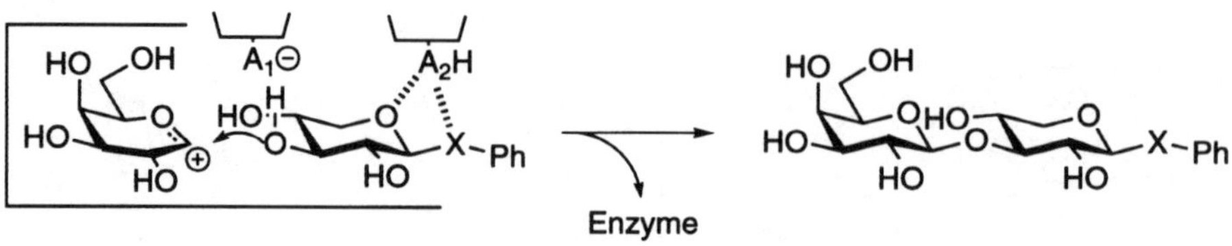

(b)

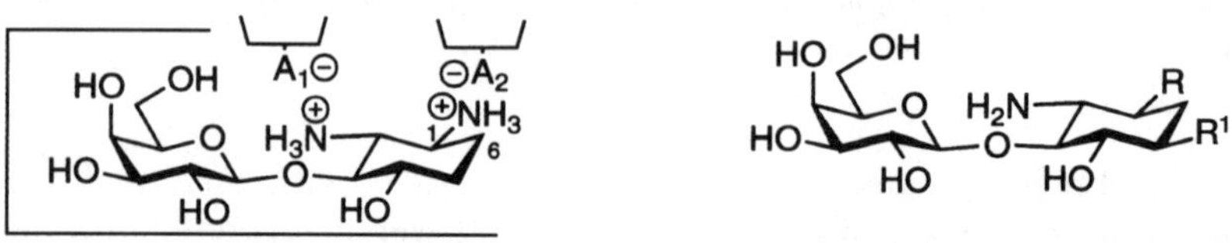

2 R=NH$_2$, R^1=H
3 R=H, R^1=NH$_2$

Fig. 3. **a** Postulated complex for the formation of β-D-Gal(1,3)-β-D-XylR showing polar interaction between O-5 of xylose and A_2H of the enzyme [24]. **b** Similar interaction with the 1-NH$_2$ in **2** may be responsible for enhanced affinity in comparison with **3** [25]

enzyme produce respectively the $\beta(1,3)$ and the $\beta(1,4)$ disaccharides. In the 1,3-complex there may be an important stabilizing interaction between the lone pair electrons of the ring oxygen in the xyloside and an electrophilic center in the enzyme, that would be affected by the substitution at the anomeric carbon, i.e. the X atom (Fig. 3a). Hence, the higher the electron density in the X atom ($O > S > CH_2$), the better the interaction between ring oxygen and electrophilic center. This interaction may involve the same enzyme residue as in Hubers model for the binding at O-5 of glucose (Fig. 2a). Recently Lehmann et al. working with some basic pseudo-disaccharides as competitive inhibitors of β-galactosidase from *E. coli* [25], postulated a similar interaction to explain the enhanced affinity of inhibitor **2** in comparison with **3** (Fig. 3b).

Changes of the regioselectivity depending on the anomeric configuration of the acceptor were first shown by Nilsson [26, 27] and, later, by other authors [28]. For instance, the reaction of α-D-Gal-OPh-p-NO$_2$ with α-D-Gal-OMe catalyzed by an α-galactosidase from coffee bean gave the $\alpha(1,3)$ disaccharide, whereas the reaction with the anomer β-D-Gal-OMe afforded the $\alpha(1,6)$ linked disaccharide as the major product [26]. These regioselectivity changes upon variation of the anomeric configuration have not been rationalized so far in terms of stabilizing or destabilizing interactions between enzyme and acceptor.

Overall, the possibility of controlling yields and regioselectivities by simple structural modifications in the sugar acceptor is specially attractive for chemists, who are able to introduce a large variety of functional groups at any position of the molecule. As an example, in the galactosylation of xylopyranosides β-D-Xyl-R using the enzyme from *E. coli*, it was observed that the regioselectivity and yield of β-D-Gal-β-D-Xyl-R disaccharide products varied drastically whether β-D-XylO-Me or β-D-XylO-Bn was used as acceptor [24]. While the methyl glycoside gave exclusively the β-(1,4) disaccharide, the benzyl glycoside provided the β-(1,3) regioisomer as the major product, the yield increasing five-fold in the latter reaction. Kinetic data indicated that the yield enhancement is a consequence of the higher specificity of the enzyme for the acceptors containing an aromatic ring. Nilsson also observed a change of the regioselectivity in the reactions of a β-galactosidase or β-D-GlcNAcOMe and β-D-GlcNAcO$CH_2CH_2SiMe_3$ as acceptors [29].

In addition to providing good selectivity and yield, these functional groups can, at the same time, act as temporary protecting groups. The 2-*O*-benzyl galactose derivative (2-OBn)-α-D-Gal-OMe was α-galactosylated with high regioselectivity at position 3 using an α-galactosidase [30]. The disaccharide product, α-D-Gal(1,3)-(2-OBn)-α-D-Gal-OMe, which is benzylated at position 2, can be used as a building block for the synthesis of blood group determinant B, α-D-Gal(1,3)-[α-L-Fuc(1,2)]-D-Gal.

2.3 Synthesis of Di- and Oligosaccharides

There are some comprehensive reviews containing results obtained with a large variety of di- and oligosaccharides prepared by either the reverse hydrolysis or the kinetic approach [1, 4]. Collecting these results here would be repetitive, and I refer to these reviews. However, some recent examples showing some interesting features from a synthetic point of view are discussed in this section.

Beau et al. have described the preparation of β-2-deoxy-D-glucosides and galactosides, using glycosidases and the corresponding glycals as donor substrates [31]. Yields of isolated products ranged from 14 to 60% depending on the structure of the acceptor. This work is of special interest since the stereoselective chemical synthesis of β-2-deoxy-D-glycosides is a difficult task. In some cases, it requires a stereoselective auxiliary group equatorially disposed at C-2, which is removed in a later step.

In this context, the stereocontrolled formation of sialyl glycosidic bonds is one of the most difficult to achieve by chemical methods; yields used to be relatively low in comparison with the formation of other glycosidic bonds. Thiem and Sauerbrei reported a stereoselective synthesis of sialyloligosaccharides by using an immobilized sialidase [32]. Yields ranged from 14 to 24%, and sialylation at the primary position of the acceptor was mainly obtained. From a practical point of view, fragments of biologically important oligosaccharides, such as the trisaccharide α-Neu5Ac(2,3)-β-D-Gal(1,4)-D-GlcNAc, were obtained from unprotected substrates by using a commercially available sialidase from *Vibrio cholerae*. The regioselectivity of the sialylation of lactose and *N*-acetyl-lactosamine using sialidases of various origins was studied recently by Ajisaka et al. [33]. When the enzyme from the Newcastle virus was used, the α(2,3)-linked isomer was regioselectively obtained.

By the sequential use of different enzymes, efficient synthesis of partially acylated disaccharides can be achieved. 6-*O*-acetyl glucal 5 (Scheme 4), prepared

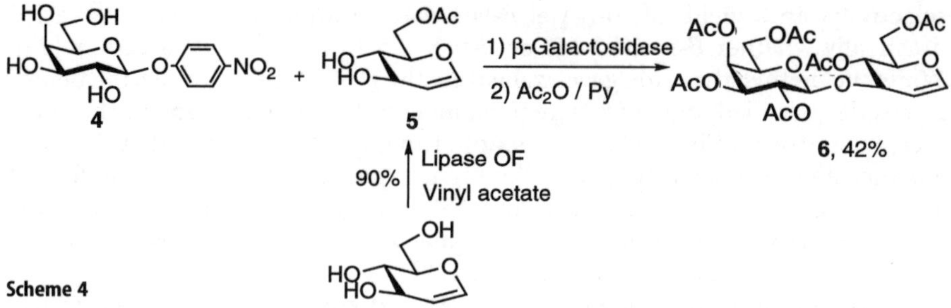

Scheme 4

through lipase or protease mediated monoacylation of glucal [34, 35], was glycosylated with **4** in the presence of *β*-galactosidase from *E. coli*. Compound **6** was the exclusive disaccharide product, obtained in 42 % yield after peracylation [36]. The acetyl group at position 6 prevented the formation of the (1,6)-regioisomer. Compound **6** is a useful building block for the synthesis of oligosaccharides possesing a *β*(1,3)-linkage such as the type I [*β*-D-Gal(1, 3)-D-GlcNAc] subunit.

The ability of most of the enzymes to work under similar conditions of pH and temperature allows the combination of different enzymes in the same reaction. Thus, glycosidases can be combined with glycosyltransferases to synthesize oligosaccharides in one-pot reactions [37, 38]. For instance, *β*-galactosidase from *Bacillus circulans* was used to synthesise *β*-D-Gal(1,4)-D-GlcNAc [37], which was in turn sialylated with an *α*-2,6-sialyltransferase (Scheme 5). The rapid synthesis of the trisaccharide avoids the *β*-galactosidase-catalyzed hydrolysis of the disaccharide intermediate, representing an additional advantage of this catalytic system.

Higher oligosaccharides can also be synthesized by use of more specific glycosidases such as cellulase, that can catalyze the formation of *β*(1,4) glycosidic bonds of glucose oligomers in a regio- and stereoselective manner. Using *β*-cellobiosyl fluoride as an activated glycosyl donor in the presence of a cellu-

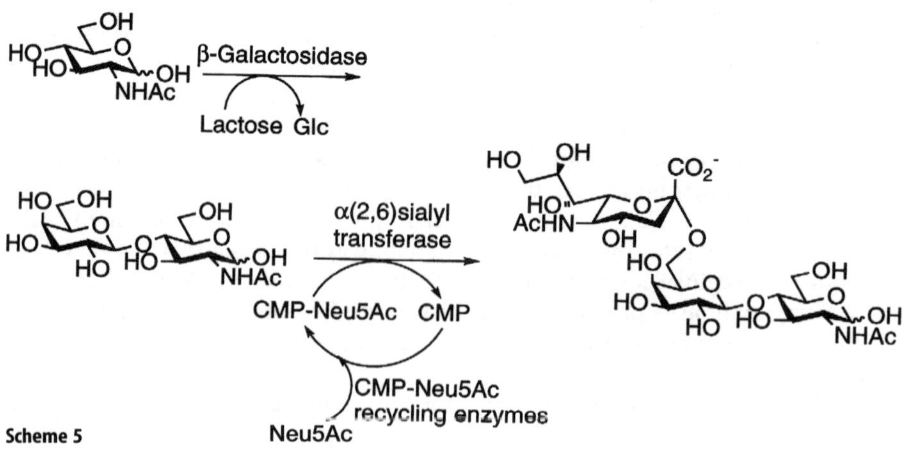

Scheme 5

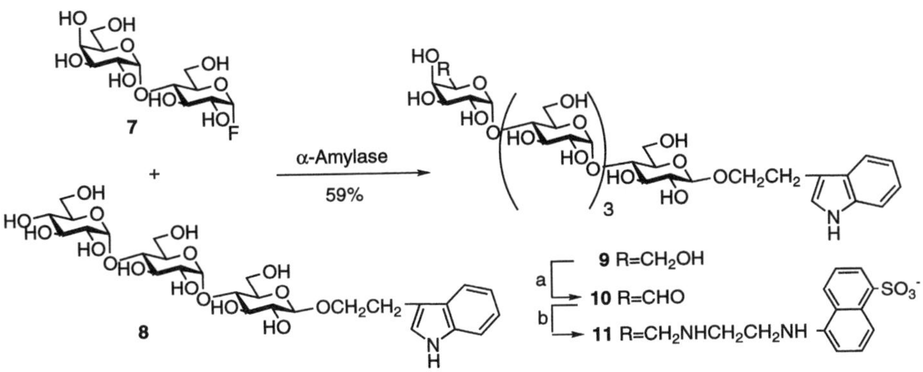

Scheme 6. Preparation of pentasaccharide **11** [42]: **(a)** galactose oxidase (*Dactylium dendroides*), catalase (bovine liver), phosphate buffer, 37 °C; **(b)** 5-(2-aminoethylamino)-1-naphthalene-sulfonate (EDANS), MeOH, reflux, then $NaBH_3CN$

lase, β(1,4)-D-glucopyranoside oligomers (cellulose) with a degree of polymerization >22 were obtained [39]. Interestingly, β-lactosyl fluoride is also recognized by the cellulase, and capable of forming a reactive glycosyl-enzyme intermediate, but it cannot be polymerized owing to the axial HO-group at the 4′-position [40, 41]. However, in the presence of appropriate acceptors, cellulase-catalyzed lactosylation takes place giving oligosaccharides with complete regio- and stereoselectivity [40]. Furthermore, the resulting lactosyl-oligosaccharide can be treated with a β-galactosidase to hydrolyze the terminal galactose unit, giving rise to a new method of elongating oligosaccharides with a glucose unit [41]. In a recent comunnication, Driguez et al. [42] reported the preparation of pentasaccharide **11** through a chemoenzymatic synthesis involving an α-amylase-catalyzed coupling as one of the key steps (Scheme 6). Thus, 4′-epimer maltosyl fluoride donor **7**, obtained by chemical synthesis, was condensed with the maltotrioside derivative **8** in the presence of α-amylase from *A. oryzae* to obtain pentasaccharide **9** in 59% isolated yield. Compound **9** was oxidized to **10** by galactose oxidase and further transformed into **11**. As in the cellulase case, **7** is accepted by the α-amylase as a donor, but neither **7** nor **9** are acceptors because of the axial 4-OH at the nonreducing end unit. This is a nice example to illustrate that by combining chemical and enzymatic methods, short syntheses of complex molecules can be achieved. Besides, compound **11** is used as a substrate for the in vitro assays of human salivary and pancreatic α-amylases and displays intramolecular fluorescence quenching [42].

2.4
Synthesis of Non-Sugar Glycosides

Glycosides of monohydroxylated molecules have been synthesized using both the reverse and the kinetic approach (transglycosylation) [12, 43, 44]. Yields can be high since the alcohol is normally used in large excess. The main benefits when using this synthetic methodology are the total stereocontrol in the glycosidic bond. Further, the reactions are easily carried out and this allows preparations on a larger scale. For instance, the preparation of allyl, benzyl, and trimethylsilylethyl β-D-galactopyranosides on a 1–20 g scale was achieved from lactose and allyl, benzyl, and trimethylsilylethyl alcohols, respectively, using a β-galactosidase [43]. With racemic and *meso* alcoholic acceptors, some degree of diastereoselectivity is obtained [45]. This reaction may be of interest if one of the glycosides formed is the desired product. However, enzymatic glycosylation is not a practical reaction for resolution of alcohols as compared, for example, with enzymatic acylations or deacylations.

Glycosylation of peptides and amino acids having a hydroxyl group by a glycosidase-catalyzed reaction has been reported [46]. In general, yields are low (10–30%), although dependent on the substrate structure and the choice of the protecting groups at the amino and acid functional groups. For acid and base labile amino acids these reactions offer the advantage over the chemical methods that mild conditions are used.

2.5 Concluding Remarks

Chemists are reluctant to use glycosidases for the synthesis of glycosides. Indeed, these enzymes do not offer the selectivity and the efficiency in terms of yield that one would expect from a biocatalyst. However, the advantages of using glycosidases may be found where limitations and inconveniences exist for other methods. Glycosidases are cheaper and more readily available than glycosyltransferases, additionally, they accept a wider range of sugar acceptors. In contrast to the classical chemical methods, glycosidases use unprotected substrates and the reactions are performed under conditions that neither require toxic or hazardous solvents nor chemical reagents. Considering these advantages there are certain applications in which their use results in a very convenient methodology. They can be summarized as follows:

1. Glycosidases can be used to prepare glycosides of simple alcohols as starting materials in a synthetic route. For example, benzyl and allyl glycosides can be prepared on a multigram scale in a single step to protect the anomeric position of the sugar [43]. Glycosides bearing a spacer arm, which can be used to make glycoconjugates or to be coupled to a solid support for affinity chromatography, can also be synthesized [47].
2. In industrial and pharmaceutical applications, non-sugar glycosides can be obtained under very mild conditions, as for instance, alkyl glycosides as non-ionic surfactants. Additionally, glycosylation of pharmaceuticals can be achieved to make compounds with enhanced activity. As an example, ergot alkaloids, which decompose at high pH and easily undergo oxidation, were galactosylated under mild conditions by use of β-galactosidases in yields up to 40% [48]; some of the products showed a significant enhancement of immunomodulatory activity.

3. In certain applications in which a pure oligosaccharide is not required. For example, by galactosylation of unprotected D-xylose using a β-galactosidase, a mixture of galactosyl-xylose disaccharides was prepared in 50% yield [49a]. This mixture, without the need of separating regioisomers, can be used to evaluate the activity of the intestinal lactase in vivo, giving rise to a new diagnostic method for lactase deficiency. Compared to the previous chemical synthesis [49b], this one-step enzymatic procedure is cheaper and less hazardous. As an additional example, it is possible to produce bread containing galactooligosaccharides which promote the growth of Bifidobacteria, that live in the human intestine and are beneficial, based on the use of lactose and β-galactosidases [50].
4. In chemoenzymatic synthesis of oligosaccharides, the use of glycosidase-catalyzed reactions can provide useful di- and oligosaccharide intermediates for transformations. For this purpose, reactions must be driven with a satisfactory degree of yield and regioselectivity to facilitate the isolation of products and the scale-up of the process. The possibility of introducing functional groups in the sugar acceptor, acting as temporary protection, makes this approach more versatile. Chemoenzymatic synthesis combining the use of cellulases or amylases allows the preparation of higher oligosaccharides in yields and regioselectivities comparable to those obtained using glycosyltransferases.

Finally, it should be mentioned that there is also a field of research which studies how functional groups introduced at different positions of a sugar acceptor can affect selectivity. This should provide valuable information concerning the nature of the active site and binding characteristics of the enzyme.

3
Lipases in Selective Acylations and Deacylations of Carbohydrates

Lipases are widely distributed among animals, plants, and microorganisms. They catalyze the hydrolysis of ester bonds at the fat/water interface. The mechanism of the lipase-catalyzed hydrolysis of esters involves two consecutive steps [51]. First, the ester is attacked by a nucleophilic group of the enzyme – normally an OH of a serine or a SH of a cysteine – to form an acyl-enzyme intermediate with concomitant liberation of the alcohol moiety of the ester. The acyl-enzyme is subsequently attacked by a water molecule to give the product.

Due to their low cost and availability from different sources, lipases have been frequently employed for the hydrolysis of acylated compounds. The reaction is easy to perform: enzyme and substrate are mixed in buffered water, and the pH is normally adjusted by addition of NaOH to neutralize the acid released during the reaction. In addition, lipases are able to work in solvents other than water and, thus, acylations and deacylations catalyzed by lipases have been performed in organic solvents. Some previous accounts [2, 52, 53] of enzyme transformations include comprehensive data of lipase-catalyzed acylations and deacylations of pyranose and furanose monosaccharides, disaccharides, nucleosides and further aglycon glycosides. I refer to these articles in order to give a good overview of the work done in this field of research. In the present article, general features about the selectivity obtained in these reactions will be discussed, with special emphasis given to show the factors that influence this selectivity. A brief description of some characteristics of lipase-catalyzed reactions in organic solvents follows.

3.1 Lipases in Organic Solvents

Lipases, and many other enzymes as well, have been shown to be active in organic solvents with low water content [54] ($<1\%$ v/v of water). From this finding some synthetic applications are derived [55], the most important are the following:

1. The possibility of shifting the thermodynamic equilibrium to favor synthesis over hydrolysis. In this way, esters are formed from acids and alcohols. Transesterifications between ester and alcohol are also possible. In water, these reactions are suppressed by hydrolysis and, therefore, do not occur to any appreciable extent.
2. The enzyme can be recovered by simple filtration since it is insoluble in most organic solvents.
3. Some enzymatic properties – such as the substrate specificity, and the regio- and stereoselectivity – can be modified by the solvents.

However, to remain active in an organic solvent, the enzyme must maintain a small and essential amount of water which is tightly bound to the enzyme. Since sugars are highly polar compounds the use of polar solvents, such as pyridine, DMSO and DMF, is required; they, however, inactivate enzymes by stripping away the essential water.

Different solutions can be applied to avoid the use of polar solvents. The most usual is to make the sugar more lipophilic by introducing hydrophobic groups by simple chemical transformations [56]. In addition, sugars can be solubilized in hydrophobic solvents by in situ complexation with phenylboronic acid [57].

Most of lipase-catalyzed acylations of sugars in organic solvents have been reported as transesterification rather than esterification reactions. The displacement of the equilibrium towards products has been accomplished by using activated acyl donors [58] such as 2,2,2-trichloroethyl esters and, more often, enol esters. The use of enol esters, such as a vinyl or an isopropenyl ester, was, in fact, first reported in lipase-catalyzed reactions with sugars [59]. In the reaction, an unstable enol is liberated which instantaneously tautomerizes to the corresponding aldehyde or ketone, making the reaction irreversible.

3.2 Regioselective Deacylations

Lipase-catalyzed deacylations of sugars can be carried out either by hydrolysis in water or by alcoholysis in organic solvents in the presence of a nucleophilic alcohol, such as *n*-propanol or *n*-butanol. In peracylated sugars the anomeric ester is the most reactive of the molecule – probably due to the intrinsic reactivity of the anomeric center – and, thus, enzymatic deacylation at this position is generally achieved in good yields [60]. The product is a useful intermediate in oligosaccharide synthesis, since it can be subsequently activated to give glycosyl donors. If the anomeric hydroxyl group is not esterified, steric effects in the modes of binding of the substrate within the enzyme's active center often govern the selectivity in the deacylation. For this reason, selective hydrolysis of the ester at the primary position of pyranoses or furanoses is normally achieved in high yield [60].

Among secondary esters, selectivity in the hydrolysis can be obtained although somehow unpredictably and in a way very dependent on the origin of the lipase. In fact, the basic modus operandi when seeking a given selectivity is, firstly, to carry out a screening with a variety of lipases. For example, by regioselective hydrolysis of octa-*O*-acetylsucrose using a variety of lipases as well as proteases, five different sucrose hepta-acetates can be prepared, two of them through a selective deacylation of secondary esters [61].

With a given lipase and different sugars, the regioselectivity may be dependent on the relative configuration of the stereogenic centers in the substrate. For example, the 1,6-anhydro-2,3,4-tri-*O*-butanoyl-β-D-galactopyranose was selectively deacylated at position 2 to give the 3,4-di-*O*-butanoyl derivative using lipase from *Candida cylindracea* in a buffer solution [62]. With the glucopyranose derivative, epimeric substrate at C-4, and the same enzyme, deacylation first takes place at position 4 after which the acyl ester at C-2 is hydrolyzed [63].

In an attempt to evaluate how other factors of the reaction can affect regioselectivity, the deacylation of some alkyl 2,3,4-tri-*O*-acyl-β-D-xylopyranosides (**12**) by lipase PS and six different solvents with varying degree of hydrophobicity has been studied [64]. With all of the substrates and in all the solvents used, deacylation at the C-4 position was observed (Scheme 7). From a synthetic point of view, this high regioselectivity allows the preparation of xylose derivatives having only the HO-4 free in high yield. For example, methyl 2,3,4-tri-*O*-acetyl-β-D-xylopyranoside (**12a**) in *tert*-amyl alcohol gave exclusively methyl 2,3-di-*O*-acetyl-β-D-xylopyranoside (**13a**) in 93% yield. The persistent

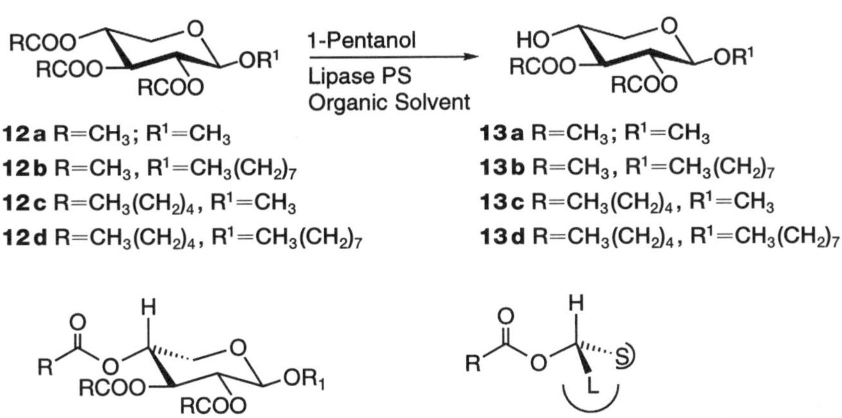

Scheme 7 **12** **13**

regioselectivity was ascribed to the fact that the C-4 position is less sterically hindered than C-2 and C-3 positions. Besides, the stereochemistry and the size difference of the substituents at C-4 in **12** (Scheme 7) fits well with the proposed model **13** to predict which enantiomer of a secondary alcohol reacts faster in enzymatic resolutions with lipase PS [65].

3.3 Regioselective Acylations

In unprotected sugars, highly regioselective acylations of the primary hydroxy group are achieved. In these cases the main problem is the selection of a suitable solvent, since highly polar solvents inactivate most of the enzymes. However, some enzymes maintain their activity in polar solvents. For example, by use of porcine pancreatic lipase in pyridine, Klibanov and Therisod [66] reported the transesterification of several unprotected monosaccharides with 2,2,2-trichloroethyl esters to give regioselective acylation of the primary hydroxyl group in yields ranging from 50 to 91%. Differentation between secondary hydroxyl groups is also possible, as in the case of deacylation, by screening different lipases [67]. Thus, the butyrylation of 6-*O*-butyrylglucose in pyridine in the presence of lipase from *Cromobacterium viscosum* resulted exclusively in acylation at the C-3 position [67a]. However, the reaction with porcine pancreatic lipase (PPL) afforded preferentially the acylated product at the C-2 position.

In the acylations of monosaccharides having several secondary hydroxyl functions, the regioselectivity can be significantly altered by different factors, such as the nature of substituents in the substrate, the organic solvent or the acylating substrate used. This is in contrast to the almost exclusive dependence on lipase origin and intrinsic steric factors in the substrate of the lipase-catalyzed deacylations. For instance, it was found that both the nature of the aglycon structure of the sugar and the organic solvent can influence the regioselectivity of the acetylation of alkyl β-D-xylopyranosides catalyzed by lipase PS [68]. The acetylation of methyl β-D-xylopyranoside with vinyl acetate in acetonitrile gave the 3,4-diacetate exclusively, which was isolated in 85% yield, whereas the reaction of octyl β-D-xylopyranoside in the same solvent gave a mixture of the 2,4- and 3,4-diacetates in almost equimolar amount. Furthermore, when the reaction on the octyl glycoside was performed in hexane, the ratio between the 2,4- and 3,4-diacetates changed to 3.6:1.0. In order to study the origin of the regioselectivity change promoted by the solvent in the acetylation of the octyl derivative, the effect of several solvents on the formation of the initial products, i.e. the monoacetates, was determined. A selective formation of the 2-acetate in hydrophobic solvents could be observed, while the formation of the 4-acetate is preferred in polar solvents. A correlation between the ratio of initial rates of formation of these monoacetates (V_2/V_4) and solvent hydrophobicity was found. It is noteworthy that in changing from acetonitrile to hexane, an eleven-fold increase in the V_2/V_4 ratio was observed.

The choice of the acylating reagent used can be important for the selectivity of the acylation [69]. With porcine pancreatic lipase, butyl α-D-glucopyranoside gave preferential acylation at C-6 or C-3 depending on whether acetic or succinic anhydride, respectively, was used as the acylating reagent [69].

An interesting observation is that the stereochemistry of the anomeric center can greatly influence the selectivity of acylations [70]. For instance [70b], the acetylation of methyl 4,6-*O*-benzylidene-α-D-glucopyranoside with lipase from *Pseudomonas fluorescens* yielded the 2-*O*-acetyl derivative exclusively, isolated in 94% yield. The β anomer, however, acetylated under the same conditions, gave mainly the 3-*O*-acetyl derivative, isolated in 86% yield.

3.4 Stereoselective Acylations

Due to their chiral nature, enzymes are capable of distinguishing between enantiomers. In this way, lipases have been extensively used to resolve racemic mixture of alcohols.

Ronchetti and Toma [71] have shown that lipases from *Pseudomonas fluorescens* (PFL) and from *Candida cylindracea* are able to acylate sugars of the D-series in a different position than their enantiomeric L-sugars. For instance, using PFL, methyl α-D-fucopyranoside was butyrylated at the C-2 position in 88% overall yield and 96% regioselectivity, and methyl α-L-fucopyranoside gave butyrylation at C-4 in 45% overall yield and 97% regioselectivity.

The size of substituents in the substrate can affect the stereoselectivity of the acylation. In work aimed at the total synthesis of enantiomerically pure carbocyclic 2′-deoxynucleosides as building blocks for oligonucleotide analogs [72], the resolution of the racemic *cis*-1,3-cyclopentanediol derivatives **14** (Scheme 8) was investigated by lipase-catalyzed acetylation. It was found that the acetylation catalyzed by *Pseudomonas fluorescens* lipase in vinyl acetate gave the two monoacetyl derivatives **15** and **16** with a degree of enantioselectivity dependent on the size of the R-protecting group in the substrate. As the size of

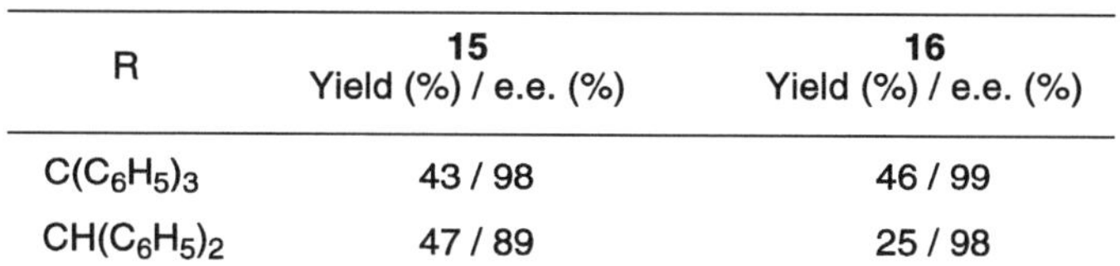

14 (±) **15** **16**

R	15 Yield (%) / e.e. (%)	16 Yield (%) / e.e. (%)
$C(C_6H_5)_3$	43 / 98	46 / 99
$CH(C_6H_5)_2$	47 / 89	25 / 98
$CH_2C_6H_5$	39 / 73	21 / --
CH_3	22 / 15	22 / --

Scheme 8

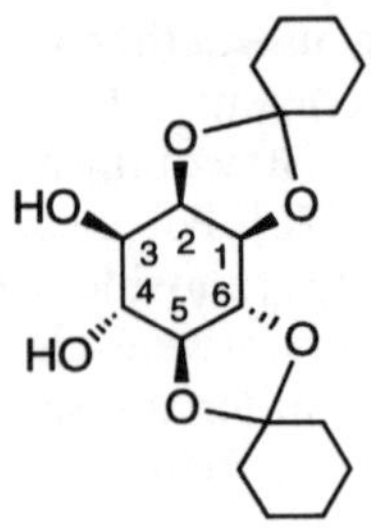

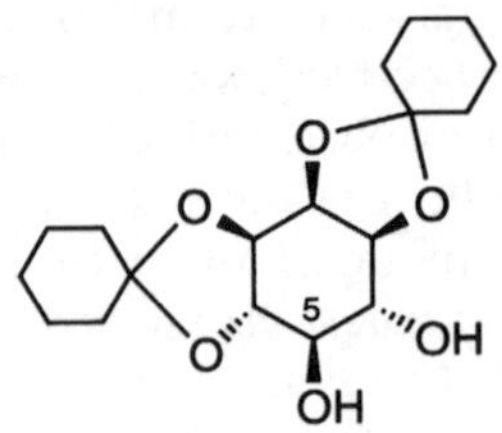

Structures *dl*-**17** *dl*-**18**

R decreased, there was a remarkable decrease in the enantiomeric excess. The highest regio- and enantioselectivities were obtained using the diol substrate protected with the bulky triphenylmethyl (trityl) group.

Optically pure inositol intermediates are very useful for synthesising D-*myo*-inositol 1,4,5-trisphosphate, which was found to be a second messenger. Ozaki et al [73] studied the resolution of racemic di-*O*-cylohexylidene-*myo*-inositol derivatives **17** and **18** by lipase-catalyzed esterification in organic solvents. Lipase from *Candida cylindracea* exclusively acetylates the hydroxyl group at C-4 or at C-5 of the D-enantiomer of **17** or **18**, respectively. Around 100% e.e's of monoacetate products and unreacted starting compounds were obtained. The efficiency of the resolution is affected by the solvent, the most hydrophobic solvents ethyl ether and benzene being more effective than the water miscible solvents such as acetone, THF, and dioxane.

In conclusion, the use of lipases for selective acylations and deacylations of carbohydrates has been shown to be an effective tool, and they deserve a place in the laboratory along with the classical chemical reagents. Lipases offer a simple and easy procedure that may provide regio- and stereoselectivities difficult to obtain under classical conditions.

Acknowledgments. I would like to thank Dr R. López for her contribution to the work carried out at our laboratory in this area, and Dr C. Jaramillo for helpful discussions. Work in the author's laboratory was funded by DGICYT (Grants PB90-0076 and PB93-0127).

References

1. (a) Toone EJ, Simon ES, Bednarski MD, Whitesides GM (1989) Tetrahedron 45:5365 (b) Wong CH, Halcomb RL, Ichikawa Y, Kajimoto T (1995) Angew Chem Int Ed Engl 34:412, 521
2. Drueckhammer DG, Hennen WJ, Pederson RL, Barbas III CF, Gautheron CM, Krach T, Wong CH (1991) Synthesis 499
3. Nilsson KGI (1988) Trends Biotechnol 6:256
4. Cote GL, Tao BY (1990) Glycoconjugate J 7:145
5. David S, Augé C, Gautheron C (1991) Adv Carbohydr Chem Biochem 49:175
6. (a) Thiem J (1995) FEMS Microbiol Rev 16:193 (b) Monsan, P, Paul F (1995) FEMS Microbiol Rev 16:187

7. (a) Kornfeld R, Kornfeld S (1985) Ann Rev Biochem 54:631 (b) Moremen KW, Trimble RB, Herscovics A (1994) Glycobiology 4:113
8. Sinnott ML (1990) Chem Rev 90: 1171
9. Johansson E, Hedbys L, Larsson PO, Mosbach K, Gunnarsson A, Svensson S (1986) Biotechnol Lett 8:421
10. Ajisaka K, Nishida H, Fujimoto H (1987) Biotechnol Lett 9:387
11. Ajisaka K, Fujimoto H (1989) Carbohydr Res 185:139
12. (a) Marek M, Novotná Z, Jary J, Kocikova V (1989) Biocatalysis 2:239 (b) Vulfson EN, Patel R, Beecher JE, Andrews AT, Law BA (1990) Enzyme Microb Technol 12:950 (c) Vic G, Crout DHG (1995) Carbohydr Res 279:315
13. López R, Fernández-Mayoralas A, Martín-Lomas M, Guisán JM (1991) Biotechnol Lett 13:705
14. Baker A, Turner NJ, Webberley MC (1994) Tetrahedron: Asymm 5:2517
15. Hedbys L, Johansson E, Mosbach K, Larsson PO, Gunnarsson A, Svensson S (1989) Carbohydr Res 186:217
16. Nilsson KGI (1987) In: Laane C, Tramper J, Lilly MD (eds) Biocatalysis in organic media. Elsevier Science Publishers BV, Amsterdam, p 369
17. Nilsson KGI (1988) Ann N Y Acad Sc 384
18. Fortun Y, Colas B (1991) Biotechnol Lett 13:863
19. Sauerbrei B, Thiem J (1992) Tetrahedron Lett 33:201
20. Wallenfels K, Weil R (1972) The Enzymes 7:658
21. Huber RE, Gaunt MT, Hurlburt KL (1984) Arch Biochem Biophys 234:151
22. Crout DHG, MacManus DA, Critchley P (1990) J Chem Soc Perkin Trans 1 1865
23. López R, Fernández-Mayoralas A (1992) Tetrahedron Lett 33:5449
24. López R, Fernández-Mayoralas A (1994) J Org Chem 59:737
25. Lehmann J, Rob B (1995) Carbohydr Res 276:199
26. Nilsson KGI (1987) Carbohydr Res 167:95
27. Nilsson KGI (1990) Carbohydr Res 204:79
28. Crout DHG, Howarth OW, Singh S, Swoboda BEP, Critchley P, Gibson WT (1991) J Chem Soc, Chem Commun 1550
29. Nilsson KGI (1989) Carbohydr Res 188:9
30. Nilsson KGI, Fernández-Mayoralas A (1991) Biotechnol Lett 13:715
31. Petit JM, Paquet F, Beau JM (1991) Tetrahedron Lett 32:6125
32. Thiem J, Sauerbrei B (1991) Angew Chem Int Ed Engl 30:1503
33. Ajisaka K, Fujimoto H, Isomura M (1994) Carbohydr Res 259:103
34. Holla EW (1989) Angew Chem Int Ed Engl 28:220
35. Look GC, Ichikawa Y, Shen GJ, Cheng PW, Wong CH (1993) J Org Chem 58:4326
36. Look GC, Wong CH (1992) Tetrahedron Lett 33:4253
37. Herrmann GF, Ichikawa Y, Wandrey C, Gaeta FCA, Paulson JC, Wong CH (1993) Tetrahedron Lett 34:3091
38. Kren V, Thiem J (1995) Angew Chem Int Ed Engl 34:893
39. Kobayashi S, Kashiwa K, Kawasaki T, Shoda S (1991) J Am Chem Soc 113:3079
40. Shoda S, Obata K, Karthaus O, Kobayashi S (1993) J Chem Soc, Chem Commun 1402
41. Kobayashi S, Kawasaki T, Obata K, Shoda S (1993) Chem Lett 685
42. Payre N, Cottaz S, Driguez H (1995) Angew Chem Int Ed Engl 34:1239
43. Nilsson KGI (1988) Carbohydr Res 180:53
44. (a) Mitsuo N, Takeichi H, Satoh T (1984) Chem Pharm Bull 32: 1183 (b) Matsumura S, Kubokawa H, Yoshikawa S (1991) Chem Lett 945 (c) Trincone A, Nicolaus B, Lama L, Morzillo P, De Rosa M, Gambacorta A (1991) Biotechnol Lett 13:235
45. (a) Gais HJ, Zeissler A, Maidonis P (1988) Tetrahedron Lett. 29:5743 (b) Bjorkling F, Godtfredsen SE (1988) Tetrahedron 44:2957 (c) Trincone A, Nicolaus B, Lama L, Gambacorta A (1991) J Chem Soc Perkin Trans 1 2841
46. (a) Cantacuzène D, Attal S, Bay S (1991) Bioorg Med Chem Lett 1:197 (b) Turner NJ, Webberley MC (1991) J Chem Soc, Chem Commun 1349 (c) Holla EW, Schudok M, Weber A, Zulauf M (1992) J Carbohydr Chem 11:659

47. Vic G, Crout DHG (1994) Tetrahedron: Asymm 5:2513
48. Kren V, Sedmera P, Havlicek V, Fiserova A (1992) Tetrahedron Lett 33:7233
49. (a) Aragón JJ, Cañada FJ, Fernández-Mayoralas A, López R, Martín-Lomas M, Villanueva D (1996) Carbohydr Res 290:209 (b) Rivera-Sagredo A, Fernández-Mayoralas A, Jiménez-Barbero J, Martín-Lomas M, Villanueva D, Aragón JJ (1992) Carbohydr Res 228:129
50. Sonoike Y, Kobayashi Y, Kato H, Kan T (1992) US Patent 5,118,521
51. Kazlauskas RJ (1994) Trends Biotechnol 12:464
52. Waldmann H, Sebastian D (1994) Chem Rev 94:911
53. Bashir NB, Phythian SJ, Reason AJ, Roberts SM (1995) J Chem Soc Perkin Trans 1 2203
54. Klibanov AM (1986) Chemtech 16:354
55. Klibanov AM (1990) Acc Chem Res 23:114
56. Fregapane G, Sarney DB, Vulfson EN (1991) Enzyme Microb Technol 13:796
57. (a) Schlotterbeck A, Lang S, Wray V, Wagner F (1993) Biotechnol Lett 15:61 (b) Oguntimein GB, Erdmann H, Schmid RD (1993) Biotechnol Lett 15:175 (c) Ikeda I, Klibanov AM (1993) Biotechnol Bioeng 42:788
58. Faber K, Riva S (1992) Synthesis 895
59. Sweers HM, Wong CH (1986) J Am Chem Soc 108:6421
60. Hennen WJ, Sweers HM, Wang YF, Wong CH (1988) J Org Chem 53:4939
61. Chang KY, Wu SH, Wang KT (1991) Carbohydr Res 222:121
62. Ballesteros A, Bernabé M, Cruzado C, Martín-Lomas M, Otero C (1989) Tetrahedron 45:7077
63. Kloosterman M, de Nijs MP, Weijnen JGJ, Schoemaker HE, Meijer EM (1989) J Carbohydr Chem 8:333
64. López R, Pérez C, Fernández-Mayoralas A, Conde S (1993) J Carbohydr Chem 12:165
65. (a) Kazlauskas RJ, Weissfloch ANE, Rappaport AT, Cuccia LA (1991) J Org Chem 56:2656 (b) Johnson CR, Miller MW (1995) J Org Chem 60:6674
66. Therisod M, Klibanov AM (1986) J Am Chem Soc 108:5638
67. (a) Therisod M, Klibanov AM (1987) J Am Chem Soc 109: 3977 (b) Nicotra F, Riva S, Secundo F, Zucchelli L (1989) Tetrahedron Lett 30:1703
68. López R, Montero E, Sánchez F, Cañada J, Fernández-Mayoralas A (1994) J Org Chem 59:7027
69. Fabre J, Betbeder D, Paul F, Monsan P, Perie J (1993) Tetrahedron 49:10877
70. (a) Carpani G, Orsini F, Sisti M, Verotta L (1989) Gazz Chim Ital 119: 463 (b) Iacazio G, Roberts SM (1993) J Chem Soc Perkin Trans I 1099 (c) Panza L, Luisetti M, Crociati E, Riva S (1993) J Carbohydr Chem 12:125
71. Ciuffreda P, Colombo D, Ronchetti F, Toma L (1990) J Org Chem 55:4187
72. Henly R, Elie CJJ, Buser HP, Ramos G, Moser HE (1993) Tetrahedron Lett 34:2923
73. Ling L, Watanabe Y, Akiyama T, Ozaki S (1992) Tetrahedron Lett 33:1911

Chemical Transformations Employing Glycosyltransferases

Ulrike Gambert · Joachim Thiem

Institut für Organische Chemie, Universität Hamburg, Martin-Luther-King-Platz 6, D-20146 Hamburg, Germany

Early on, Emil Fischer proposed the use and technological application of enzymes in natural product chemistry, in particular in peptide and carbohydrate synthesis. After extensive contributions of biochemistry to this field over several decades, bioorganic applications of enzymes for synthetic purposes have increased significantly only over recent years. This chapter will highlight the rather complex task of stereospecific and regiospecific glycosylation by use of glycosyltransferases. The application of enzymes, including cofactor regeneration cycles, are discussed. Reactions, including modified acceptor and donor substrates which give access to heterooligosaccharides in preparative quantities, are described. The following contribution focuses on modern developments and presents some selected recent examples from the author's laboratory and others.

Table of Contents

Topics in Current Chemistry, Vol. 186

1 Introduction

The amazing diversity of carbohydrate structures is intimately related to their large variety of biological functions. Apart from their importance as energy storage and supporting substances, carbohydrates are involved in diverse dynamic physiological processes. Their essential role in cell-cell recognition phenomena and cell differentation processes [1], as antigenic determinants in malignant tissues [2, 3] and as components for receptor-mediated import of bacteria, viruses, toxins and hormones [4–6] has been reported.

All these different functions of heterooligosaccharides can be ascribed to the enormous structural variation due to the polyfunctionality of sugars. Thus, biological information is inherent not only in the sequence but also in the type of interglycosidic linkages, configuration, and branching. This very variety makes the straightforward synthesis of a defined oligosaccharide a difficult project.

Even though in recent years some elegant methods for glycosylation have been developed and previously existing methods have been improved [7–10], approaches toward such structures following classical organic synthetic means are often restrained by the complex protective group chemistry, leading to time-consuming multistep syntheses with low overall yields. As most of the hetero-oligosaccharides of interest are produced in nature by enzymes, it is entirely plausible to use as well the synthetic potential of enzymes in vitro. In contrast to multistep chemical reactions, such synthetic processes should work regio- and stereopecifically, and the products should become available more easily. Unfortunately, only about 300 enzymes are commercially available. Some of them are cheap, but others are extremely expensive. Therefore, many enzymes have to be isolated by the synthetic chemists themselves, who must be familiar with many biochemical techniques.

Of special interest is the elaboration of conditions for large scale preparations, reactions in semi-aqueous systems, and synthetic applications of less specific enzymes for unnatural substrates.

Although reaction routes are influenced by organic solvents, some enzymes retain their activity [11]. For example, in organic solvents, lipases [12] esterify the saccharides nearly quantitatively, but in aqueous solutions ester hydrolysis prevails.

Superior to lipases with regard to activity at higher solvent polarity are other hydrolases such as proteases. For example, subtilisin from *Bacillus subtilis* can be used to acetylate sucrose in dimethylformamide to the 1-mono-acetylated derivative [13], whereas the chemical acetylation reaction regioselectively leads to the 6-mono-acetylated product [14]. Thus, the two procedures nicely complement each other.

Enzymes turned out to be very helpful in the de novo synthesis of certain monosaccharides. Generally, two chiral carbonyl compounds are combined in an aldol-type reaction. In carbohydrate metabolism, aldolases catalyze the condensation of dihydroxyacetone phosphate (DHAP) and aldehydes to higher sugar components. To date, about thirty aldolases have been classified, but only a few of them have been used in a preparative way. Well known and often applied is fructose diphosphate aldolase from rabbit muscle which needs DHAP and glyceraldehyde 3-phosphate to synthesize fructose 1,6-diphosphate. FDP aldolase is highly specific for DHAP [15], but the aldehyde component can be widely varied in order to design a particular sugar derivative. In addition to the above mentioned D-fructose 1,6-diphosphate aldolase (FruA), L-rhamnulose 1-phosphate aldolase, D-tagatose 1,6-diphosphate aldolase, and L-fuculose 1-phosphate aldolase can be used with a variety of aldehydes to generate all possible stereoisomeres of ketohexoses [16].

In addition, there have been some recent applications of FruA in the field of pheromones [17]. The two independent stereogenic centers of the condensation product were stereospecifically generated by FruA-catalyzed aldol addition of 5-oxohexanal to DHAP. Further, enzymatic dephosphorylation and a series of straightforward chemical transformations yielded (+)-exo-brevicomin, the sex aggregation pheromone of the bark beetle.

One of the most important sugars in glycoconjugates with regard to molecular recognition is *N*-acetyl neuraminic acid (Neu5Ac). Originally, this had to be isolated from biological sources, but a recent approach makes use of the enzyme *N*-acetyl neuraminate lyase from *Clostridium*, which by an aldol-type reaction between *N*-acetyl mannosamine (ManNAc) and pyruvate leads to the formation of Neu5Ac [18]. Further, starting with the cheaper *N*-acetyl glucosamine (GlcNAc), which in turn is no substrate for the aldolase, this can be epimerized to ManNAc with GlcNAc-2-epimerase [19].

Principally, two groups of enzymes can be used in oligosaccharide synthesis. Glycohydrolases [20] usually cleave the terminal non-reducing saccharide unit of an oligosaccharide, but, if the equilibrium is shifted, they can also form interglycosidic linkages under appropriate reaction conditions. Thermodynamically controlled reactions can be regulated by temperature or concentration of the starting materials. Alternatively, reactions under kinetic control involve activated glycosyl donor components carrying a good anomeric leaving group. In contrast to reverse hydrolysis, yields and regioselectivities in transglycosylation are higher and therefore preferred. Glycosidases have the advantage that they convert sugars without prior activation, so they do not need expensive precursors or cofactors. Furthermore, many of them are available at low or reasonable prices. However, glycosidases show a lack of regioselectivity, and it is difficult to separate the product from the starting materials and even the isomer formed.

2 Glycosylation by Glycosyltransferases

It follows from the above that, if highly specific oligosaccharides need to be synthesized, glycosyltransferases [21] of the Leloir type [22] should be the enzymes of choice. These enzymes connect sugars via the activated nucleotide components with high stereo- and regioselectivity. They are substrate-specific, but, in vitro, with solubilized enzymes, it becomes possible to transfer modified donors to modified acceptors and thus broaden the scope of the synthetic applicability.

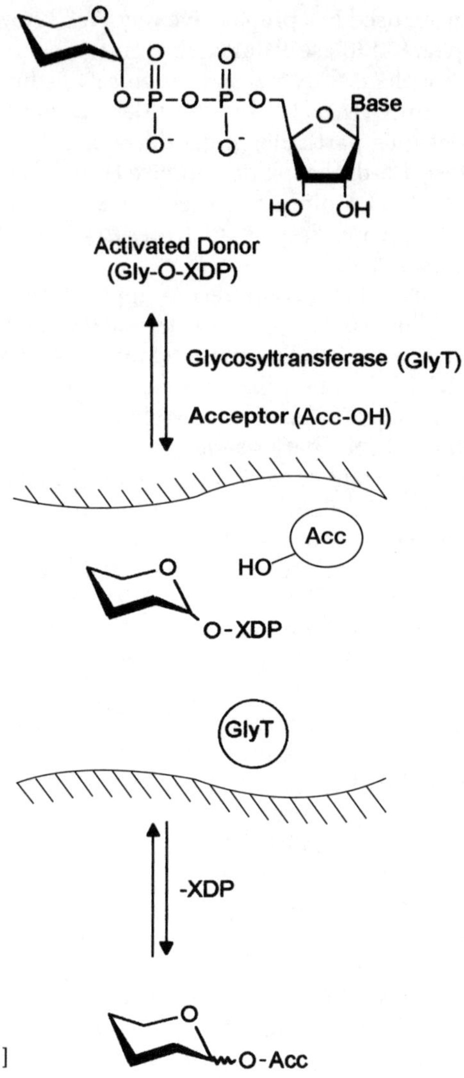

Fig. 1. Glycosylation by glycosyltransferases [21]

In the past, the use of glycosyltransferases was limited because of their lack of availability and lability, whereas nowadays modern techniques such as genetic engineering and recombinant techniques can overcome these problems.

Transferases require the sugar to be transferred as a nucleotide-activated compound, which can be prepared either chemically or enzymatically in preliminary reactions.

The chemical synthesis of a sugar nucleotide diphosphate makes use of an activated nucleotide monophosphate [23] and a glycosyl monophosphate. The most efficient nucleoside monophosphates are phosphorimidazolidates

[24] and phosphormorpholidates [23]. Glycosyl phosphates are obtained by reacting activated glycosyl donors [10] with phosphates or by chemical phosphorylation of anomeric hydroxyl groups [25]. The enzymatic synthesis of α-glucose-1-phosphate is performed with glycogen phosphorylase [26] or sucrose phosphorylase [27]. Phosphomutase transforms glucose-6-phosphate into glucose-1-phosphate [28].

A typical glycosyl transfer reaction is shown in Fig. 1. The sugars employed are activated by nucleoside diphosphates (XDP) (with the exception of CMP-neuraminic acid, which is a monophosphate nucleoside) to give potent donors (Gly-O-XDP). An activated intermediate is formed between the donor and the enzyme (GlyT). Little is known about the mechanism and the transition state involved in the reaction, but it is assumed to involve an oxocarbenium ion intermediate which is internally stabilized by the anomeric leaving group, the nucleoside diphosphate [29]. Further, the acceptor (Acc-OH) in turn is recognized by the enzyme-donor complex and the glycosyl donor is transferred onto a specific hydroxyl function of the acceptor molecule, often a sugar derivative, to give the interglycosidically linked entities. With regard to the donor, this process is stereospecific, giving the anomeric linkage inverted or not inverted with respect to the sugar nucleoside diphosphate. In addition, it is regiospecific with regard to the acceptor.

2.1 Galactosyl Transfer

2.1.1 *Galactosylation with Galactosyltransferase and Integrated Cofactor Regeneration*

Because of its availability, β1-4-galactosyltransferase (GalT) is one of the most extensively studied enzymes with regard to substrate specificity and preparative synthesis. Initial studies showed the regioselective transfer of uridin-diphosphate galactose (UDP-Gal) to the 4-position of GlcNAc to give *N*-acetyl lactosamine (Galβ1-4-GlcNAc) structures [30]. Because of their marked lability and the feed-back inhibition of released uridine-diphosphate (UDP), the respective galactosyl donors themselves are formed in situ [31]. Unprotected glucose is phosphorylated to give glucose-6-phosphate (Glc-6-P). The reaction is catalyzed by hexokinase, which transfers the phosphate from adenosine triphosphate (ATP) and releases adenosine diphosphate (ADP), which is subsequently recycled by pyruvate kinase. In the subsequent step, Glc-6-P is converted into glucose-1-phosphate (Glc-1-P) by phosphoglucomutase. Uridine triphosphate (UTP) and Glc-1-P are transformed by UDP-glucose-pyrophosphorylase to pyrophosphate (PP_i) and the activated glucose (UDP-Glc). In order to drive the reaction to the product side, the pyrophosphate must be transformed into inorganic phosphate (P_i), which is performed with inorganic pyrophosphatase. The transformation of UDP-Glc into UDP-Gal is achieved with galactose-4-epimerase. This activated donor serves as a substrate for galactosyltransferase (GalT), which galactosylates the acceptor regioselectively at the 4-position and stereoselectively with regard to UDP-Gal to afford Galβ1-4-linked products. The released

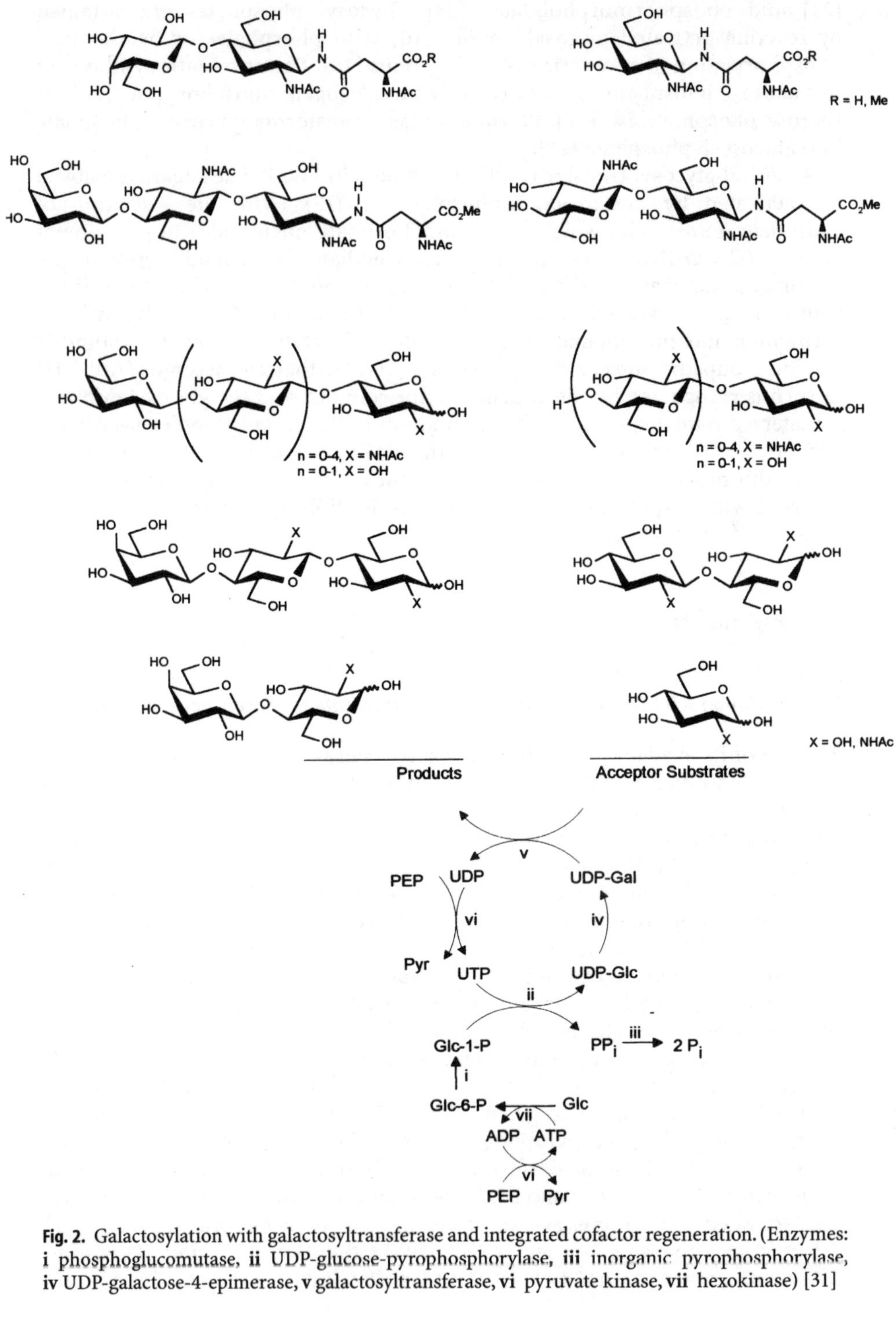

Fig. 2. Galactosylation with galactosyltransferase and integrated cofactor regeneration. (Enzymes: **i** phosphoglucomutase, **ii** UDP-glucose-pyrophosphorylase, **iii** inorganic pyrophosphorylase, **iv** UDP-galactose-4-epimerase, **v** galactosyltransferase, **vi** pyruvate kinase, **vii** hexokinase) [31]

UDP is recycled by pyruvate kinase to give UTP, employing the activated phosphate donor phosphoenol pyruvate (PEP) which gives pyruvate (Pyr) (Fig. 2).

By employing this integrated cofactor-regenerated approach, *N*-acetyl lactosamine can be formed in multigram amounts with native [30] or immobilized enzymes on silica gel [32] or on a polyamide-based carrier [33].

An extension with regard to acceptor modifications such as 4-substituted GlcNAc [34], chitobi-, tri-, tetra- and penta-ose (GlcNAcβ1-4(GlcNAc)$_{n=1\text{-}4}$) [35], *N*-acetylglucosamine-asparagine (GlcNAcβ1-NAsn), chitobiose-asparagine (GlcNAcβ1-4GlcNAcβ1-NAsn) and chitobiose-asparagine methyl ester (GlcNAcβ1-4GlcNAcβ1-NAsnOCH$_3$) [36] open further possibilities for enzymatic approaches. All glycosides that are reported to be acceptor substrates for GalT have β-glycosidic linkages. Both α- and β-glucosides are accepted.

In the presence of α-lactalbumin, however, only glucose and cellobiose (Glcβ1-4Glc) [37] are the preferred acceptors. Neither D-mannose, D-allose, D-galactose, D-ribose, nor D-xylose are substrates. Glucuronic acid and α-glucose-1-phosphate (monosaccharides with a negative charge) are also not accepted.

2.1.2
Donor-Substrate Specificity of Galactosyltransferases

Donor modifications are of great interest in glycosyltransferase-catalyzed reactions. Information about the donor substrate binding site can lead to better understanding of the topography of the active center of the enzyme and of the mechanism involved in the reaction. For the preparative chemist,

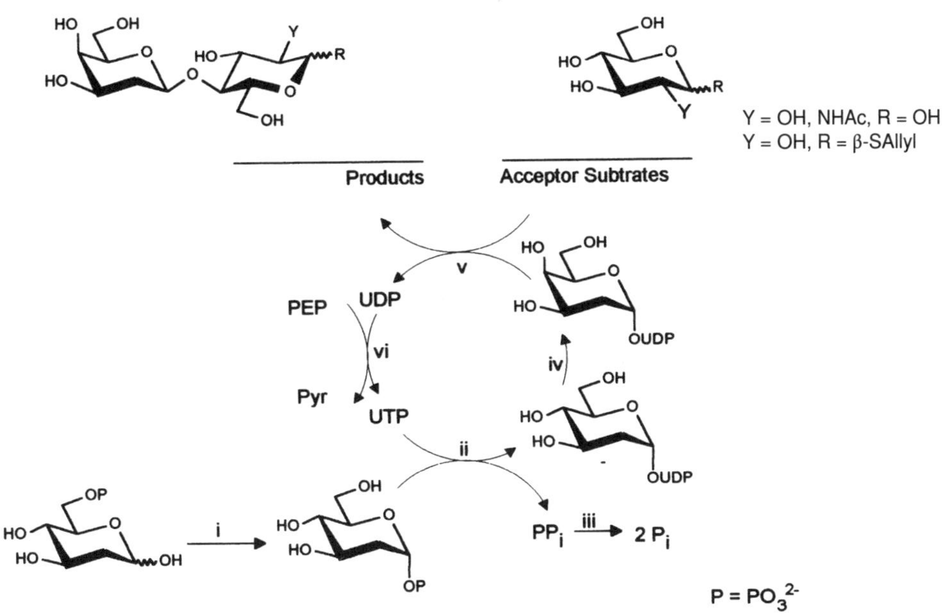

Fig. 3. Galactosylation with UDP-2d-Gal as donor (Enzymes: **i–vi** cf. Fig. 2) [38]

it is of interest to enhance the number of novel compounds by applying new donors.

The enzyme recognizes modified donors such as uridine diphosphate 2-deoxygalactose (UDP-2-d-Gal). The starting material 2-deoxyglucose-6-phosphate, is formed with hexokinase from 2-d-Glc and ATP, which in turn can be regenerated from PEP with pyruvate kinase, and this is transformed into 2-deoxyglucose-1-phosphate.

Applying a similar recycling system, UDP-2d-Gal is obtained as the activated sugar via UDP-2d-Glc, which is formed from UTP and 2-deoxyglucose-1-phosphate; UTP is regenerated by the PEP-pyruvate kinase system from UDP. The following galactosylation of *N*-acetyl glucosamine (GlcNAc), Glucose (Glc) or GlcβSAll alone or in the presence of α-lactalbumin leads to β1-4 linked products on a preparative scale [38] (Fig. 3).

New sources of biocatalysts such as bovine α1-3-galactosyltransferase produced by the baculovirus expression system are available and extend the vast field of enzymatic galactosylation. The enzyme transfers galactose with an α-linkage to the 3-position of the Gal residue in Fucα1-2Galβ1-OR [39].

2.1.3
Frame-Shifted Galactosylation with Galactosyltransferases

A surprising outcome is observed when unnatural amino sugars are galactosylated. The transfer reaction is performed in two steps: first the epimerization of UDP-Glc gives UDP-Gal catalyzed by UDP-galactose-4-epimerase, second GalT/lactalbumin catalyzes the galactosylation of *N*-acetyl kanosamine (Glc3NAc) [40], *N*-acetyl gentosamine (Xyl3NAc) [41] and *N*-acetyl-5-thio gentosamine (5SXyl3NAc) [42]. Apparently, this frame-shifted galactosylation leads to β,β-trehalose-type disaccharides. In an extension to these results, it is found that xylose is recognized by the enzyme in both its normal and its reverse orientation. Therefore, galactosylation of xylose results in a mixture of β1-4- and β1-β1-galactopyranosyl xylopyranosides in a 2:1 ratio [43] (Fig. 4).

These results may be rationalized by comparing the shapes of β-D-xylose in its normal and reverse orientation (Fig. 5). Except for the position of the ring oxygen and the isosteric 5-methylene group, both orientations are essentially similar. Concerning the enzyme, merely an exchange of the ring oxygen and the methylene group has occurred, but the exact ratio of the products might be a consequence of several factors. First, the oxygen/methylene switch might cause different binding constants of the enzyme-substrate and enzyme-product complexes. Second, since galacto-configurated sugars are not recognized as acceptor substrates, reverse orientated D-xylose will be fixed at the acceptor binding site only with the anomeric hydroxy group in equatorial orientation, i.e. in its β-form. On the other hand, galactosyltransferase accepts both the α and β form of the normally orientated sugar, provided that α-lactalbumin is present and the aglycon is small. Therefore, the apparent concentration of xylose for the formation of the β1-β1-disaccharide is probably much lower than for β1-4-galactosylation. Third, the nucleophilic reactivity of the anomeric hydroxyl group is assumed to be more pronounced than that of the 4-OH group.

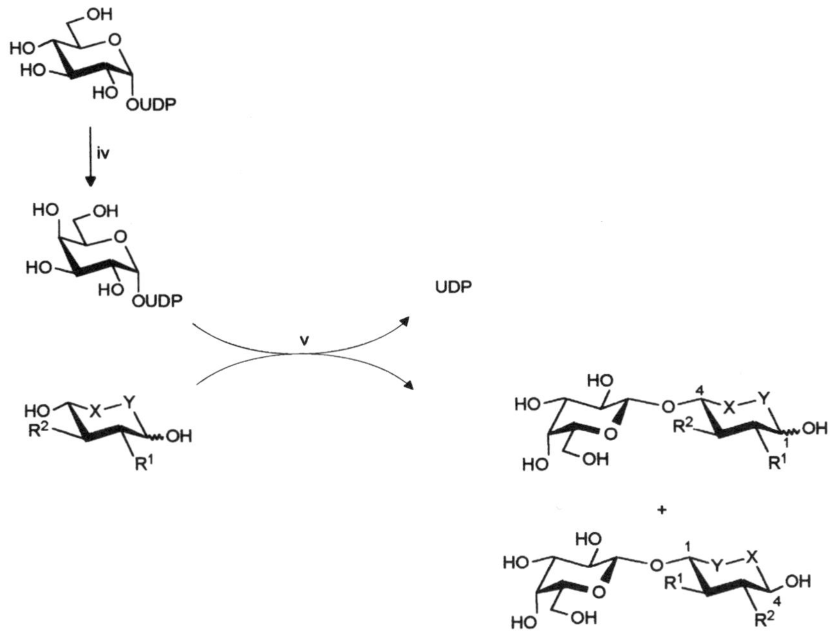

Fig. 4. Galactosylation leading to β1-4- and β1-β1-transfer products (Enzymes: **iv, v** cf. Fig. 2; R^1, R^2, X, Y cf. Table 1) [43]

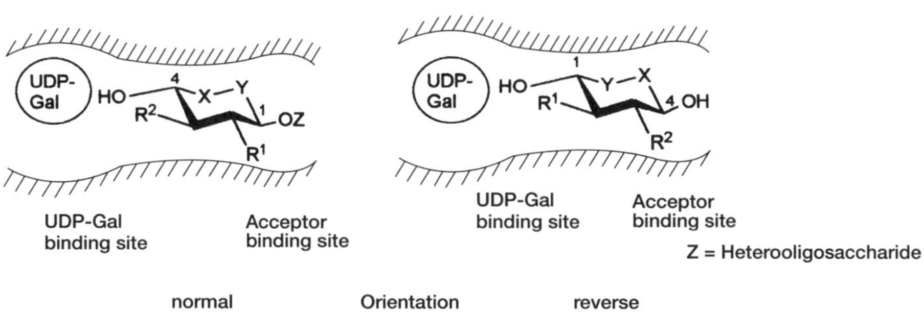

Fig. 5. Postulated structural map of the substrate binding site of galactosyltransferase (R^1, R^2 cf. Table 1) [43]

The results concerning the substituent effects on acceptor ability are summarized in Table 1. Because of its high symmetry, xylose is recognized in both the normal and the reversed orientation and therefore represents the first ambident acceptor substrate for galactosyltransferase. If the symmetry is reduced by an equatorial $CH(CH_2OH)$ group in the X or Y position, Glc is recognized by the enzyme in its normal orientation (X=$CH(CH_2OH)$, Y=O), an equatorial $CH(CH_2OH)$ group is not tolerated, and only the β1-4-linked galactosylation product is formed. But there is an exeption when R^1=NHAc, since *N*-acetyl

Table 1. Effect of substituent on acceptor ability [43]

Acceptors	R^1	R^2	X	Y	Substrate	Orien-tation	Product
Xyl	OH	OH	CH_2	O	+	n	β1-4
	OH	OH	O	CH_2	+	r	β1-β1
Glc	OH	OH	e-$CH(CH_2OH)$	O	+	n	β1-4
	OH	OH	O	CHe-(CH_2OH)	–	r	
Glc3NAc	OH	NHAc	e-$CH(CH_2OH)$	O	–	n	
	NHAc	OH	O	CHe-(CH_2OH)	+	r	β1-β1
Xyl3NAc	OH	NHAc	CH_2	O	–	n	
	NHAc	OH	O	CH_2	+	r	β1-β2
GlcNAc	NHAc	OH	e-$CH(CH_2OH)$	O	+	n	β1-4
	OH	NHAc	O	CHe-(CH_2OH)	–	r	
5SXyl3NAc	OH	NHAc	CH_2	S	–	n	
	OH	NHAc	CH_2	S	+	r	β1-β4

kanosamine (Glc3NAc) gives only the β1-β1-product (R^1=NHAc, X=O, Y=$CH(CH_2OH)$). This deviant behavior may be rationalized by assuming that binding of the NHAc group in R^1 causes a change of the three-dimensional structure of the enzyme at the Y-binding site and/or a large negative contribution to the total binding energy.

If the symmetry of xylose is reduced by an NHAc group in position 3, the corresponding *N*-acetyl gentosamine (Xyl3NAc) is recognized by the enzyme only in its reversed orientation (R^1=NHAc, X=O, Y=CH_2). An equatorial NHAc group at R^2 is not tolerated, whereas other space-requiring groups in this position are, e.g., *N*-acetyl muramic acid (R^2=CH_3–$CH(CO_2H)O$–). The enzymatic galactosylation of xylose described differs from other GalT-reactions, which gives new insights into the binding site of GalT and may provide a valuable enzymatic access to some non-reducing oligosaccharides on a preparative scale.

2.2 Sialyl Transfer

2.2.1 *Enzymatic Synthesis of CMP-Neu5Ac*

It has so far only been possible to solve the complex problem associated with chemical sialylation in a few examples in a convincing way [44]. Due to the lack of a neighboring group at C-3 there is little possibility to influence the stereochemistry at the anomeric center except by solvents. Further, the marked reverse anomeric effect of the carboxyl group leads to the electronically preferred β-configuration. Thus, this area would be ideally suited for the application of stereo- and regioselective biocatalysts.

In contrast to the diphosphosugars mentioned, sialic acids are found to be anomerically linked via a single phosphate unit to the nucleoside cytidine. The activated donor (CMP-Neu5Ac) has been synthesized enzymatically on an analytical scale from CTP and Neu5Ac by catalysis of CMP-Neu5Ac-synthase [45, 46]. An improved method for the preparative scale operates by the enzymatic formation of the costly CTP from cheap CMP using adenylate kinase and pyruvate kinase [47]. Adenylate kinase transfers one phosphate group from ATP to CMP leading to CDP, which is phosphorylated by pyruvate kinase to give CTP. In a preceding reaction, Neu5Ac is made from *N*-acetyl mannosamine and pyruvate with the corresponding aldolase. These two crude reaction solutions are mixed, and CMP-Neu5Ac is made with immobilized CMP-Neu5Ac-synthase. This enzyme is isolated from calf brain by ammonium sulfate precipitation and subsequently purified by affinity chromatography. The stationary phase is improved and consists of CNBr-activated Sepharose 4B reacted with β-[3-(2-aminoethylthio)propyl]-*N*-acetyl neuraminic acid. This in turn is synthesized by irradiating a mixture of the allyl glycoside and cysteamine to achieve radical C–S bond formation [48].

A purely chemical synthesis of CMP-Neu5Ac has also been reported by Schmidt et al. [49].

2.2.2
Sialylation with Modified Donor and Acceptor Substrates

In transferase-catalyzed reactions, the activated neuraminic acid serves as precursor for the introduction of the terminal Neu5Ac residue into various oligosaccharides. This approach is particularly attractive because the chemical synthesis with Neu5Ac glycosyl donors has two drawbacks: the low yielding glycosylation reaction and the formation of anomeric mixtures.

CMP-neuraminic acid can be used in oligosaccharide synthesis as a substrate for various α2-6-sialyltransferases isolated from rat liver [50] or bovine colostrum [51] as well as α2-3-sialyltransferases isolated from porcine liver [52] and porcine submaxillary glands [53]. Recently, the enzyme has been overexpressed in a baculovirus system [54].

Following the enzymatic synthesis of CMP-Neu5Ac, isolated α2-6-sialyltransferases could be used to introduce sialic acid to the *N*-acetyl lactosamine derivates **a**–**f** in position 6. This afforded a series of trisaccharide derivates Neu5Acα2-6 Galβ1-4GlcNAcβ1-R. The unblocked trisaccharide **a** [55], the methyl glycoside **b** [56], the β-Asn derivative **c** [57], the pentapeptide derivative **d** [58], the allyl **e** [59], and the pent-4-enyl glycoside **f** [48] can be prepared in convincing yields (Fig. 6).

Some sialyltransferases recognize modified CMP-Neu5Ac such as those in which the hydroxyl group at C-9 is replaced with an amido, fluoro, azido, acetamido, or benzamido group [60–63]. Acceptor analogs in which the acetamido function is replaced by an azide, phthalimide, carbamate, or pivaloyl functionality are also accepted by the enzyme [64].

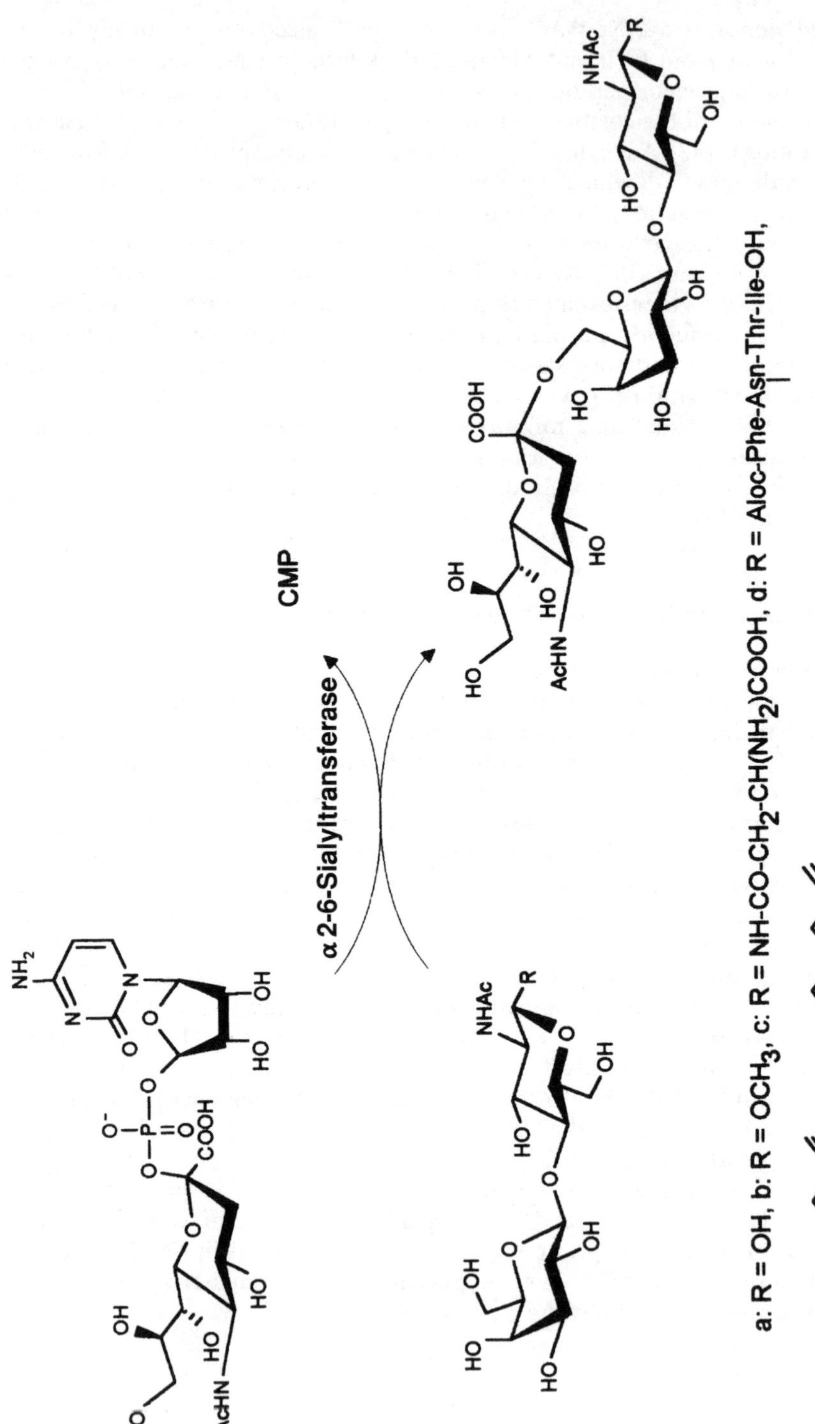

Fig. 6. Sialylation of lactosamine derivates with *α*2-6-sialyltransferase [48, 55–59]

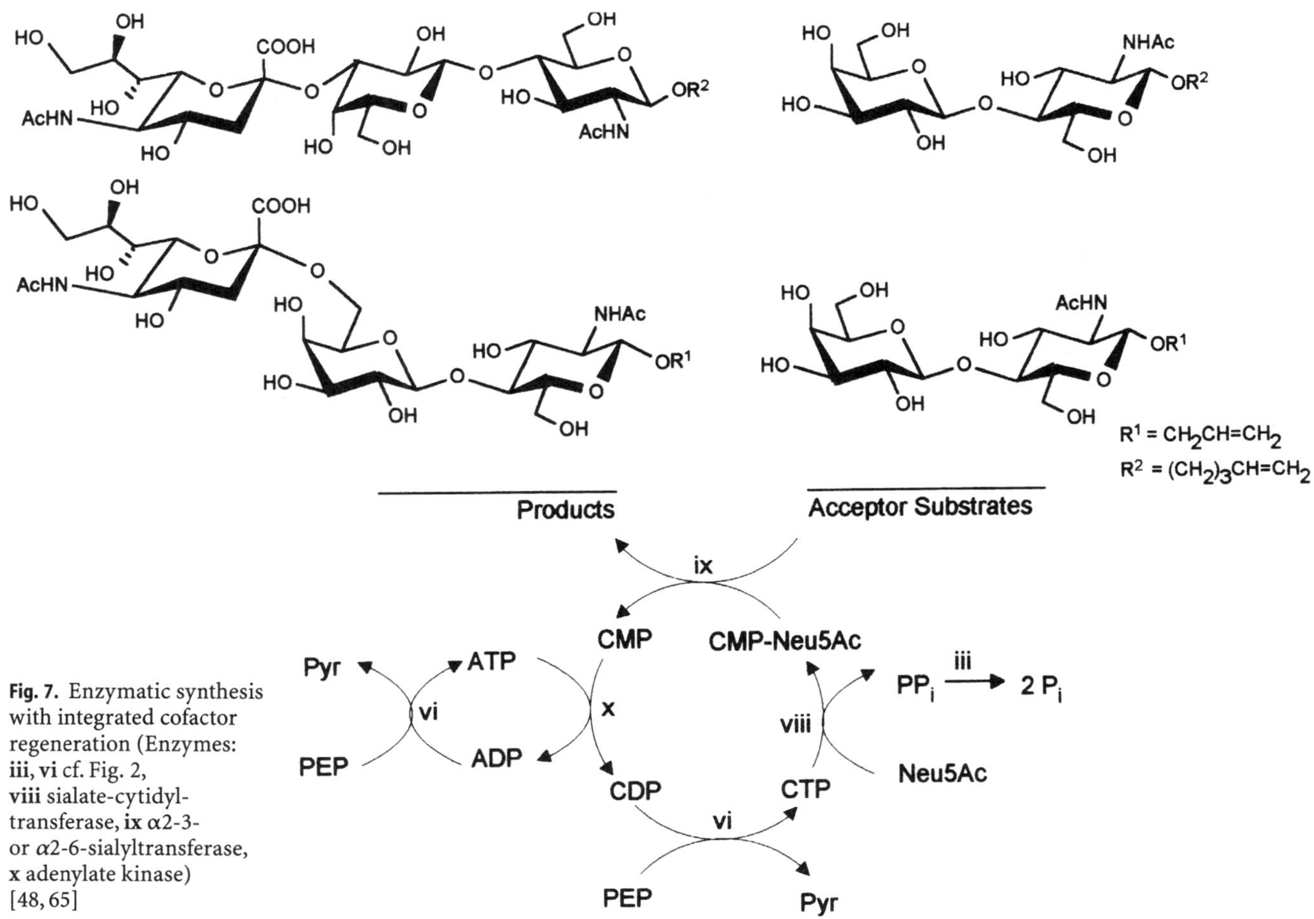

Fig. 7. Enzymatic synthesis with integrated cofactor regeneration (Enzymes: **iii**, **vi** cf. Fig. 2, **viii** sialate-cytidyltransferase, **ix** α2-3- or α2-6-sialyltransferase, **x** adenylate kinase) [48, 65]

2.2.3
Sialylation with Integrated Cofactor Regeneration

The sialyltransferase step has recently been incorporated into the above discussed recycling pathway, leading to the formation of activated neuraminic acid (CMP-Neu5Ac), which is transferred in a one-pot reaction to give the sialylated compound. The derivative **e** is synthesized by this route [59]. Parallel studies employing the allyl *N*-acetyl lactosaminide with CMP-Neu5Ac and the rare α2-3-sialyltransferase lead to the α2-3-linkage between the terminal Neu5Ac and the preterminal Gal residue [65], and sialyl Lewis x was thus obtained in acceptable yield. It was possible to perform the corresponding synthesis of the α2-6-linked trisaccharide pent-4-enyl glycoside from the corresponding *N*-acetyl lactosaminide by applying the integrated cofactor regeneration with α2-6-sialyltransferase [48] (Fig. 7).

Coupled with the above-described integrated synthesis of Neu5Ac from *N*-acetyl mannosamine and PEP using Neu5Ac-aldolase, it has been possible to prepare terminally sialylated oligosaccharides. Altogether, nine enzymes are applied with the starting monosaccharides GlcNAc, ManNAc, Glc-1-P and with phosphoenol pyruvate (PEP) to build up the complex trisaccharide [66].

2.3
Fucosyl Transfer

2.3.1
Fucosylation with Integrated Cofactor Regeneration

For a long time, the utility of fucosyl transfer reactions suffered from short cuts in the chemical synthesis of GDP-L-fucose as well as from the feedback inhibition of released GDP. Another drawback is the tedious work-up procedure for α1-3/4-fucosyltransferase from human milk. The chemical synthesis of GDP-fucose and its analogs requires the stereospecific hydrolysis of acetobromofucose to form the β-anomer of the reducing sugar. However, isomerization leads to the more stable α-anomer. Functionalization over three steps then provides the β-phosphate which can be converted to GDP-fucose [67]. Since it is not certain that the required analogs can be isolated, a simpler and potentially more general synthesis of β-pyranosyl phosphates is developed using Ag_2CO_3 catalyzed glycosylation with dibenzylphosphate [68]. Two procedures are examined for conversion of the phosphates to the required GDP derivates. Both the coupling of these phosphates using GMP-morpholidate [69] or the phosphinothioic anhydride [70] give good yields.

To overcome the synthetic and inhibition problems, an integrated recycling system is applied. GTP and Fuc-1-P are transformed by GDP-fucosyl-pyrophosphorylase to pyrophosphate and the activated donor (GDP-Fuc) [71]. This in turn in the presence of α1-3/4-fucosyltransferase fucosylates the acceptor Galβ1-4GlcNAc at the GlcNAc-3-position or the acceptor Galβ1-3GlcNAc at the GlcNAc-4-position. The released GDP is phosphorylated to GTP with pyruvate kinase employing the activated phosphate donor phosphenol pyruvate, which in

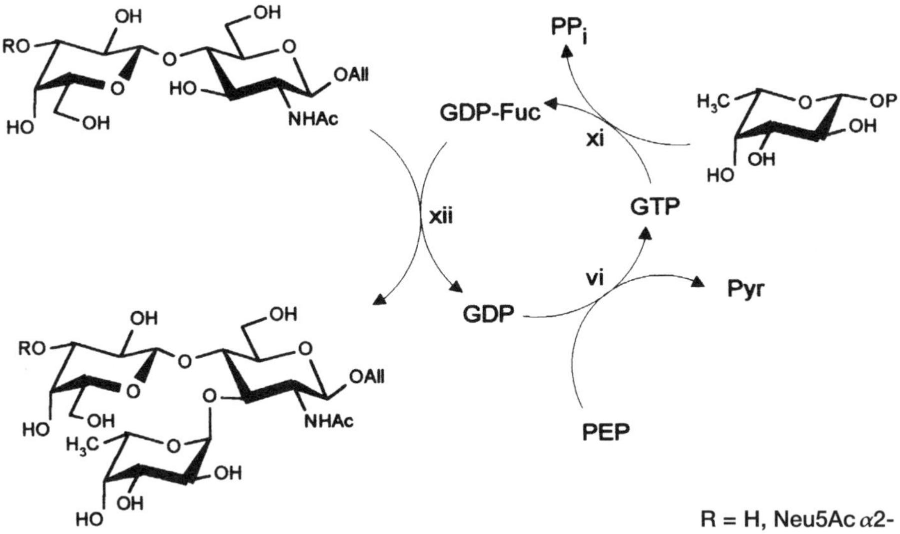

Fig. 8. Enzymatic fucosylation with integrated cofactor regeneration (Enzymes: **vi** cf. Fig. 2, **xi** GDP-fucose-pyrophosphorylase, **xii** α1-3/4-fucosyltransferase) [72]

turn gives pyruvate. The integrated cofactor regeneration has been successfully applied in the synthesis of Lewis x or Lewis a structures and the corresponding sialylated analogs [72] (Fig. 8).

An associated enzyme, α1-3-fucosyltransferase, has also been used in syntheses to provide the Lewis x and sialyl Lewis x structural motifs [65]. Increasing availability of fucosyltransferases via cloning techniques are expected to have a strong influence on oligosaccharide synthesis in the near future.

2.3.2
Fucosylation Employing Modified Donor and Acceptor Substrates

Acceptors with modification of the terminal sugar residue, such as Galβ1-4Glc, Galβ1-4Glucal and Galβ1-4(5-SGlc), open further possibilities for fucosylation [73].

Recent studies showed that fucosyltransferase reactions can also be extended to modified donors. For example, 3-deoxyfucose and L-arabinose are transferred to the Lewis c acceptor catalyzed by α1-4fucosyltransferase [67]. In addition, GDP-L-galactose, GDP-3-deoxy-L-galactose and GDP-3,6-dideoxy-L-galactose can be used to fucosylate Lewis derivates by catalysis of α1-3/4-fucosyltransferase [74] (Fig. 9). It is possible to transfer GDP-L-galactose either to the 3-position of Galβ1-4GlcNAc or to the 4-position of Galβ1-3GlcNAc, whereas GDP-3-deoxy-L-galactose and GDP-3,6-dideoxy-L-galactose are only transferred to the 4-position of Galβ1-3GlcNAc. There is no plausible explanation for this specificity at present.

Similarly, GDP-L-galactose derivatives functionalized at O-6 were prepared and studied in transfer reactions [75].

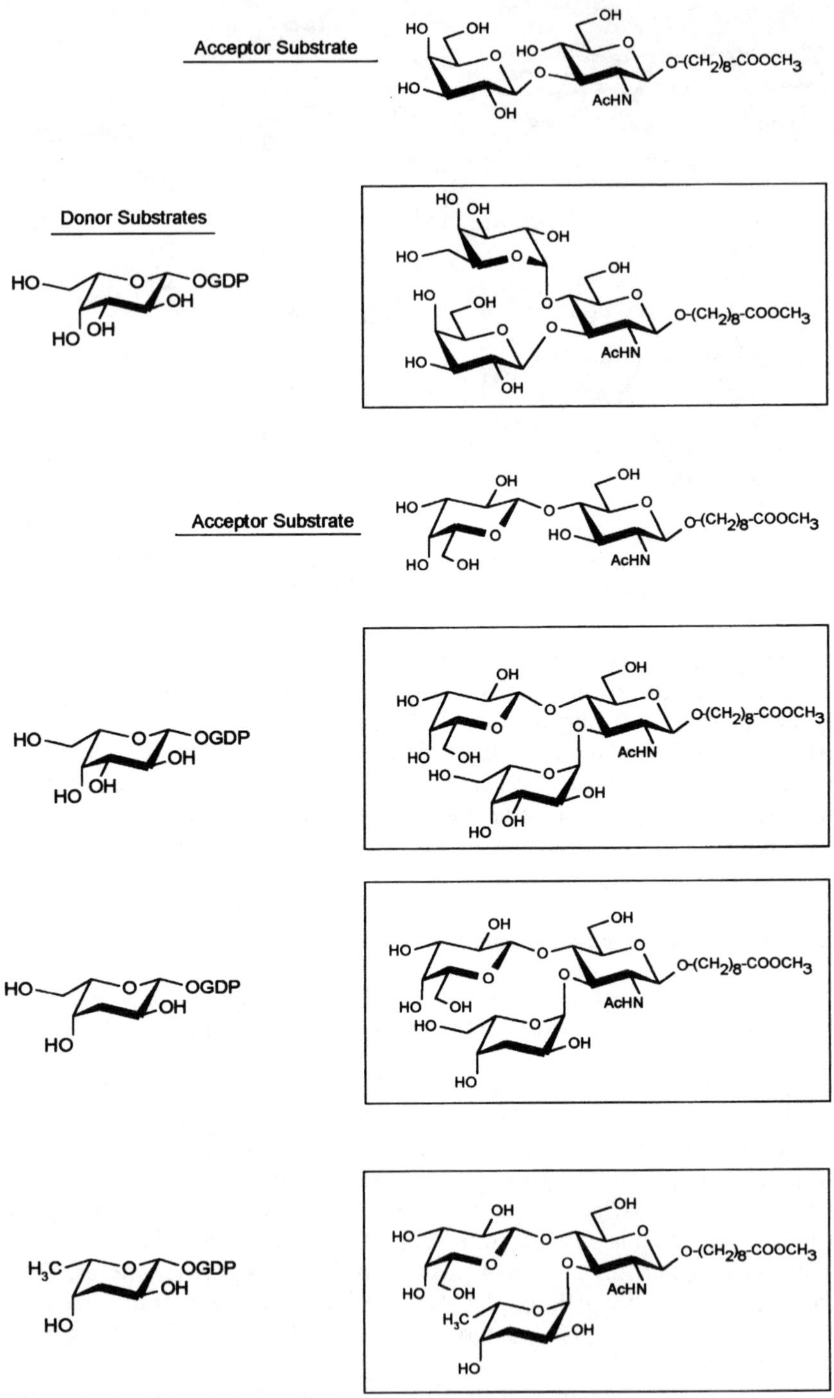

Fig. 9. Fucosylation employing modified donors and acceptors [74]

2.4 *N*-Acetyl Glucosaminyltransfer

N-Acetyl glucosaminyltransferases transfer *N*-acetyl glucosamine to various oligosaccharide acceptors, e.g the pentasaccharide

GlcNAc β1-2 Gal α1-6 \
GlcNAc β1-2 Man α1-3 / Man β1-OR

is synthesized using a mixture of *N*-acetyl glucosaminyltransferases (GnT) I and II [76]. UDP-GlcNAc serves as a GlcNAc donor for transfer by GnT-I from bovine colostrum to the α1–3-linked mannose in

Gal α1-6 \
Man α1-3 / Man β1-OR

whereas the synthesized tetrasaccharide is a substrate for the GnT-II. β1-2-Transfer occurs onto the α1-6 branch of the oligosaccharide to give the desired pentasaccharide.

Core-II GlcNAc transferase, which is involved in the biosynthesis of O-linked glycoprotein glycans catalyzes the β1-6-glycosylation of the T-antigen derivative Galβ1-3GalNAcα1-OR. The trisaccharide is then galactosylated by galactosyltransferase to yield the tetrasaccharide

Gal β1-4 GlcNAc α1-6 \
Galβ1-3 / GalNAc α1-OR

Further sialylation give a sialyl Lewis x-containing hexasaccharide [77].

2.5 Combined Synthesis Employing Several Enzymes

The use of mixed catalytic systems with several enzymes can provide multiple benefits in terms of costs, effectiveness of the particular production process, stability of the bio-catalysts, and possible structures.

A novel one-pot multienzyme system with cofactor regeneration for the synthesis of sialylated heterooligosaccharide Neu5Acα2-3Galβ1-3GalNAc is demonstrated [78]. A combined sequential use of glycosidase together with glycosyltransferase including a cofactor regeneration transforms complicated multistep reactions into a one-pot reaction, thus avoiding laborious purifications of intermediates. First step is the galactosylation of GalNAc with β-galactosidase from bovine testes, leading to a β1-3-linked disaccharide. The reverse hydrolysis of this intermediate is blocked by its α1-3-sialyltransferase-mediated conversion into the trisaccharide, which is no longer a substrate for the glycosidase. This part of the reaction sequence was coupled to a modified multienzymatic system regenerating CMP-Neu5Ac.

A corresponding approach has been employed for a chemoenzymatic synthesis of Neu5Acα2-6Galβ1-4GlcNAc using β-galactosidase from *Bacillus circulans* and α2-6-sialyltransferase from rat liver [79]. In this reaction, less complicated problems had to be solved than in the above case in which the pH optima of all enzymes were in a wider range.

In summary, the concept of multienzyme reactions with integrated cofactor regeneration has been shown to be useful for sequential synthesis of rather complicated heterooligosaccharides. This conception opens up new perspectives for the synthesis of glycosides having up to three or four different glycosyl units in one-pot reactions.

If the very demanding synthesis of the undecasaccharide-asparagine:

Neu5Ac α2-6 Gal β1-4-GlcNAc β1-2 Man α1-6 \
                                                              > Man β1-4 R \
Neu5Ac α2-6 Gal β1-4-GlcNAc β1-2 Man α1-6 /

R = GlcNAc β1-4-GlcNAc β1-NAsn

was to be performed using only classical synthetic chemistry, a large number of synthetic steps would be required, with the expectation of little product. An efficient alternative turned out to be the combination of chemical and enzymatic glycosylation to reduce the number of steps [80]. First the heptasaccharide

GlcNAc β1-2 Man α1-6 \
                              > Man β1-4 GlcNAc β1-4 GlcNAc β1-NAsn \
GlcNAc β1-2 Man α1-3 /

is synthesized chemically with a protecting group strategy designed for the subsequent chemoenzymatic elongation to form the undecasaccharide-asparagine. The terminal *N*-acetyl glucosamine residues serve as acceptors for glycosyltransferases. In an enzymatic one-pot synthesis, the GlcNAcβ1-OR heptasaccharide is prolonged, forming the trisaccharide structural element sialyl-*N*-acetyl lactosamine (Neu5Acα2-6Galβ1-4-GlcNAcβ1-OR) in high yield.

3 Enzymatic Reactions with Non-Leloir Glycosyltransferases

3.1 Cyclodextrin Glycosyltransferases

Some transferases can directly use and transform oligosaccharides as donors without further activation. The following paragraph will discuss some reactions with non-Leloir glycosyltransferases which are performed on a preparative level. For example, cyclodextrin-α1-4-glucosyltransferase (CGT) from *Klebsiella pneumonia* or from *Bacillus macerans* are rather versatile enzymes which can catalyze various reactions.

Amylose may be cyclized to give cyclodextrins, acceptors may be coupled onto cyclodextrins to give acceptor-attached open chain maltooligomer derivatives, or disproportionation reactions between amylose chains of different dp may occur [81]. The normal enzymatic synthesis of cyclodextrins from amylose with CGT gives about 17% α-, 19% β-, and 6% γ-cyclodextrin. It could be demonstrated with immobilized CGT that α-D-glucopyranosyl fluoride was accepted as an unnatural substrate to give only α- (30%), β- (38%), and no γ-cyclodextrin, and in addition maltooligomers (maltose to maltooctaose, 32%) [82].

Further, various glycosyl fluorides were tested as substrates for CGT [83]. This approach gave novel insights into the specificity of the catalytic site of CGT from *Bacillus macerans* and new routes for the preparation of regioselectively modified cyclodextrins.

3.2 Glycosyl Phosphorylases

Phosphorylase transfers glycosyl residues to inorganic phosphate. The reverse reaction, the elongation of a primer with glycosyl phosphate, is also catalyzed, but in vivo the degradation reaction prevails. As well as the natural substrate, D-glucal can be employed with phosphorylase in the presence of inorganic phosphate [84]. Then, 2-deoxy-glycosyl residues are transferred to a primer, and novel 2-deoxy-maltooligosaccharides can be synthesized [85] (Fig. 10). The minimum chain length of the primer for elongation is four, and the maltotetraose is not further degraded by phosphorolysis. The course of the reaction is mainly controlled by the phosphate concentration, and the chain length distribution is dependent on the incubation time. 2-Deoxy-maltooligosaccharides with an average chain length of 20 are obtained from glucal and maltotetraose with 0.05 equivalents of phosphate. When the synthesis is stopped after a few hours, a fraction with an average chain length of 12 can be isolated. The penta-, hexa-, and heptasaccharide, that is the maltotetraose elongated by one, two or by three 2-deoxy-glucosyl residues, have been isolated.

In the presence of equimolar amounts of phosphate in relation to glucal, the phosphorolysis prevails and 2-deoxy-α-D-glucosyl phosphate can be produced in a one-pot reaction which proceeds in two steps. First a 2-deoxy-glucosyl unit is transferred to the primer, second the 2-deoxy-glucosyl residue is phosphorolysed (Fig. 10, 11) [86].

In addition, 6-deoxy-D-glucal and D-xylal were tested as substrates for phosphorylase which would lead to 2,6-dideoxy-maltooligosaccharides and to 2-deoxy-xylooligosaccharides or to the corresponding glycosyl phosphates. However, no reaction could be observed with any of the modified glycals, which implies that the 6-OH group is essential for the substrate binding to the active site of phosphorylase [85].

The 2-deoxy-oligomer with the chain length of 6 was used for further glycosylation reactions with the aim of obtaining oligosaccharides carrying natural glucose at the end. Incubation of the hexamer with glucose-1-phosphate leads to the glucosylated heptamer [87] (Fig. 11).

Furthermore, α-D-mannose-1-phosphate represents a substrate for the enzyme, and the α-mannosyl-maltotetraose is obtained. Supplementary to this result, the α-phosphate of uronic acid was tested, but no enzymatic elongation

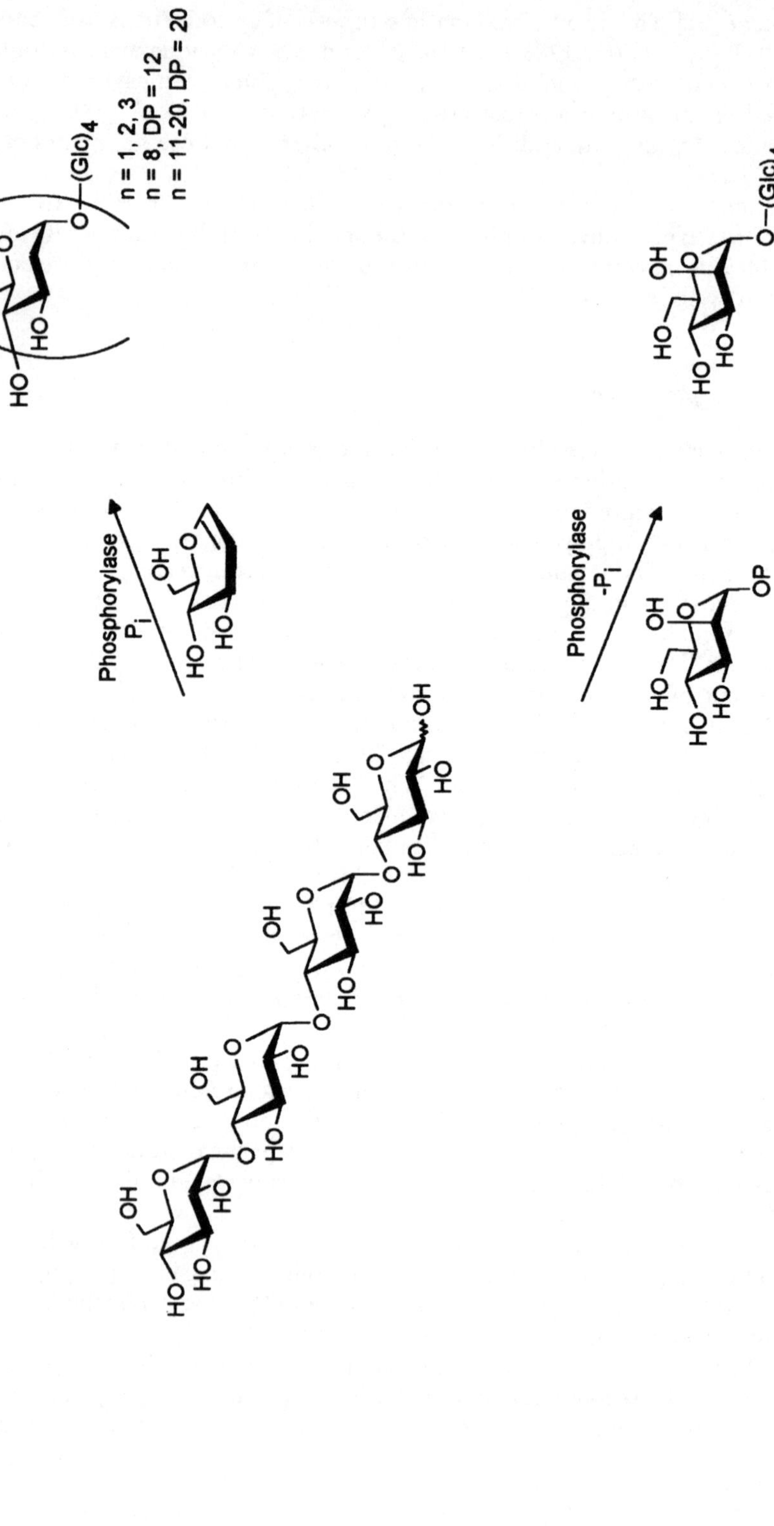

Fig. 10. Phosphorylase-catalyzed synthesis of modified maltooligosaccharides [85, 86]

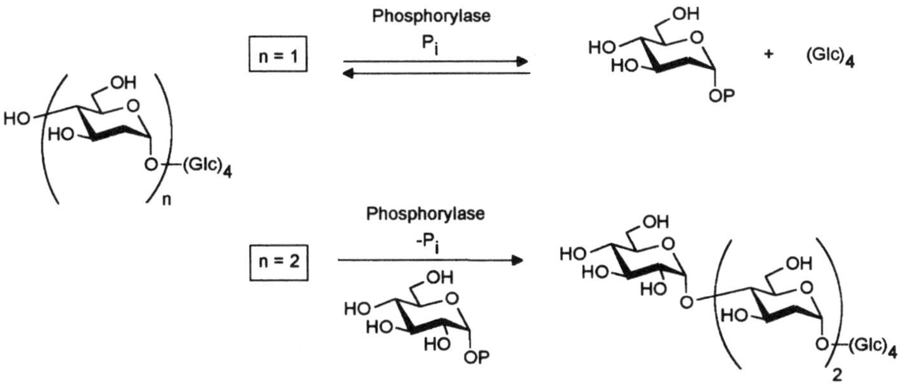

Fig. 11. Phosphorolytic formation and degradation of modified maltooligosaccharides [86, 87]

reaction was observed, which confirms the assumption that there is a high specificity of the enzyme's active site towards the 6-OH group in contrast to the more flexible 2-OH group [87].

4 Future Outlook

Chemoenzymatic techniques provide relatively fast routes to complex oligosaccharide structures. Nevertheless, they always combine biochemical and synthetic elements and therefore require the cooperation of both fields. To date, few glycosyltransferases have been cloned, expressed, and produced in quantities sufficient for preparative enzymatic syntheses. Given the advantages of enzymatic synthesis of oligosaccharides over traditional schemes, research into overexpression of glycosyltransferases will continue to flourish and will offer the benefit of preparative enzymology techniques for carbohydrate chemists in the future.

References

1. Edelmann GM (1983) Science 219:450
2. Feizi T (1985) Nature 314:53
3. Feizi T, Childs RA (1985) Trends Biochem Sci 10:24
4. Sharon N, Lis H (1981) Chem Eng News 58(13):21
5. Hakomori S-I (1986) Spektrum Wiss 7:90
6. Wiley DC, Wilson JA, Skeiiel JJ (1981) Nature 289:373
7. Paulsen H (1982) Angew Chem Int Ed Engl 21:155
8. Paulsen H (1984) Chem Soc Rev 13:15
9. Ogawa T, Yamamoto H, Nukada T (1984) Pure Appl Chem 56:779
10. Schmidt RR (1986) Angew Chem Int Ed Engl 25:212
11. Klibanov AM, Samokhin GP, Martinek K, Berezin IV (1977) Biotechnol Bioeng 19:1351
12. Erdmann H, Fritsche K, Kordel M, Lang S, Lokotsch W, Markweg M, Schneider M, Syldatk C, Wagner F (1992) Biotechnology Focus 3:333
13. Riva S, Chopineau J, Kieboom APG, Klibanov AM (1988) J Am Chem Soc 110:584

14. Kahn R (1984) Pure Appl Chem 56:833
15. Bischofberger N, Waldmann H, Saito T, Simon ES, Lees W, Bednarski MD, Whitesides GM (1988) J Org Chem 53:3457
16. Fessner W-D, Sinerius G, Schneider A, Drever M, Schulz GE, Badia J, Aguilar J (1991) Angew Chem Int Ed Engl 30:555
17. Schultz M, Waldmann H, Kunz H, Vogt W (1990) Liebigs Ann Chem 1019
18. Auge C, David S, Gautheron C (1984) Tetrahedron Lett 25:4663
19. Auge C, David S, Malleron A (1989) Carbohydr Res 188:201
20. Nilsson KGI (1988) Trends Biochem Sci 6:257
21. Toone EJ, Simon ES, Bednarski MD, Whitesides GM (1989) Tetrahedron 45:365
22. Leloir LF (1971) Science 172:1299
23. Clark VM, Hutchinson DW, Kirby AJ, Warren SG (1984) Angew Chem Int Ed Engl 76:704
24. Simon ES, Grabowski S, Whitesides GM, (1990) J Org Chem 55:1834
25. Gokhale UB, Hindsgaul O, Palcic MM (1990) Can J Chem 68:1063
26. Pfannemuller B, (1968) Staerke 11:341
27. Waldmann H, Gygax D, Bednarski MD, Shangraw WR, Whitesides GM, (1986) Carbohydr Res 157:C4
28. Wong CH, Haynie SL, Whitesides GM, (1982) J Org Chem 47:5416
29. Sinnot JM, (1990) Chem Rev 90:1171
30. Nunez HA, Barker R (1980) Biochemistry 19:489
31. Wong C-H, Haynie SL, Whitesides GM (1982) J Org Chem 47:5416
32. Thiem J, Treder W (1986) Angew Chem Int Ed Engl 25:1096
33. Pollak A, Blumenfeld H, Wax M, Baughn RL, Whitesides GM (1980) J Am Chem Soc 102:6324
34. Palcic MM, Srivastava OP, Hindsgaul O (1987) Carbohydr Res 159:315
35. Streicher H (1996) PhD thesis, University of Hamburg, Germany
36. Thiem J, Wiemann T (1990) Angew Chem Int Ed Engl 29:80
37. Thiem J, Treder W, Wiemann T (1988) DECHEMA Biotechnology Conferences, VCH, Weinheim 2:189
38. Thiem J, Wiemann T (1991) Angew Chem Int Ed Engl 30:1163
39. Lowary TL, Hindsgaul O (1993) Carbohydr Res 249:163
40. Nishida Y, Wiemann T, Sinnwell V, Thiem J (1993) J Am Chem Soc 115:2536
41. Nishida Y, Wiemann T, Thiem J (1992) Tetrahedron Lett 33:8043
42. Nishida Y, Wiemann T, Thiem J (1993) Tetrahedron Lett 34:2905
43. Wiemann T, Nishida Y, Sinnwell V, Thiem J (1994) J Org Chem 59:6744
44. Okamoto K, Goto T (1990) Tetrahedron 46:5835
45. Kean EL, Roseman S (1966) Meth Enzymol 8:208
46. Haverkamp J, Beau JM, Schauer R (1979) Hoppe Seyler's Z Physiol Chem 360:159
47. Simon ES, Bednarski MD, Whitesides GM (1988) J Am Chem Soc 110:7159
48. Stangier P, Treder W, Thiem J (1993) Glycoconjugate J 10:20
49. Martin TJ, Schmidt RR (1993) Tetrahedron Lett 34:1765
50. Weinstein J, de Souza-e-Silva U, Paulson JC (1982) J Biol Chem 257:13835
51. Paulson JC, Beranek WE, Hill RL (1977) J Biol Chem 252:2356
52. Lubineau A, Auge C, Francois P (1992) Carbohydr Res 228:137
53. Sadler JE, Rearick JA, Paulson JC, Hill RL (1979) J Biol Chem 254:4434
54. Gillespie W, Kelm S, Paulson JC (1992), J Biol Chem 267:21004
55. Thiem J, Treder W (1986) Angew Chem Int Ed Engl 25:1096
56. Sabesan S, Paulson JC (1986) J Am Chem Soc 108:2068
57. Auge C, Gautheron C, Pora H (1989) Carbohydr Res 193:288
58. Unverzagt C, Kunz H, Paulson JC (1990) J Am Chem Soc 112:9308
59. Ichikawa Y, Shen G-J, Wong C-H (1991) J Am Chem Soc 113:4698
60. Gross HJ, Brossmer R (1988) Eur J Biochem 177:583
61. Schauer R, Wember M, do Amaral CF (1972) Hoppe-Seyler's Z Physiol Chem 353:883
62. Conradt HS, Bunsch A, Brossmer R (1984) FEBS Lett 170:295
63. Petrie CR, Sharma M, Simmons OD, Korytnyk W (1989) Carbohydr Res 186:326

64. Ito Y, Gaudino JJ, Paulson JC (1993) Pure Appl Chem 65:753
65. Ichikawa Y, Lin Y-C, Dumas DP, Shen G-J, Garcia-Junceda E, Williams MA, Bayer R, Ketcham C, Walker LE, Paulson JC, Wong C-H (1992) J Am Chem Soc 114:9283
66. Ichikawa Y, Liu JL, Shen G-J, Wong C-H (1991) J Am Chem Soc 113:6300
67. Gokhale UB, Hindsgaul O, Palcic MM (1990) Can J Chem 68:1063
68. Lindhorst TK, Thiem J (1991) Carbohydr Res 209:119
69. Moffatt JG, Khorana HG (1961) J Am Chem Soc 83:649
70. Nunez HA, O'Connor JV, Rosevear PR, Barker R (1981) Can J Chem 59:2086
71. Stiller R, Thiem J (1992) Liebigs Ann Chem 467
72. Palcic MM, Venot AP, Ratcliff RM, Hindsgaul O (1989) Carbohydr Res 190:1
73. Wong C-H, Dumas DP, Ichikawa Y, Koseki K, Danishefsky SJ, Weston BW, Lowe JB (1992) J Am Chem Soc 114:7321
74. Stangier K (1994) PhD thesis, University of Hamburg, Germany
75. Hällgren C, Hindsgaul O (1995) J Carbohydr Chem 14:453
76. Kaur KJ, Alton G, Hindsgaul O (1991), Carbohydr Res 210:145
77. Öhrlein R, Hindsgaul O, Palcic MM (1993) Carbohydr Res 244:149
78. Kren V, Thiem J (1995) Angew Chem Int Ed Engl 34:893
79. Herrmann GF, Ichikawa Y, Wandrey C, Gaeta FCA, Paulson JC, Wong C-H (1993) Tetrahedron Lett 34:3091
80. Unverzagt C (1994) Angew Chem Int Ed Engl 33:1102
81. Bender ML, Komiyama M (1978) Cyclodextrin chemistry. Springer Verlag, Berlin Heidelberg New York
82. Treder W, Thiem J, Schlingmann M. (1986) Tetrahedron Lett. 27:5605
83. Cottaz S, Apparn C, Driguez H (1991) J Chem Soc Perkin Trans 1:2235
84. Klein HW, Palm D, Helmreich EJM (1982) Biochemistry 21:6675
85. Evers B, Thiem J (1995) Starch/Staerke 47:434
86. Evers B, Mischnick P, Thiem J (1994) Carbohydr Res 262:335
87. Evers B (1996) PhD thesis, University of Hamburg, Germany

Enzymatic and Chemical Glycosylations of Ergot Alkaloids and Biological Aspects of New Compounds

Vladimír Křen

Institute of Microbiology, Academy of Sciences of the Czech Republic,
Laboratory of Biotransformation, Vídeňská 1083, CZ-142 20 Prague 4, Czech Republic

The primary aim of this chapter is to demonstrate different strategies, chemical and chemoenzymatic, for the preparation of glycosides of pharmaceutically active and broadly used compounds – ergot alkaloids. New glycosides prepared recently have interesting immunomodulatory, neuroactive and cytostatic activity.

Table of Contents

Introduction

This paper, which deals with a rather specialized topic of carbohydrate chemistry, aims to illustrate possibilities and different strategies for the preparation of glycosides of complex molecules such as ergot alkaloids and to show pharmacological perspectives of such work.

Topics in Current Chemistry, Vol. 186

Ergot alkaloids (EA) are broadly used in many fields of medicine. At present there exist more than 50 formulations in clinical use containing natural or semisynthetic ergot alkaloids.

Ergot alkaloids cover a large field of therapeutic use as drugs of high potency in the treatment of uterine atonia, postpartum bleeding, migraine, orthostatic circulatory disturbances, senile cerebral insufficiency, hypertension, hyperprolactinemia, acromegaly, and parkinsonism [1, 2]. Recently, new therapeutic uses have emerged, e.g., against schizophrenia [3] and in applications based on newly discovered antibacterial [4] and cytostatic effects [5], immunomodulatory [6, 7] and hypolipemic activity [8].

The broad physiological effects of EA are based mostly on their interaction with the neurotransmitter receptors on the cells. The presence of "hidden structures" resembling some important neurohumoral mediators (e.g., noradrenaline, serotonin, dopamine) (Scheme 1) in the molecules of EA could explain their interactions (agonistic or antagonistic) with these receptors [2].

EA are produced by filamentous fungi of the genus *Claviceps* (e.g., *Claviceps purpurea* – Ergot, Mutterkorn) and by some other filamentous fungi, and can also be found infrequently in the plant kingdom. These alkaloids are produced on the industrial scale either by parasitic cultivation (field production of the ergot) or by submerged fermentation of selected strains [9] (for biosynthesis see Scheme 8). Although other methods for the total synthesis of EA have been developed [10], they are not practicable.

Most of the ergot alkaloids used in therapy are semisynthetic derivatives. There exists a continuous search for new, more potent or more selective EA derivatives. Because of alkaloid complexity and the sensitivity of the EA skeleton to harsh conditions, biotransformation methods are more often used for their derivatization [11].

Glycosides of EA were isolated as naturally occurring products, and a large number of them were recently prepared by chemical and enzymatic methods. Their promising physiological effects stimulate future research in this field.

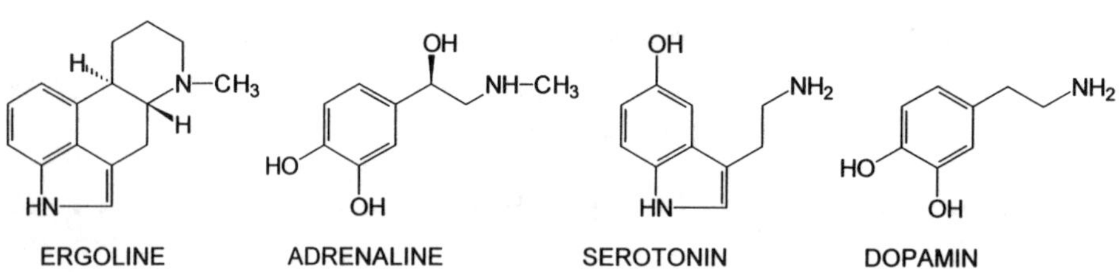

Scheme 1

2
Ergot Alkaloid *O*-Glycosides

2.1
Ergot Alkaloid Monoglycosides

2.1.1
Naturally-occuring Ergot Alkaloid Glycosides

The first natural EA glycoside, elymoclavine-*O*-β-D-fructofuranoside (**2**) (Scheme **2**) was isolated from a saprophytic culture of *Claviceps* sp. strain SD-58 by Floss et al. [12]. This glycoside was formed from elymoclavine produced by the microorganism by the action of the enzyme invertase present in the fungal mycelium. Invertase transferred the fructose moiety from sucrose present as a nutrient to the hydroxyl group of elymoclavine. Later it was found that also higher oligofructosides of elymoclavine are formed by the same enzymatic mechanism, e.g., elymoclavine-*O*-β-D-fructofuranosyl$(2 \rightarrow 1)$-*O*-β-D-fructofuranoside [13] and tri- and tetrafructoside (**3**) (Scheme 3) [14]. No other EA glycosides have been found up to now in nature.

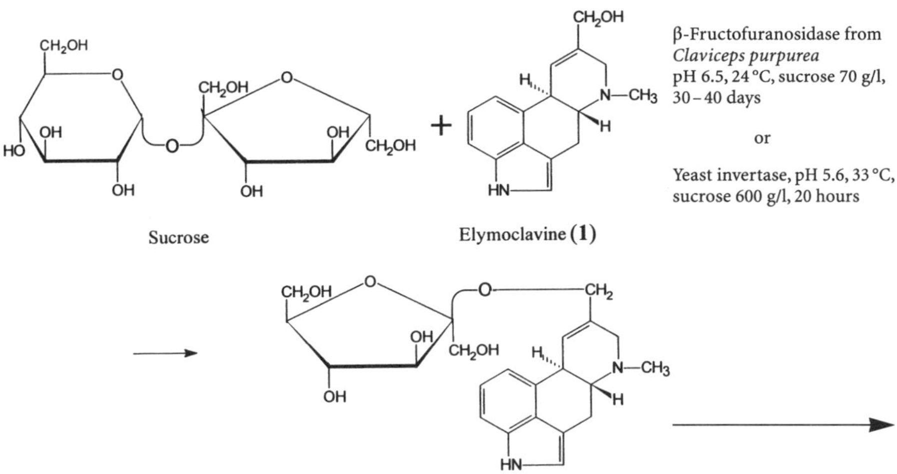

Scheme 2 **2**

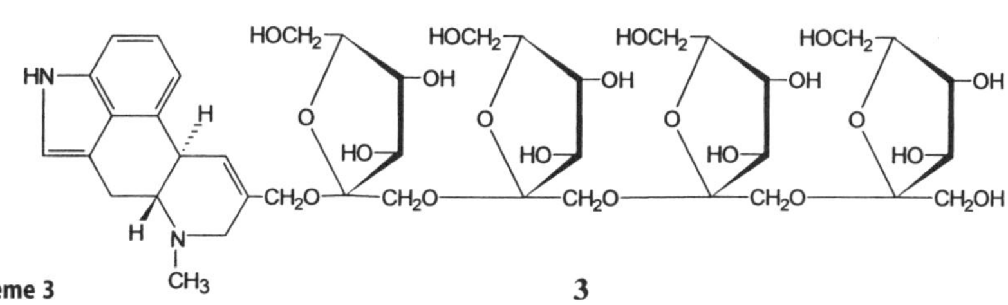

Scheme 3 **3**

2.1.2
Chemical Synthesis of O-Glycosides of Ergot Alkaloids

The first attempt to synthesize glycosides of 9,10-dihydrolysergol was undertaken by Seifert and Johne [15]. They used Koenigs-Knorr conditions (peracetylglycosyl bromide, Fétizon reagent as promoter and THF as a solvent). This method afforded rather low yields of the β-glucopyranoside, β-galactopyranoside, β-xylopyranoside and α-arabinopyranoside of 9,10-dihydrolysergol. Large amounts of the respective ortho esters were always formed; in case of mannose only the ortho ester was obtained in low yield. Variation of the reaction conditions performed in our laboratory and even the use of more potent promoters (e.g., $AgClO_4$, AgOTf, $Hg(CN)_2$) gave mostly negative results. Some other alkaloids, especially those having a double bond adjacent to a primary hydroxyl group gave under the above conditions almost exclusively ortho esters with the glucose and galactose donors. Use of trichloroacetimidates gave inseparable mixtures. Finally, moderate yields (approx. 40%) of β-glucosides and β-galactosides of elymoclavine and 9,10-lysergol were obtained using peracetylated sugar as a donor and trimethylsilyl-triflate as a catalyst. A small amount (1–4%) of α-glycoside was formed and part of the aglycon (30%) was acetylated or degraded [16]. This method is, however, not suitable for glycosylation of alkaloids with a peptidic bond in the molecule, e.g., ergometrine. Also, partial degradation of the rather valuable raw material is a considerable drawback.

Glycosyl esters of xylopyranose, arabinopyranose, glucopyranose and galactopyranose were prepared from lysergic and 9,10-dihydrolysergic acid [17] under Koenigs-Knorr conditions (peracetylbromo sugars, Ag_2O, THF, r.t., 2–10 d) in rather low yields and as anomeric mixtures. The authors were not able to prepare pure and deblocked lysergyl- and 9,10-dihydrolysergyl-1-*O*-glycopyranosides.

2.1.3
Preparation of Ergot Alkaloid Glycosides by the Use of Glycosidases

The problematic chemical synthesis of the ergot alkaloid glycosides prompted us to test enzymatic glycosylation methods.

β-Galactosylation of elymoclavine (**1**), chanoclavine (**7**), lysergol, 9,10-dihydrolysergol and ergometrine was accomplished by β-galactosidase from *Aspergillus oryzae* using *p*-nitrophenyl-β-D-galactopyranoside or lactose as β-galactosyl donor (Scheme 4). Transglycosylation yields ranged from 13 to 40%. β-Galactosidase from *E. coli* gave somewhat lower yields [18].

The enzymatic method made it possible for the first time to glycosylate ergometrin bearing in the molecule a peptidic bond (for the structure of ergometrin glycosides, see, e.g., Scheme 7). Galactosylation also proceeded to a higher degree, affording e.g. β-D-galactopyranosyl$(1 \rightarrow 6)$-β-D-galactopyranosyl-$(1 \rightarrow O)$-elymoclavine (**5**).

The introduction of aminosugars into the EA molecules would be rather complicated by chemical means. However, aminosugars bearing EA had been

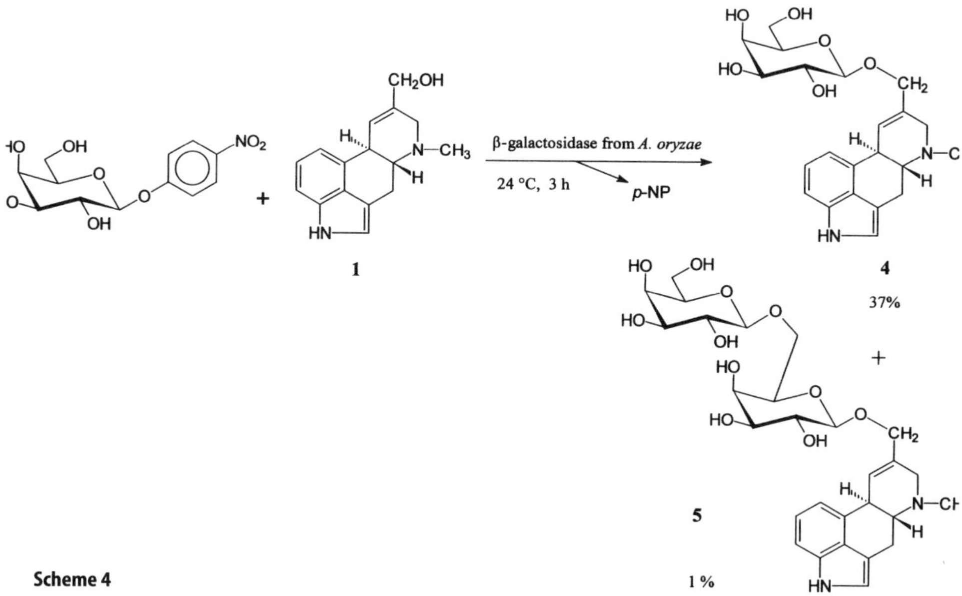

Scheme 4

expected to have immunomodulatory activities, and also this glycosylation would create a basis for further extension of the carbohydrate chain (introduction of LacNAc or sialyl residue – see below).

This task was accomplished by tranglycosylation using β-hexosaminidase from *A. oryzae* [19]. Representatives of each class of the EA, e.g. clavines [elymoclavine (**1**)], secoclavines (chanoclavine) and lysergic acid derivatives (ergometrine) were chosen to demonstrate the wide applicability of this method (Scheme 5). As a donor, *p*-nitrophenyl-*N*-acetylglucosaminide or galactosaminide were used, the yields ranging from 5 to 15%. Interestingly, higher glycosides were also formed, and one of them, 2-acetamido-2-deoxy-β-D-glucopyranosyl-(1 $\rightarrow$ 3)-2-acetamido-2-deoxy-β-D-glucopyranosyl-(1 $\rightarrow$ *O*)-elymoclavine, (**6**) was isolated and spectrally characterized. β-Hexosaminidase from *A. oryzae* is generally known to form (1 $\rightarrow$ 4)chitobiosyl bonds [20].

Enzymatic mannosylation of EA by α-mannosidase from *Canavalia ensiformis* (Jack beans) was accomplished by two different strategies – transglycosylation using *p*-nitrophenyl-α-D-mannopyranoside or reversed glycosylation using a high concentration of mannose [21]. In the case of chanoclavine (**7**), a higher yield of the respective α-mannoside (**8**) was obtained in a shorter time by use of the transglycosylation concept (Scheme 6). Lower yields in reversed glycosylation are, however, compensated for by the considerably lower cost of the mannosyl donor (mannose). Unreacted aglycone can be almost quantitatively recuperated.

The yield of the reversed glycosylation is controlled by the equilibrium constants of the respective glycosides. For the various alkaloids, elymoclavin (**1**), chanoclavine (**7**) and ergometrine (Scheme 7), the yields of the respective mannosides after 2 days were 16%, 11% and 10%, respectively (Fig. 1).

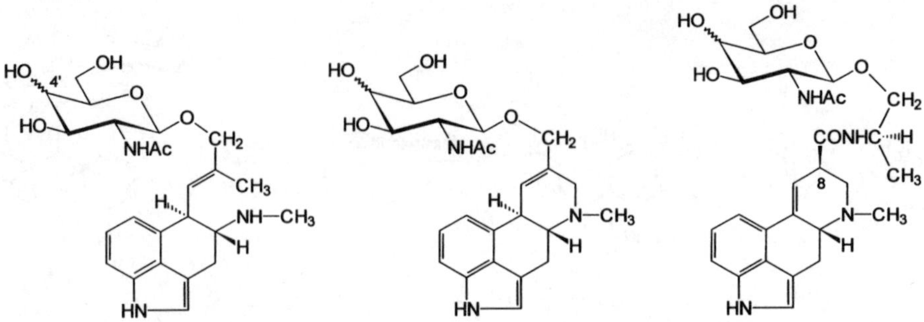

4′-α-OH Chanoclavine O-β-GlcNAc
4′-β-OH Chanoclavine O-β-GalNAc

4′-α-OH Elymoclavine O-β-GlcNAc
4′-β-OH Elymoclavine O-β-GalNAc

4′-α-OH Ergometrine O-β-GlcNAc
4′-β-OH Ergometrine O-β-GalNAc

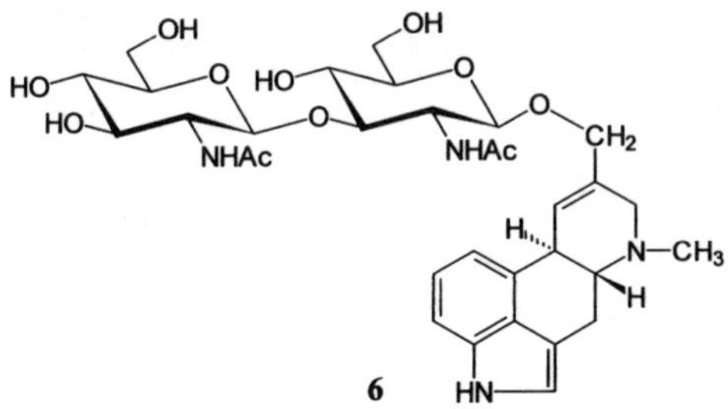

Scheme 5 6

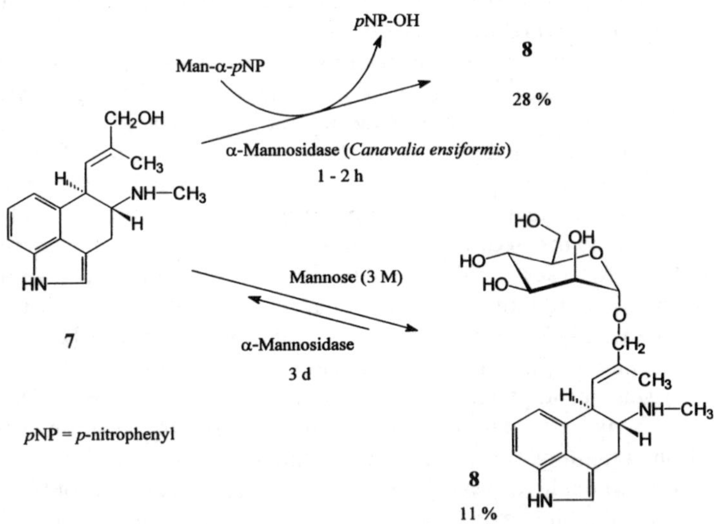

Scheme 6

Series of other glycosides of EA was prepared by enzymatic transglycosylations using activated *p*-nitrophenylglycosides as donors. *β*-Glucosides of, e.g., elymoclavine and DH-lysergol, were prepared with *β*-glucosidase from *A. oryzae* [16], *α*-glucosides of, e.g., elymoclavine and chanoclavine were prepared with *α*-glucosidase from *Bacillus stearothermophilus* or with *α*-glucosidase from rice

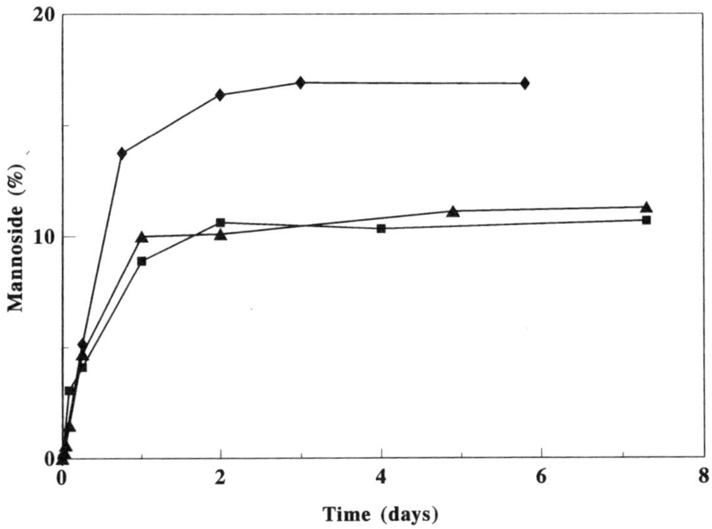

Fig. 1. α-Mannosidase-catalyzed synthesis of elymoclavine α-mannoside (-◆-), ergometrine α-mannoside (-▲-) and chanoclavine α-mannoside (-■-). The initial concentration of mannose and respective alkaloid were 3 and 0.2 M, respectively. The enzyme concentration was 10 U/ml [21]

[22]. α-Galactosides of the same alkaloids were prepared with α-galactosidase from *Coffea arabica* (green coffee beans) or from *A. niger* [22] (Scheme 7).

2.1.4
Fructosylation of Ergot Alkaloids by Microbial Biotransformations

The discovery of elymoclavine-O-β-fructofuranoside (**2**) [12] and higher $2 \rightarrow 1$ fructofuranosides of elymoclavine [13, 14] and chanoclavine [22] revealed a new group of naturally occurring ergot alkaloid glycosides. All these compounds were obtained from the cultures of the fungi *Claviceps purpurea*. These glycosides were formed from their aglycones (produced by the fungus *per se*) by β-D-fructofuranosidase present in the periplasmic space of *Claviceps* cells [23]. This enzyme primarily splits sucrose into glucose and fructose, which is then transferred to a hydroxymethyl group either on another sucrose molecule [24] or on some other acceptor (Scheme 2). This enzyme is even capable of the resynthesis of sucrose [25].

Fructosylation of elymoclavine and chanoclavine was performed by the intact mycelium of *C. purpurea* in a cultivation medium under sterile conditions [26]. The respective alkaloids were fed as their citrate salts. The bioconversion was completed after 25–45 days. The reaction was optimized with respect to pH and substrate concentrations.

The transfructosylation activity of this enzyme is strongly dependent on pH and the substrate concentration. Some glycosidases are known to be more hydrolytically active under acidic conditions, whereas their transglycosylation activity is higher at pH values above their pH optimum. Transfructosylation of

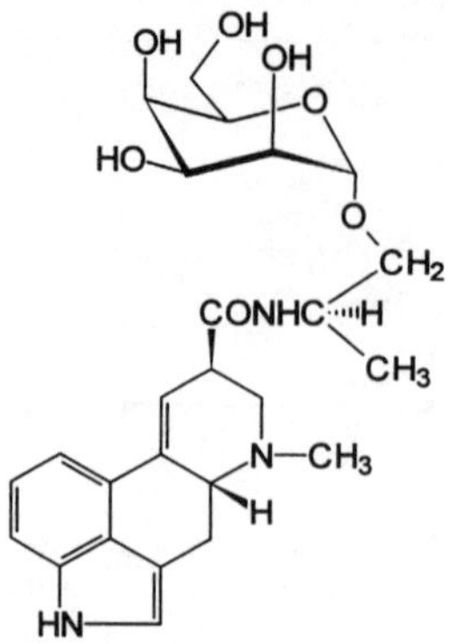

Ergometrine *O*-α-mannoside
Enzyme: α-mannosidase from *Canavalia ensiformis*

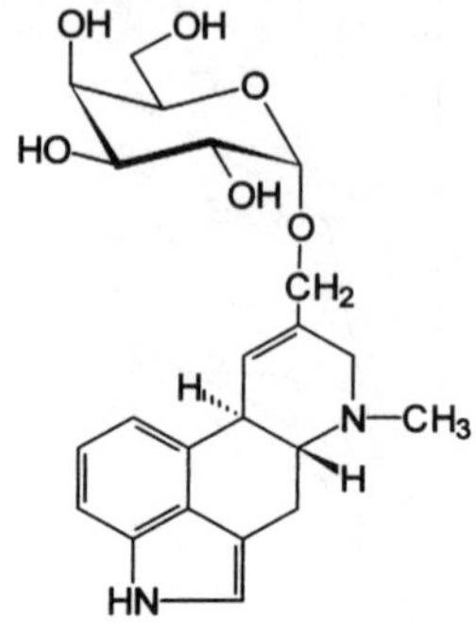

Elymoclavine *O*-α-galactoside
Enzyme: α-galactosidase from *Coffea arabica* or from *Aspergillus niger*

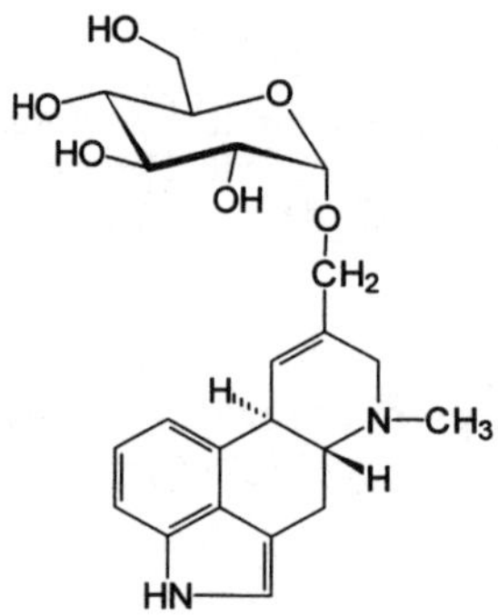

Elymoclavine *O*-α-glucoside
Enzyme: α-glucosidase from *Bacillus stearothermophilus* or that from rice

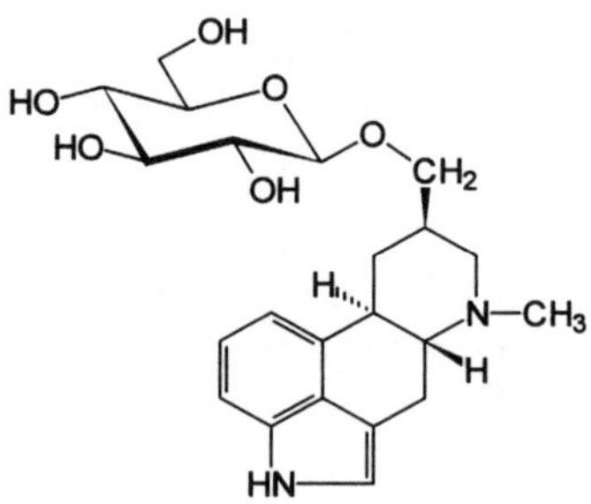

DH-Lysergol *O*-β-glucoside
Enzyme: β-glucosidase from *A. oryzae*

Scheme 7

elymoclavine was enhanced fivefold by a pH shift from 5.0 to 6.5 (Fig. 2) [26]. Concentration of sucrose (fructosyl donor) is another crucial parameter influencing the yield of alkaloid fructosides. At lower sucrose concentrations the substrate did not saturate the enzyme, whereas at higher concentrations the sucrose itself competed with the alkaloids as an acceptor of the fructose (Fig. 3) [26]. Immobilized cells of *C. purpurea* were repeatedly used for elymoclavin and chanoclavin fructosylation [26].

Fructosylation of other alkaloids than elymoclavine is complicated by the fact that (living) cells of *C. purpurea* also produce elymoclavine, and this alkaloid competes with the introduced alkaloids for the fructosyl transfer. It was found that elymoclavine has a higher affinity for the fructosyl transfer than other alkaloids (Table 1). Both the lowered yield and separation problems prevented us from obtaining lysergol and DH-lysergol fructosides. Less reactive alkaloids

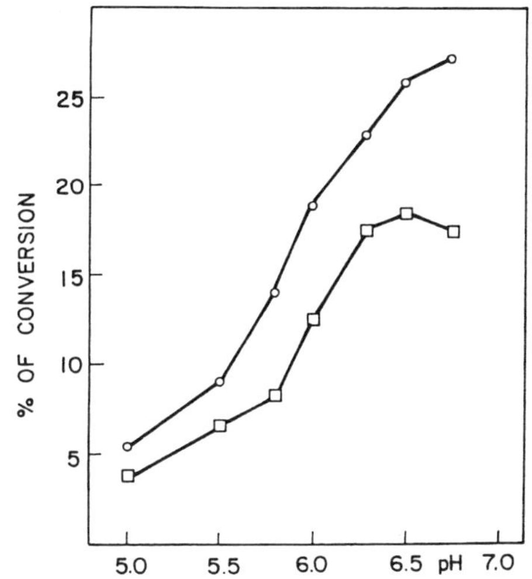

Fig. 2. Influence of pH on elymoclavine fructosylation by *C. purpurea* cells, showing conversion to elymoclavine monofructoside (-□-) and total elymoclavine fructosides (-○-) (incl. the higher fructosides) [26]

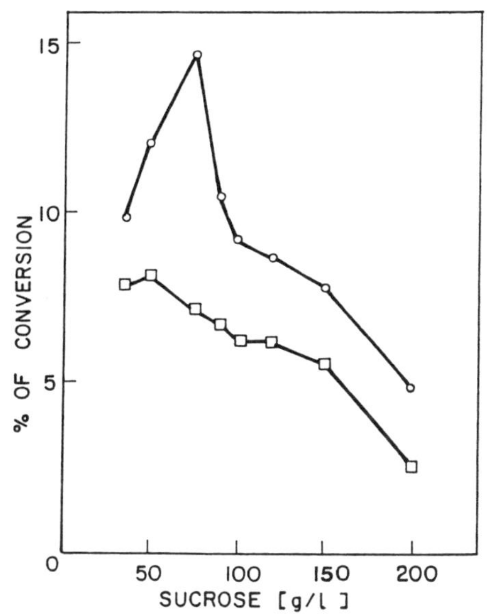

Fig. 3. Influence of sucrose concentration on elymoclavine fructosylation by *C. purpurea* cells, showing conversion to elymoclavine monofructoside (-□-) and total elymoclavine fructosides (-○-) (incl. the higher fructosides) [26]

than *iso*-lysergoles gave no fructosylation products under elymoclavine competition.

We have therefore selectively blocked the first enzyme in the biosynthesis pathway of EA (DMAT-synthase) by 5-fluorotryptophan (Scheme 8). This treatment did not influence other biochemical processes in the living cells of *C. purpurea*, thus enabling the fructosylation of introduced alkaloids to be performed [27].

Yields of alkaloid fructosides differed considerably (Table 1). Both *iso*-lysergols gave considerably lower yields of fructosides. This was probably caused by

Table 1. Ergot alkaloid fructosylation by alkaloid-blocked culture of *C. purpurea*

Alkaloid	Yield of fructoside [%]
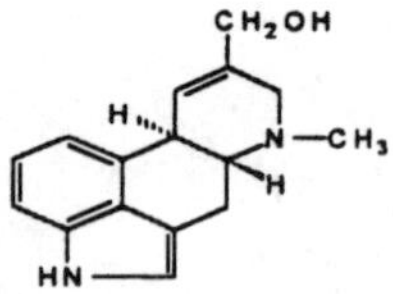 Elymoclavine	36.6
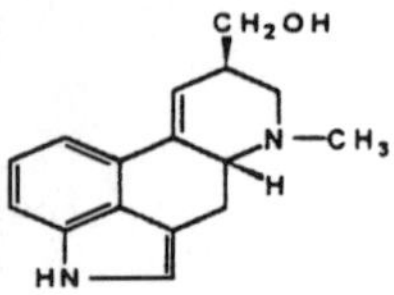 Lysergol	14.7
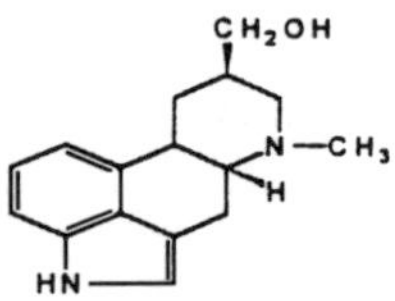 9,10-Dihydrolysergol	16.7
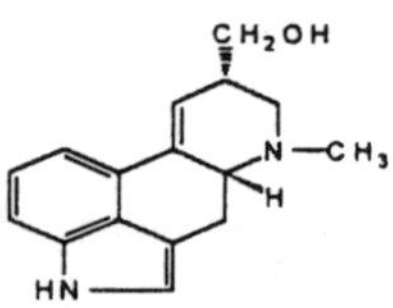 *iso*-Lysergol	5.7
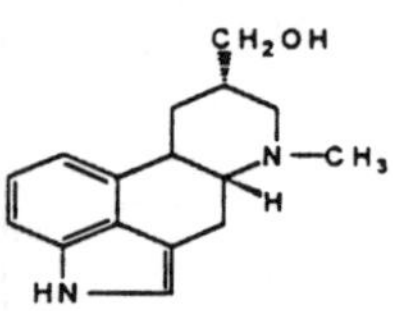 *iso*-9,10-Dihydrolysergol	5.1

stabilization of their acceptor hydroxymethyl groups by an H-bond to the N-6 atom of the ergoline system (confirmed by molecular modelling). The high yield of elymoclavine fructoside could be explained by the allylic configuration in the D-ring activating the OH-group.

2.2 Enzymatic Preparation of Complex Ergot Alkaloid Glycosides

Complex alkaloid glycosides bearing, e.g., lactosyl (Lac), lactosaminidyl (LacNAc) or sialyl (Neu5Ac) moieties were required for immunomodulation

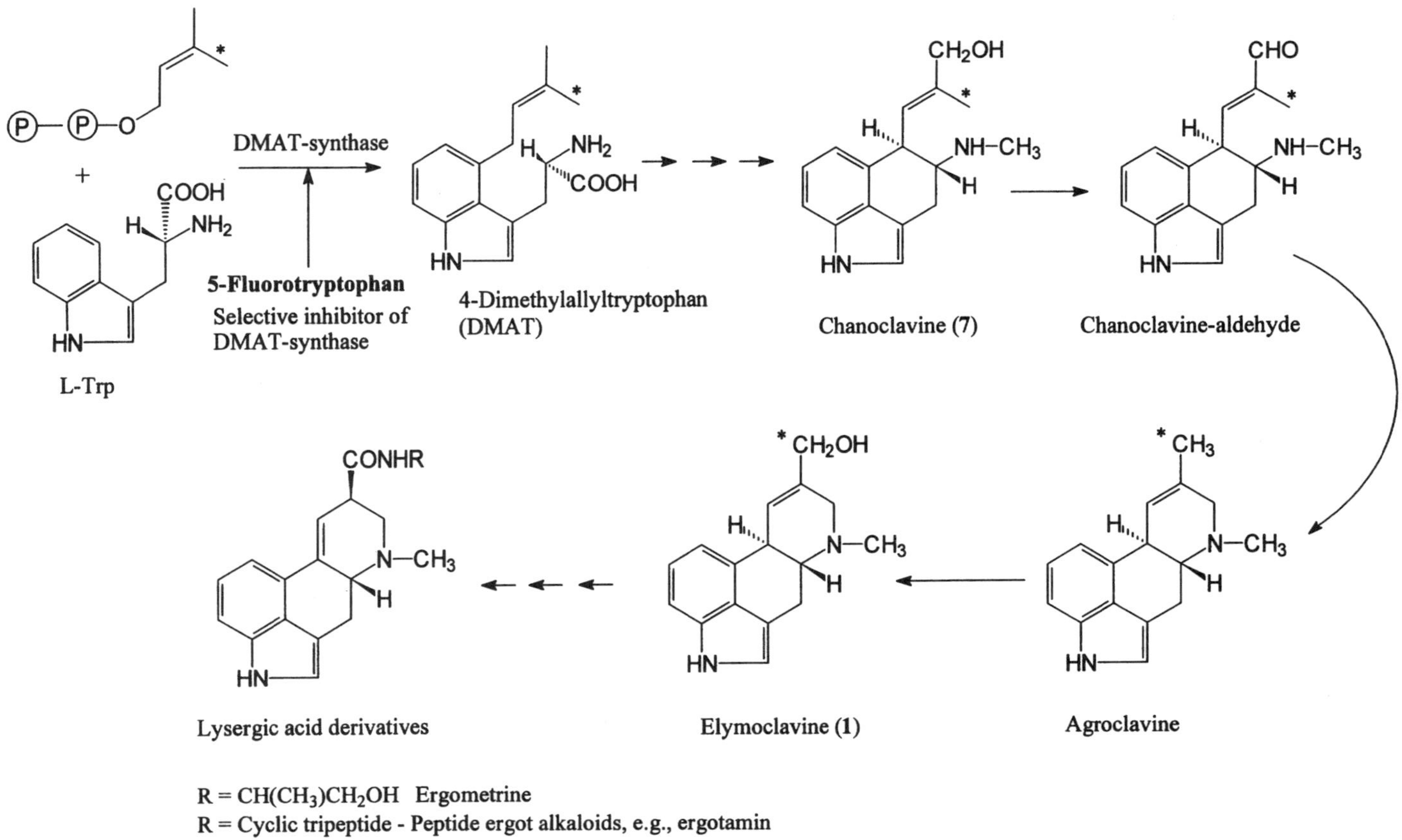

Scheme 8. Biosynthesis of ergot alkaloids. Selective inhibition of DMAT-synthase in the biosynthesis pathway by 5-fluorotryptophan blocks synthesis of alkaloids *de novo* but does not influence other biochemical processes in the cells, e.g., transfructosylating enzymes [27]

tests (see below). Because of the paucity of starting material (alkaloid monoglycosides) and the need of regioselective glycosylation, only enzymatic reactions were practicable.

For the preparation of β-D-glucopyranosyl (1 → 4)-2-acetamido-2-deoxy-β-D-glucopyranosyl-(1 → *O*)-elymoclavine (**11**), the extension of previously prepared 2-acetamido-2-deoxy-β-D-glucopyranosyl-(1 → *O*)-elymoclavine (**10**) by the use of bovine β-1,4-galactosyltranferase was chosen. Uridine 5′-diphosphogalactose (UDP-Gal) served as a substrate, and alkaline phosphatase from calf intestine (EC 3.1.3.1) was used to remove feedback inhibition caused by the UDP produced [28].

For the generation of UDP-Gal in situ, UDP-Glc and UDP-Gal 4′-epimerase were used. It was found, however, that β-1,4-galactosyltransferase was able to transfer glucose, forming in parallel also β-D-glucopyranosyl (1 → 4)-2-acetamido-2-deoxy-β-D-glucopyranosyl-(1 → *O*)-elymoclavine (**12**) (Scheme 9). This was later confirmed on a semipreparative scale using UDP-Glc without the epimerase. The transfer of glucose was confirmed fully by spectral methods [28]. This is the first example of concomitant transfer of glucose and galactose by galactosyltransferase.

β-Lactosyl elymoclavine was prepared from the respective β-glucoside by the use of bovine β-1,4-galactosyltransferase in the presence of α-lactalbumine [16]. Analogously, β-Lac and β-LacNAc derivatives of other ergot alkaloids, e.g., 9,10-dihydrolysergene were prepared [16].

Attachment of neuraminic acid to β-LacNAc-elymoclavine (**11**) yielding **19** was accomplished by the use of α-2,6-sialyltransferase from rat liver (EC 2.4.99.1) (Scheme 10). It was interesting that the affinity of the sialyltransferase towards **11** was approximately 20% higher than that towards *N*-acetyllactosamine, the natural substrate of this enzyme.

2.3 Glycosylation of Agroclavine

Agroclavine (**13**) displays interesting biological effects. It causes vasoconstriction through its effect on α-adrenoreceptors or 5-hydroxytryptamine receptors. Cytostatic activity of agroclavine and its derivatives is comparable with clinically used cytostatics, e.g., adriamycin [29]. Also, the antibiotic activity of agroclavine was documented. Agroclavine also stimulates the killing capability of NK-leukocytes towards the NK-resistant target cells and enhances the production of interleukines (IFN-γ, IL-2) [31]. From previous studies, it was known that glycosylation of some alkaloids augments, e.g., immunomodulatory effects [16, 30]. Glycosylation of agroclavin (having no free OH group) would be possible either after derivatization or by preparation of N-glycosides (see below).

By condensation with formaldehyde, *N*-1-hydroxymethyl agroclavine was prepared. This compound was β-galactosylated in good yield (30–40%) by β-galactosidase from *A. oryzae* (Scheme 11) [31].

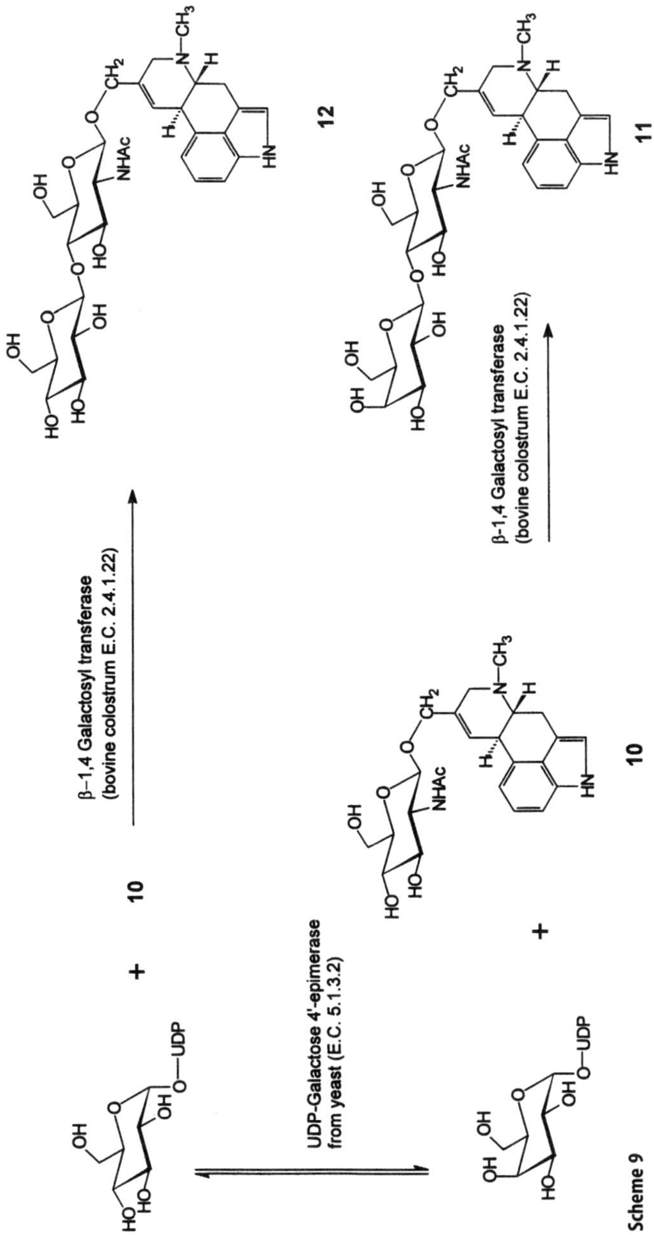
UDP-Galactose 4'-epimerase
from yeast (E.C. 5.1.3.2)
β-1,4 Galactosyl transferase
(bovine colostrum E.C. 2.4.1.22)
β-1,4 Galactosyl transferase
(bovine colostrum E.C. 2.4.1.22)
10
10
11
12

Scheme 9

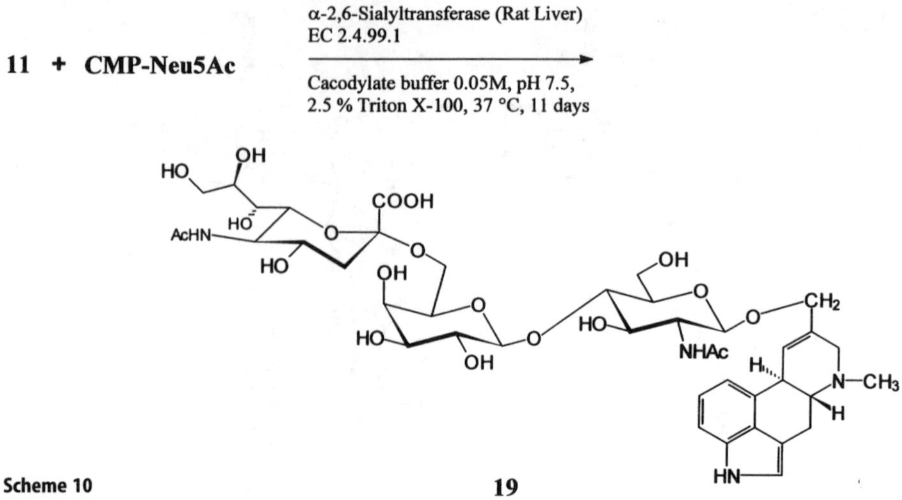

Scheme 10

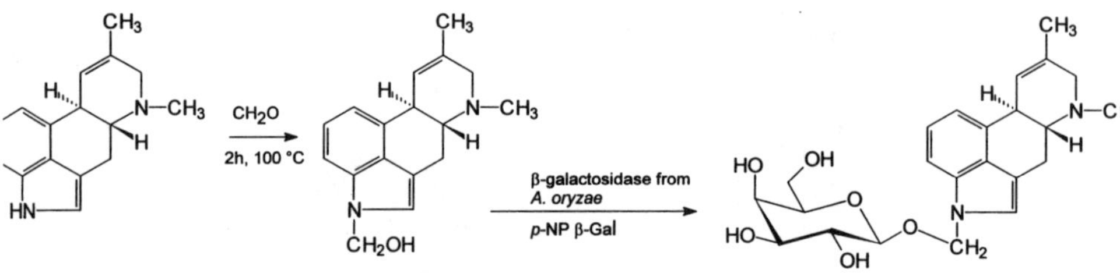

Scheme 11

3
Ergot Alkaloid *N*-Glycosides

3.1
Ergot Alkaloid *β*-*N*-Ribosides

Besides their activity mediated by neurotransmitter receptors, clavine alkaloids also possess antibiotic and cytostatic activities. The antibiotic activity of EA was ascribed to inhibition of nucleic acid replicatory processes [32]. The antineoplastic and antiviral activity of various heterocycles could be augmented by their *N*-ribosylation. The preparation of *N*-ribosides of EA could form analogous compounds to nucleosides with the aglycon possessing both neurohumoral and cytostatic activity.

The synthesis of these nucleosides was achieved using a silylation procedure. Trimethylsilyl (TMS) derivatives of agroclavine, lysergene, elymoclavine-*O*-acetate, lysergol-*O*-acetate and 9,10-dihydrolysergol-*O*-acetate were prepared from *N*-methyl-*N*-(trimethylsilyl)-trifluoroacetamide in MeCN [33]. The TMS derivatives were then treated with 1-*O*-acetyl-2,3,5-tri-*O*-benzoyl-*β*-D ribofuranose. From the different reaction conditions tested, the best involved

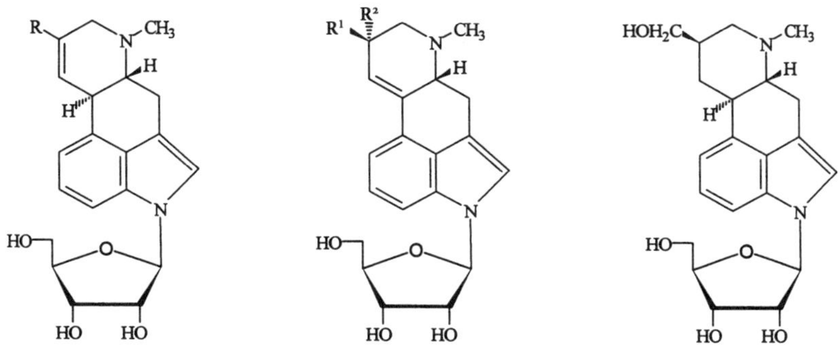

14 R = CH_3
15 R = CH_2OH

16 $R^1 = CH_3OH$, $R^2 = H$
17 $R^1, R^2 = CH_2$

18

Scheme 12

1,2-dichloroethane as the solvent and $SnCl_4$ as the catalyst. Protected ribosides were obtained in 20–40% yields. By deprotection with NEt_3: MeOH:H_2O 1:8:1 (r.t., overnight), 1-(β-D-ribofuranosyl)-agroclavine (**14**), 1-(β-D-ribofuranosyl)-elymoclavine (**15**), 1-(β-D-ribofuranosyl)-lysergol (**16**), 1-(β-D-ribofuranosyl)-lysergene (**17**), and 1-(β-D-ribofuranosyl)-9,10-dihydrolysergol (**18**) were obtained (Scheme 12). A small amount of α-anomer (6–19%) that was not separable was formed in all cases [34].

3.2 Ergot Alkaloid *N*-Deoxyribosides

All tested enzymatic (microbial ribosylnucleotide transferase) methods for the transfer of the ribosyl and 2-deoxyribosyl moiety to the alkaloid molecule (agroclavine) failed (e.g.[35]), presumably due to the very narrow specificity of the particular enzymes for certain aglycone molecules.

Therefore, a synthetic approach to the preparation of β-*N*-deoxyribosyl ergolines was tested. As a glycosyl donor, 2-deoxy-3,5-di-*O*-*p*-toluoyl-α-D-ribofuranosylchloride [36] was chosen.

For glycosylation, in the first approach, analogous conditions to those used for β-*N*-ribosylation were tested, i.e., $SnCl_4$ as catalyst, 1,2-dichloroethane as solvent, and the TMS derivative as acceptor. Only traces of the desired product were formed. We have therefore tested another possibility. The attempt to prepare the *N*-Na derivative of the ergolines (NaH, acetonitrile) failed. The use of other catalysts, e.g., TMS-*O*-Tf (1,2-dichloroethane) gave negligible yields of product. Finally, the use of acetonitrile and $SnCl_4$ as catalyst proved to be effective. The TMS-alkaloid (agroclavine) derivative reacted nearly quantitatively. However, in the case of deoxy-ribosides a mixture of α/β anomers was formed. This mixture was separable in protected form. After deprotection and purification, NMR and MS spectra proved the formation of a glycosidic bond. Contrary to the situation in natural nucleosides, it was not possible to use the NMR rules developed for the nucleosides of pyrimidine or purine bases. We have therefore

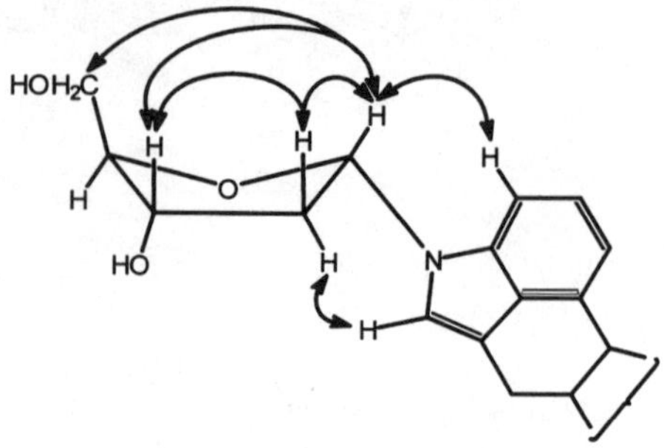

α-Deoxyriboside of Ergoline

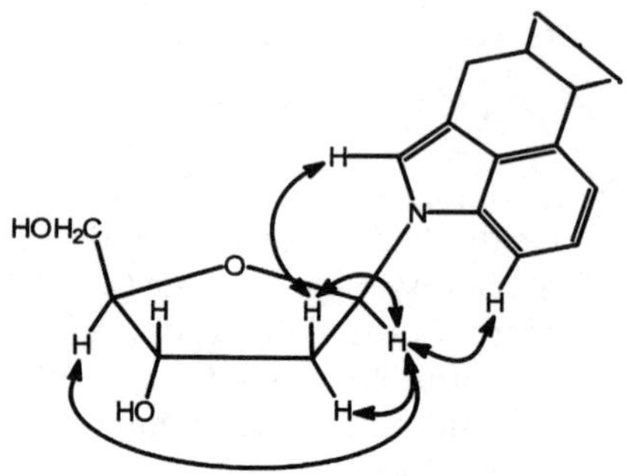

Scheme 13 β-Deoxyriboside of Ergoline

used a kinetic approach stemming from the fact that *β*-deoxyriboside is formed primarily, and it later isomerizes (mostly by the action of the HCl released) to the *α*-anomer. We have found that during the first few minutes of the reaction, one anomer was formed preferentially, and this was assumed to be the *β*-anomer.

The anomeric configuration was eventually determined by NOE (studied using ROESY) experiments, establishing both inter-sugar and sugar-aglycon relations [37] (Scheme 13).

New *β*-*N*-deoxyribosides (and also *α*-anomers) of agroclavine, lysergol, DH-lysergol and lysergene were prepared. The average yield of the *β*-deoxyribosides isolated was approximately 20–30% [38].

4 Biological Activity of New Ergot Alkaloid Glycosides

Most glycosides of EA have been prepared recently, and therefore a systematic study of their biological activity is not yet complete. However, preliminary results obtained indicate that some of these of derivatives could have very interesting types of activity compared with their aglycones.

Some glycosides of 9,10-dihydrolysergol and elymoclavine were tested for their inhibitory activity towards prolactin secretion [15, 39] (Table 2).

Table 2. Inhibition of prolactin secretion in rats

Compound	Prolactin inhibition [%]
9,10-Dihydrolysegol	63
9,10-Dihydrolysegol-*O*–D-glucopyranoside	86
9,10-Dihydrolysegol-*O*–D-xylopyranoside	38
9,10-Dihydrolysegol-*O*–D-arabinopyranoside	60
Elymoclavine	71
Elymoclavine-*O*–D-ribofuranoside	42

Only 9,10-dihydrolysegol-*O*-β-D-glucopyranoside exhibited significantly higher ($p < 0.001$) inhibitory activity.

A more systematic study was performed on the immunomodulatory activity of new alkaloid glycosides. A large range of the glycosides, e.g., of elymoclavine and 9,10-dihydrolysergol, was tested for their stimulatory activity on cytotoxic lymphocytes. These lymphocytes form the effector arm of cell-mediated immune responses to infection and tumors.

The effect of the alkaloid glycosides was tested on natural killer (NK) cell-mediated cytotoxicity of the resting and activated human peripheral blood mononuclear cells (PBMC) against MOLT4 T lymphoma cells (resistant to lysis by fresh PBMC cells and sensitive to activated cells) (Fig. 4) [7, 16]. All compounds were tested in the concentration range from 10^{-6} to 10^{-15} M. Maximum immunomodulatory effect was obtained at 10^{-10} M. These effects are mediated by cell surface receptors. After addition of free saccharides to effector-target cell mixture, the cytotoxic activity of resting fresh lymphocytes was enhanced in all cases. The most potent stimulation was observed with glucose. The glycosylation of elymoclavine does not influence its cytotoxic activity. However, the cytotoxic-potentiating activity of DH-lysergol was strongly increased, especially in the the case of β-glucoside and β-lactoside. The lytic capacity of activated killer cells was not influenced by any preparation tested.

The effects of elymoclavine and DH-lysergol glycosides were also tested on the cytotoxicity of resting PBMC cells towards the NK-sensitive tumor cell line K562 and the NK-resistant RAJI tumor cell line (Fig. 5) [7, 16]. Stimulation of NK cells against the sensitive K 562 was best with DH-lysergol itself; its glycosylation lowered the stimulatory effects. Galactosylation of elymoclavine potentiated stimulation activity compared to the aglycone.

Interesting results were obtained in the stimulation of the activity of NK cells against the resistant tumor cell line RAJI. The attachment of β-Glc and mainly Neu5Acα (2-6)Galβ(1-4)GlcNAcβ-*O*- sugar moieties to elymoclavine had a strong stimulatory effect (Fig. 5). Galactosylation of DH-lysergol also potentiated its effects.

These and some other glycosides were further tested for their immunomodulatory activity on mouse splenocyte models (Balb/c and athymic nude Nu/Nu mice). Here, mostly elymoclavine galactoside and lactoside had the highest activity [7]. More detailed studies attempting to resolve effect-structure relations are now under way.

Fig. 4. Comparison of immunomodulative effects of single saccharides, EA aglycones, and their glycosylated derivatives on human NK cell-mediated cytotoxicity towards MOLT4 target cells. All compounds were used at a concentration of 10^{-10} M. Average values from six experiments are shown. Values are presented as % of control [7, 16]

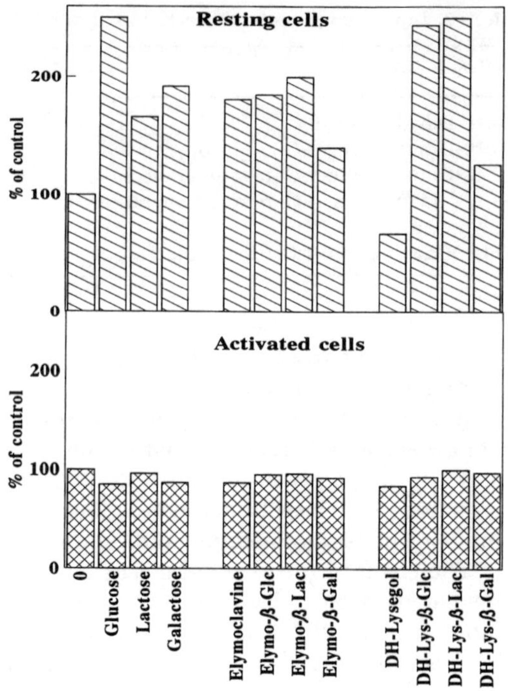

Fig. 5. Influence of EA and EA glycosides on human NK cell-mediated cytotoxicity towards K562 and RAJI target cells. Values are presented as % of control. Average values from six experiments are shown [7, 16]; Glc*β*(1-4)GlcNAc*β*-*O*-elymoclavine (**12**) and Neu5Ac*α*(2-6)Gal*β*-(1-4)GlcNAc*β*-*O*-elymoclavine (**19**)

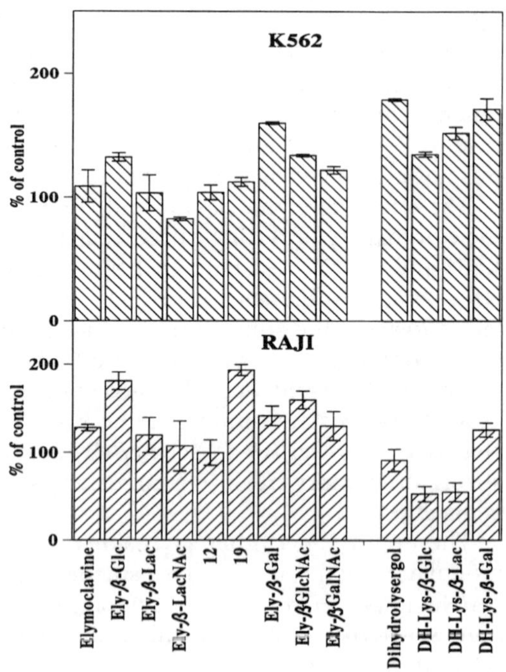

A large range of EA glycosides and their aglycons were tested for the antiviral activity including anti-HIV activity against the replication of HIV-1(IIIB) and HIV-2(ROD) in acutely infected MT-4 cells, for their activity in persistently infected HUT-78/IIIB cells and for broad spectrum antiviral activity in E6SM cells cultures against *Herpes simplex* virus-1 (KOS), *Herpes simplex* virus-2 (G), *Vaccinia* virus, *Vesicular stomatitis* virus, *Herpes simplex* virus-1 TK- B2006 and *Herpes simplex* virus-1 TK- VMW1837, in HeLa cell cultures against *Vesicular stomatitis* virus, *Coxsackie* virus B4 and *Respiratory syncytial* virus and in Vero cell cultures against *Parainfluenza*-3 virus, *Reovirus*, *Sindbis* virus, *Coxsackie* virus B4 and *Punta Toro* virus. It was found, however, that virtually all alkaloids and their glycosides tested have cytotoxic concentration below the threshold of their antiviral acivity. In some alkaloids, e.g., dihydrolysergol, their cytotoxity was considerably lowered by *N*-ribosylation [34].

Testing of the cytostatic activity of mainly *N*-glycosides of EA (ribosides and deoxyribosides) will be neccessary.

Some selected *β*-galactosides of ergot alkaloids and respective aglycons (elymoclavine and chanoclavine) were tested for their capability of influencing basal- and forskoline-stimulated adenylate-cyclase in ciliary protrusion (*recessus ciliaris*) of the rabbit eye. No significant effects were found. However, both EA-galactosides strongly diminished intraocular pressure in rabbits, and their effect was considarably higher than in the respective aglycones (J Čepelík and V Křen, unpublished results). This suggests their potential anti-glaucoma activity and a more detailed study is now in progress.

Acknowledgements. This work was supported by a grant No. 203/96/1267 from the Grant Agency of the Czech Republic and by an EU grant PECO ERBCIPDCT 930194.

References

1. Berde B, Schild HO (1978) Ergot alkaloids and related compounds. Springer, Berlin Heidelberg New York
2. Eich E, Pertz H (1994) Pharmazie 42:867
3. Markstein R, Seiler MP, Jaton A, Briner U (1992) Neurochem Int 20:211 S
4. Eich E, Eichberg D, Schwarz G, Clas F, Loos M (1985) Arzneim-Forsch/Drug Res 35(II): 1760
5. Eich E, Becker C, Sieben R, Maidhof A, Müller, WEG (1986) J Antibiot 39:804
6. Šterzl J, Řeháček Z, Cudlín J (1987) Czech Med 1:90
7. Fišerová A, Křen V, Augé C, Šíma P, Pospíšil M (1995) Ergot alkaloid derivatives with immunomodulatory activities. Proceedings of the International Conference on Experimental, Therapeutic and Toxic Manipulations of Host Defence Systems. June 12–15, 1995 Hradec Královѐ, Czech Rep. p 67
8. Cincotta AH, Meier AH (1989) Life Sci. 45:2247
9. Křen V, Harazim P, Malinka Z (1994) *Claviceps purpurea* (Ergot): Cultivation and ergot alkaloid bioproduction. In: Bajaj YPS (ed) Medicinal and aromatic plants VII, Biotechnology in Agriculture and Forestry (vol 28). Springer, Berlin Heidelberg New York, p 139
10. Ninomiya I, Kiguchi T (1990) Ergot alkaloids. In: Brossi A (ed) The alkaloids, Academic Press, New York, vol 38, p 142
11. Křen V (1991) Bioconversions of ergot alkaloids. In: Fiechter A (ed) Advances in biochemical engineering (vol 44). Springer, Berlin, Heidelberg, New York, p 123
12. Floss HG, Günther H, Mothes U, Becker I (1966) Z Naturforsch 22b:399

13. Flieger M, Zelenkova NF, Sedmera P, Křen V, Novák J, Rylko V, Sajdl P, Řeháček Z (1989) J Nat Prod 52:506
14. Havlíček V, Flieger M, Křen V, Ryska M (1994) Biol Mass Spectrom 23:57
15. Seifert K, Johne S (1984) Arch Pharm (Weinheim) 317:577
16. Křen V, Fišerová A, Augé C, Sedmera P, Havlíček V, Šíma P (1996) Bioorg Med Chem 4:869
17. Seifert K, Johne S, Meyer H (1979) J Prakt Chem 321:171
18. Křen V, Sedmera P, Havlíček V, Fišerová A, Šíma P (1992) Tetrahedron Lett 33:7233
19. Křen V, Ščigelová M, Přikrylová V, Havlíček V, Sedmera P (1994) Biocatalysis 10:181
20. Sing S, Gallagher R, Derrick PJ, Crout DHG (1995) Tetrahedron: Asymmx 6:2803
21. Ščigelová M, Křen V, Nilsson KGI (1994) Biotechnol Lett 16:683
22. Flieger M, Křen V, Zelenkova NF, Sedmera P, Novák J, Sajdl P (1989) J Nat Prod 53:171
23. Dickerson AG (1972) Biochem J 129:263
24. Arcamone F, Barbieri W, Cassinelli G, Pot C (1970) Carbohydr Res 14:65
25. Křen V, Pažoutová S, Rylko V, Sajdl P, Wurst M, Řeháček Z (1984) Appl Environ Microbiol 48:826
26. Křen V, Flieger M, Sajdl P (1990) Appl Microbiol Biotechnol 32:645
27. Křen V, Svatoš A, Vaisar T, Havlíček V, Sedmera P, Pažoutová S, Šaman D (1993) J Chem Res (*S*) 1993: 89; J Chem Res (*M*) 1993:0652
28. Křen V, Augé C, Sedmera P, Havlíček V (1994) J Chem Soc, Perkin Trans 1 1994:2481
29. Eich E, Eichberg D, Muller WEG (1984) Biochem Pharmacol 33:523
30. Fišerová A, Trinchieri G, Chan S, Bezouška K, Flieger M, Pospíšil M (1995) Ergot alkaloid-induced proliferation, cytotoxicity, and lymphokine production. In: J. Mestecky et al. (ed) Adv Mucosal Immunol. Plenum, New York, p 163
31. Křen V (manuscript in preparation)
32. Eich E, Becker C., Mayer K, Maidhof A, Müller WEG (1986) Planta Med 1986:290
33. Křen V, Sedmera P (1996) Coll Czech Chem Commun 61:1248
34. Křen V, Pískala A, Sedmera P, Havlíček V, Přikrylová V, Witvrouw M, de Clercq E (1997) Nucleosides & Nucleotides 15:(in press)
35. Mikhailopulo IA, Zinchenko AI, Kazimierczuk Z, Barai VN, Bokut SB, Kalinichenko EN (1993) Nucleosides and Nucleotides 12:417
36. Hoffer M (1960) Chem Ber 93:2777
37. Olšovský P, Pískala A, Křen (1996) NMR differentiation of α- and β-deoxyribosides of ergot alkaloids. 11th Joint Conference of Czech and Austrian Spectral Societies, April 15–17, Valtice, Czech Republic
38. Křen V, Olšovský P, Havlíček V, Sedmera P, Witvrouw P, de Clercq E Tetrahedron (submitted)
39. Cassady JM, Li GS, Spitzner EB, Floss HG (1974) J Med Chem 17:300

Enzymatic Synthesis of Peptide Conjugates – Tools for the Study of Biological Signal Transduction

B. Sauerbrei · T. Kappes · H. Waldmann

Universität Karlsruhe, Institut für Organische Chemie, Richard-Willstätter-Allee 2, D-76128 Karlsruhe, Germany

Because of their key position in signal transduction cascades, passing chemical signals across the cell membrane and beyond to the cell nucleus, protein conjugates such as glyco-, lipo-, phospho- and nucleoproteins are at the focal point of intense biochemical research. They are involved, for instance, in the regulation of cell growth and, in consequence, also in malignant proliferation processes which can lead to cancer. This biological importance has created a high demand for these complex macromolecules and for fragments thereof carrying characteristic structural features, especially the linkage region of the peptide backbone with the carbohydrate, the lipid or the phosphoric acid ester. However, the synthesis of peptide conjugates by means of classical chemical methods has come up against difficulties and limitations, prompting the development of new synthetic techniques and strategies. Thus, in recent years, combinations of enzymatic and classical methods proved to offer rewarding alternatives. This chapter highlights the potential of hydrolytic enzymes for the mild and selective introduction of protecting groups as well as for the cleavage of such groups, and later presents examples of their application as central steps in synthetic routes for the preparation of peptide conjugates.

Table of Contents

Topics in Current Chemistry, Vol. 186

1 Introduction

The transduction of chemical signals from the outside of a cell, across the cell membrane, further on into the interior of the cell, and ultimately to the cell nucleus is decisively influenced by proteins carrying covalently linked additional structural units, whose presence is vital to fulfill the designated biological functions of the biomacromolecules. To these so-called protein conjugates (Fig. 1) belong

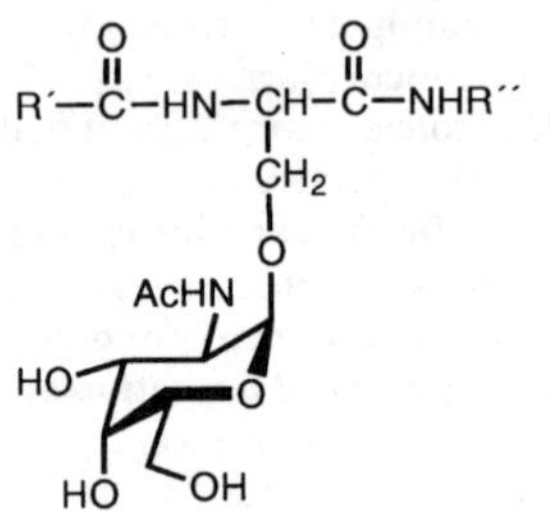

glycopeptide
1

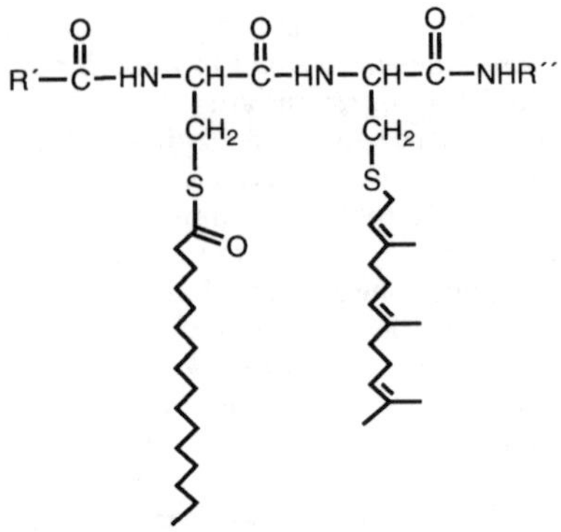

lipopeptide
2

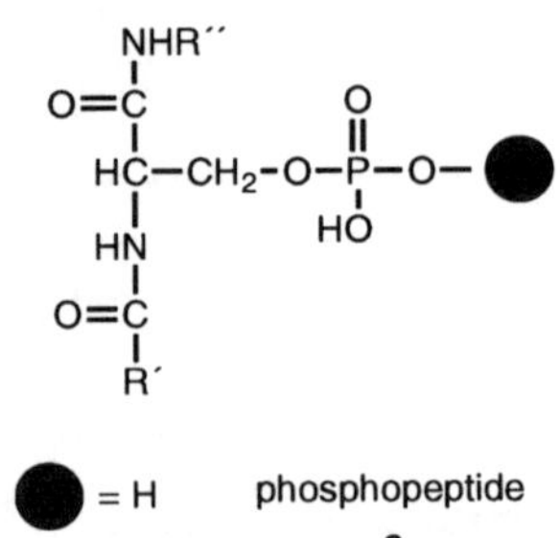

phosphopeptide
3

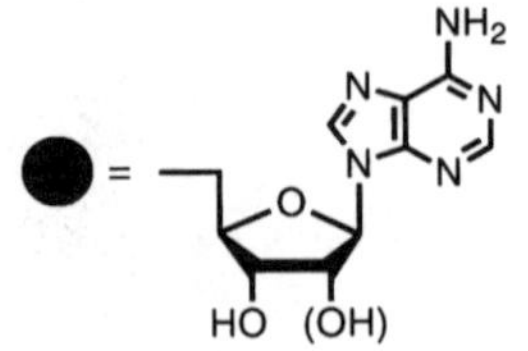

nucleopeptide
4

Fig. 1

(1) the glycoproteins, in which oligosaccharides are glycosidically attached to serine, threonine or asparagine residues,
(2) the lipoproteins, in which cysteine residues may be acylated by palmitic acid or alkylated with long unsaturated alkyl chains, in particular farnesyl residues,
(3) the phosphoproteins, characterized by the phosphorylation of serine, threonine and tyrosine residues, and
(4) the nucleoproteins, in which additional nucleoside units are attached to these phosphoric acid groups.

The extracellular messenger substrates and the corresponding receptors located on the outside of the cell membrane are often glycoproteins. On binding of suitable ligands, these receptors often transfer the signals to lipoproteins, which are anchored by means of their lipophilic residues at the interior face of the cell membrane. In many cases, this class of protein conjugates triggers a cascade of protein phosphorylation reactions mediated by kinases. Finally, the signal is passed into the cell nucleus via phosphorylation of transcription factors that interact with DNA, which, in turn, may be linked as a nucleoprotein to the matrix of the cell nucleus (Fig. 2).

By means of such signal transduction cascades, cell growth and proliferation, among other physiological events, are regulated. If this regulation is disturbed or interrupted, uncontrolled cell proliferation and in its wake the development of cancer might occur. Several of such signal transduction pathways have been identified and studied in detail [1]. A particularly important one is the so-called *Ras*-pathway [2, 3]. The *Ras*-proteins belong to the class of the lipoproteins, and are expressed in organisms as diverse as mammals, flies, worms and yeast. As shown in Fig. 2, they are activated by growth factors, and, as a central switch station, pass growth signals to the cell nucleus. If in the corresponding genes, the

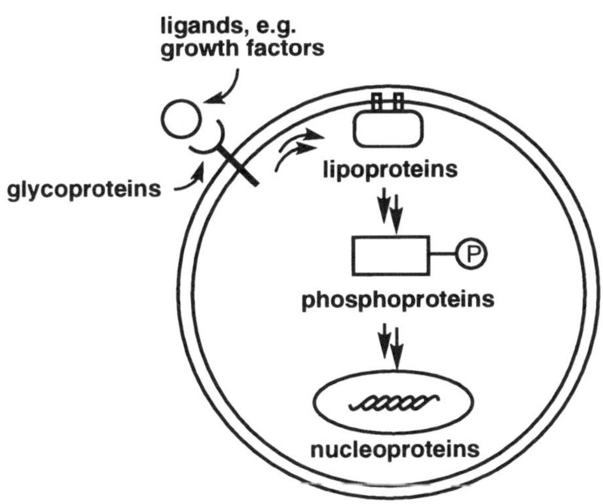

Fig. 2

ras-oncogenes, a point mutation occurs, *Ras*-mutants are generated which cannot be reset from the activated back to the inactivated state. As a consequence, they permanently stimulate cell growth, possibly leading to malignant cell transformation. Especially impressive is the finding that in ca. 40% of all cancer patients a point mutation in *ras*-genes can be detected. In some of the major malignancies like lung, colon and pancreas cancer, this figure rises to as much as 85% [4].

In view of the enormous biological importance of lipo-, phospho-, nucleo- and glycoproteins, they are, of course, the subject of intense bioorganic studies. Furthermore, the relevance of signal transduction cascades is increasingly recognized by medicinal chemistry. Thus, research is directed toward the development of substances which specifically intercept these pathways.The scope of signal transduction therapies is thereby investigated [4]. For these studies, well-defined low molecular weight peptides which contain the characteristic structural units of the parent natural products, i.e. the linkage regions between peptide chain and lipid portion, phosphoric acid ester or carbohydrate, are often required. However, the synthesis of these so-called peptide conjugates is not an easy exercise. For instance, the chemical synthesis of nucleopeptides (**4**) requires that the protecting group strategies employed for the nucleobases, the sugar units, the phosphates and the *C*- and *N*-termini of the peptide parts are compatible with each other, i.e. they must be orthogonally stable [5]. Moreover, for all three classes of peptide conjugates the situation is even more complicated, because in addition to their structural complexity they are characterized by a pronounced chemical lability. Thus, from the serine glycosides (**1**) at a pH higher than nine, the entire carbohydrate unit is split off by a β-elimination sequence. Similarly, the lipopeptides (**2**), the phosphopeptides (**3**) and the nucleopeptides (**4**) are subject to an even more easily occurring β-elimination, since in these cases acyl groups, which are even better leaving groups, are in place to be cleaved off. In addition, the lipopeptides (**2**) contain thio-esters, which are readily hydrolyzed under basic conditions. In acidic media, an anomerization or even a rupture of the *O*- or the *N*-glycosidic bonds in **1** and **4** has to be feared. Also, the olefins present in the farnesyl residues of **2** are sensitive to acids. Consequently, in the construction of these substrates, not only various orthogonally stable blocking groups are required, but furthermore all protecting group manipulations have to be carried out under almost neutral conditions.

For the solution of these problems, enzymatic methods should offer advantageous alternatives and should complement established classical chemical procedures [6–8]. Enzymes often operate at neutral, weakly acidic or weakly basic pH values and in many cases combine a high selectivity for the reactions they catalyze and the structures they recognize with a broad substrate tolerance. In this review, first an overview of the recently developed individual tools for the construction of polyfunctional peptide conjugates is given, i.e. enzymatically removable blocking groups for the OH-groups of carbohydrates, for the amino groups of peptides and nucleosides, and for the carboxy groups of peptides [9–11]. Successful applications of these methods for the synthesis of lipo- and glycopeptides are then presented.

2
Enzymatic Deprotection of Carbohydrates

As well as lipases and esterases (see the relevant chapter of this book by A. Ferandez-Mayoralas), penicillin G acylase from *Escherichia coli* proved to be an interesting biocatalyst for the selective deprotection of carbohydrates [12–15]. This enzyme selectively hydrolyzes phenylacetic acid (PhAc) esters and amides. For example, the acylase liberates with complete selectivity the 3-OH groups of diacetone glucofuranose (**5**) [12, 14] and of the glucal **6** [15], as well as the 2-OH group of the acylated glucopyranose **7** [12–14] (Fig. 3). The conditions are so mild that both base-labile acetic acid esters and acid-labile acetal groups remain completely unaffected. In the case of the peracylated glucose, moreover, the ester of a secondary hydroxyl function is quantitatively and chemoselectively hydrolyzed in the presence of a more reactive ester of a primary alcohol. In the case of analogous non-enzymatic deacylations, the ester at C-2 is actually the least reactive. Thus, by means of this enzymatic method, a complete reversal of the regioselectivity is accomplished.

A further remarkable biocatalyst for carrying out protecting group manipulations is the acetyl esterase which can be isolated from the *flavedo* of oranges. This enzyme preferably hydrolyzes acetic acid esters, longer acyl groups being attacked only slowly or not at all. The acylase can be advantageously employed to split off acetates – the standard acyl protecting group of carbohydrate chemistry – from various sugars in a chemo- and regioselective manner [16, 17]. For

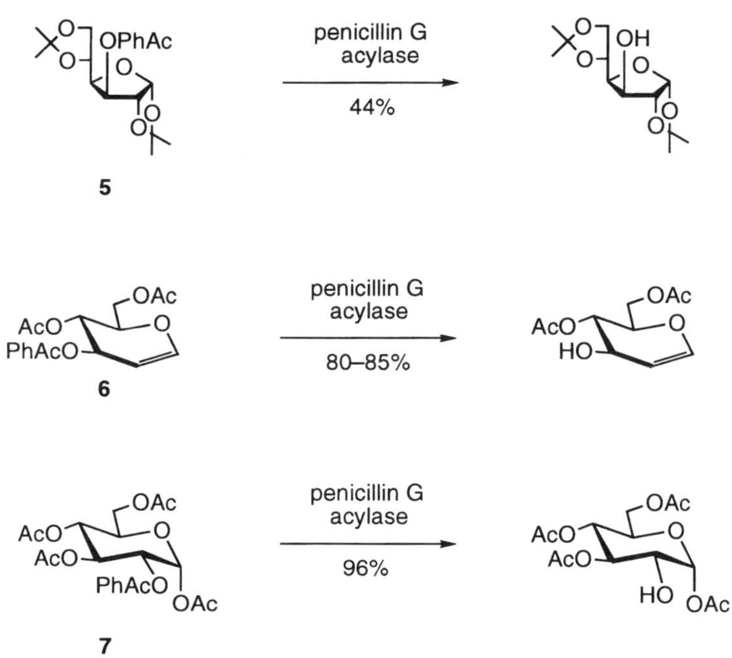

Fig. 3

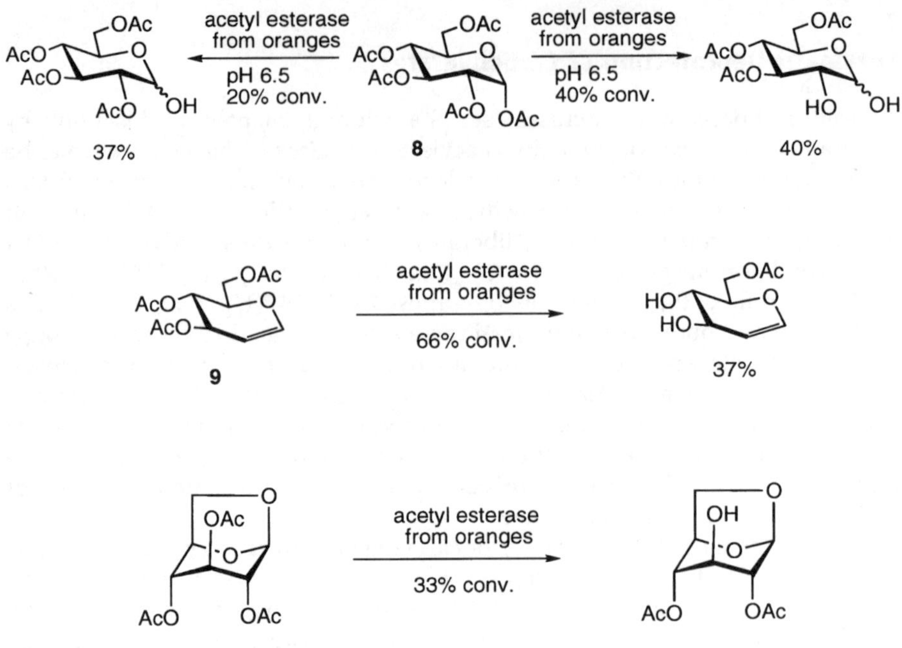

Fig. 4 **10**

instance, it chemoselectively removes the base-labile acetates from various isopropylidene-protected furanoses and pyranoses without attacking the acid-labile acetal groups. Issues of regioselectivity are addressed in the deblocking of peracetylated pyranoses and further functionalized carbohydrates. In these cases, acetyl esterase displays an unusual behavior which differs from the results recorded for classical chemical deprotections [16, 17] (Fig. 4). Thus, if the transformation of pentaacetyl glucose (**8**) with the enzyme is terminated after 20% conversion, i.e. corresponding to the cleavage of one of the five acetates present, the compound deblocked at the 1-position is formed predominantly. If the reaction is allowed to proceed further, the primary ester at C-6 is not attacked, but the 2-OH group is liberated instead, leading to 3,4,5-tri-*O*-acetyl-glucopyranose. If tri-*O*-acetyl glucal (**9**) is subjected to the enzymatic reaction, the ester in the 4-position is removed first, unexpectedly, and after 66% conversion the hexose carrying the remaining acyl function at the 6-position is obtained as the major product in useful yield. Finally, in 1,6-anhydro glucose (**10**) the acetyl esterase preferably attacks the acetate in the 3-position, which, owing to steric hindrance, is less reactive in this molecule. Therefore, in this case also, the enzymatic procedure makes possible the complete reversal of the regioselectivity.

3
Enzymatic Deprotection of Nucleosides

Acetyl esterase is a useful enzyme for the removal of acetyl groups from furanoses also [17]. In all cases, the reaction conditions are so mild (pH 6.5, room temperature, 0.15 N NaCl buffer) that other functionalities are not affected. However, besides the deprotection of simple furanoses, the enzyme shows a particularly interesting behavior towards per-*O*-acetylated 2-deoxynucleosides [16]. Thus, the adenosine and guanosine derivatives **11** and **12**, which do not carry *N*-protecting groups in the nucleobases, are deprotected at the secondary 3-OH to give **13** and **14** respectively (Fig. 5). However, if the amino groups are blocked as phenylacetamides as in **15** and **16**, the enzyme preferably saponifies the 5-acetate, generating **17** and **18**, respectively [16, 17]. Consequently, a modification of the substrate at a site relatively remote from the enzyme-labile functional group changes the regioselectivity of the enzymatic transformation entirely.

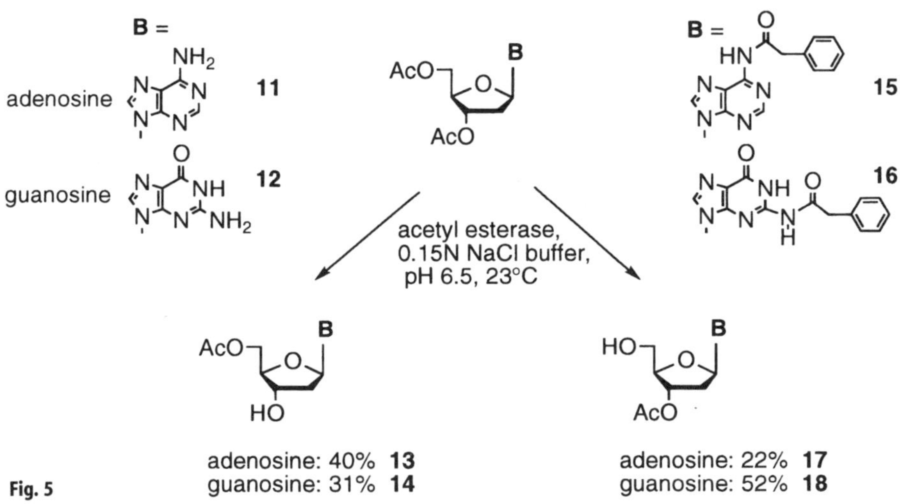

Fig. 5

In addition to the selective functionalization of the hydroxyl groups, oligonucleotide synthesis also requires suitable protecting groups for the nucleophilic amino functions of the nucleobases adenosine, guanosine, and cytidine. For this purpose, the amino groups were masked as phenylacetamides, and the resulting compounds **19, 20,** and **21** were examined as substrates for the penicillin G acylase mentioned above. It turned out that the enzyme cleaves the phenylacetamides in all cases without any detectable side reaction [16] (Fig. 6). Again, the enzymatic transformation is fully chemoselective, i.e. the acetates are left intact, and the reaction conditions are so mild that, for example, the acid-labile *N*-glycosides are not attacked.

This enzymatic deprotection technique can also be successfully applied in oligonucleotide synthesis [18]. Thus, for instance, the pentanucleotide **22** was built up from PhAc-protected guanosine employing the well-established phosphor-

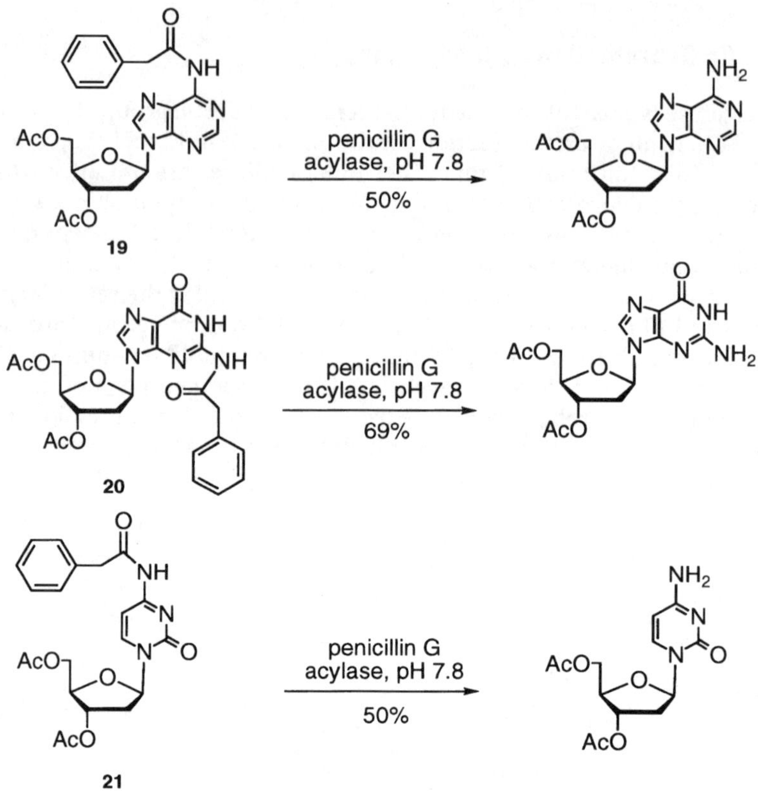

Fig. 6

amidite method for the coupling steps and controlled pore glass beads (CPG) as solid support (Fig. 7). Upon treatment of **22** with ammonia at room temperature for 40 minutes, the nucleotide was released, and simultaneously the cyanoethyl phosphates were saponified. Subsequently, all phenylacetamides were cleaved from **23** by treatment with penicillin G acylase under neutral conditions to give the desired pentanucleotide **24** quantitatively. In contrast, in the classical chemical deprotection method, the removal of the amide-protecting groups of the nucleobases is carried out by prolonged heating with concentrated ammonia.

All the transformations carried out with penicillin acylase and employing phenylacetates or -amides as substrates are hampered by the very limited solubility of these esters in aqueous environments. Although the enzyme tolerates considerable amounts of organic cosolvents, generally their application results in at least a partial deactivation of the enzyme. Since penicillin acylase accepts variations in the phenylacetic acid part of its substrates, pyridyl acetic acid esters were employed to enhance the solubility of the substrates in aqueous solution. In fact, several simple 4-pyridylacetates turned out to be fairly soluble in aqueous media and were attacked at very acceptable rates by the enzyme [19]. It is interesting to note that the velocity of the enzymatic transformations depends

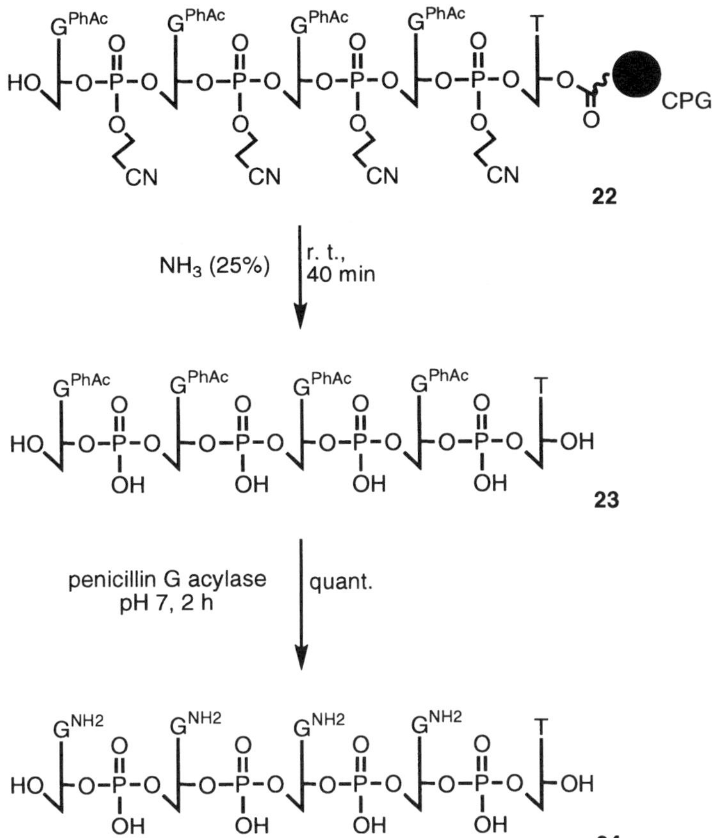

Fig. 7

on the position of the nitrogen in the ring. Thus, 3-pyridyl acetates are attacked considerably more slowly than the 4-*N* derivatives, whereas 2-pyridyl esters are even worse substrates.

Surprisingly, the introduction of the pyridine ring not only influences the velocity of the enzymatic transformations, but also induces promising stereochemical effects (Table 1). For instance, at 40% conversion (R)-phenylethanol is obtained from the pyridyl acetate **25** with 73% ee, whereas the value for the corresponding phenylacetate is only 28%. Also, the secondary alcohol liberated from the ester **26** displays 98% ee at 40% conversion, whereas the respective phenylacetate leads to 1-phenylpropanol with 94% ee but at a conversion rate of 12% only [19, 20]. These results demonstrate that the stereoselecting properties of penicillin acylase may be enhanced by appropriate engineering of the substrate. This is of particular interest since this enzyme has already been used for the kinetic resolution of various chiral alcohols [21–24], e.g. furyl alkyl carbinols [24], which are valuable precursors for the de novo synthesis, with moderate to high ee values, of carbohydrates.

Table 1. Penicillin G acylase-catalyzed hydrolysis of pyridyl acetates

No.	substrate	relative velocity [%]*	conversion [%]	ee [%][#]
	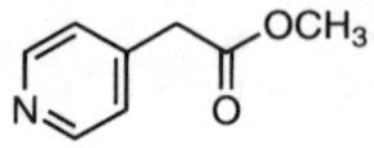	97		
	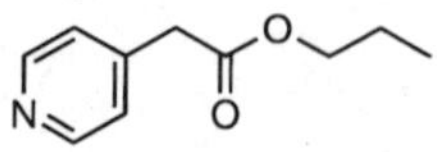	76		
	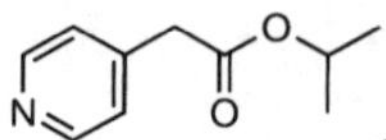	44		
25	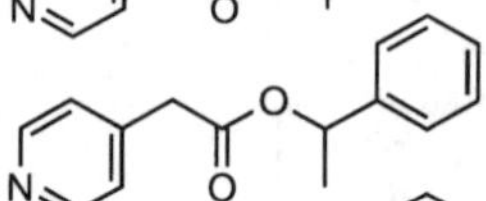	4	40	73 (R)
26	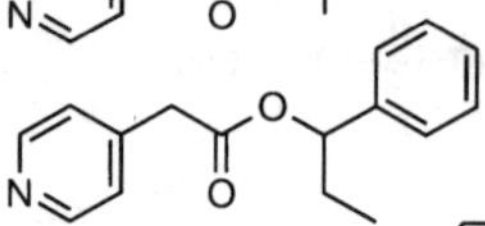	1.1	40	98 (R)
	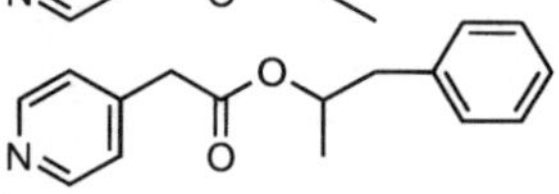	12	40	64 (S)
	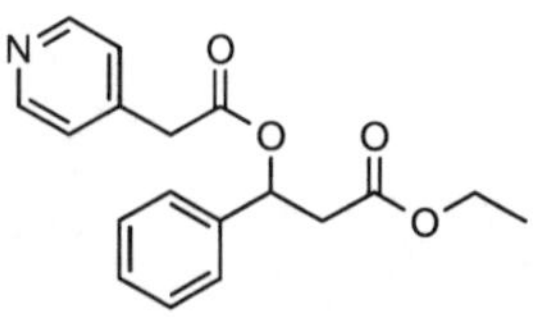	2.5	25	54 (R)

* 100% = hydrolysis of benzylpenicillin at pH = 8.
[#] enantiomeric excess and absolute configuration of the liberated alcohol.

4 Enzymatic Deprotection of Peptides

The selective protection and liberation of the α-amino function, the carboxy group, and the various side-chain functional groups of polyfunctional amino acids constitute some of the most fundamental problems in peptide chemistry.

For the *N*-terminal deprotection of peptides, the enzyme penicillin G acylase from *E. coli* has been applied. This attacks phenylacetic acid (PhAc) amides and esters but does not hydrolyze peptide bonds [12–14, 25]. The danger of a competitive cleavage of the peptide backbone at an undesired site, which always exists when proteases like trypsin and chymotrypsin are used, is overcome by using the acylase. The penicillin G acylase accepts a broad range of protected dipeptides (**27**) as substrates, and selectively liberates the *N*-terminal amino group under almost neutral conditions (pH 7–8, room temperature), leaving the peptide bonds as well as the *C*-terminal methyl-, allyl-, benzyl-, and *tert*-butyl ester unaffected (Fig. 8) [25a, b]. On the other hand, the phenylacetamide

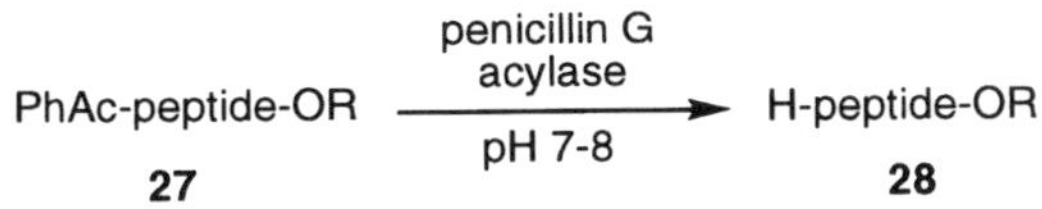

peptide	R	yield [%]
Gly-Phe	*t* Bu	65
Thr-Ala	*t* Bu	75
Ala-Ala	*t* Bu	90
Ser-Leu	*t* Bu	95
Asp(OH)-Phe	Me	87

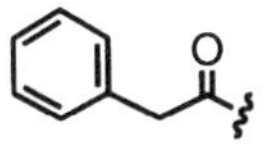

= PhAc

Fig. 8

moiety is stable to acids, bases, and hydrogenolysis and is therefore not attacked during the cleavage of the *C*-terminal ester functions.

The application of the non-urethane PhAc blocking group results in ca. 6% racemization during the construction of the phenylacetamido-protected dipeptides by chemical activation of the phenylacetamido amino acids. This disadvantage can be overcome by forming the peptide bonds enzymatically, e.g. with trypsin [26, 27], chymotrypsin [26, 28], or carboxypeptidase Y [26, 29]. An interesting example is the biocatalyzed synthesis of leucine-enkephalin *tert*-butyl ester [25e], in which phenylacetamides are introduced and cleaved by means of penicillin G acylase, and the elongation of the peptide chain is carried out with papain or α-chymotrypsin.

To overcome the drawbacks associated with the application of non-urethane *N*-terminal blocking groups, a new strategy had to be developed which enables a biocatalyst to cleave a urethane structure without the need for a direct attack on the urethane carbonyl group. The principle was successfully realized by the introduction of the *p*-acetoxybenzyloxycarbonyl (AcOZ) [30] and the *p*-phenylacetoxybenzyloxycarbonyl (PhAcOZ) [31] groups (Fig. 9).

These urethanes embody (1) a functional group (e.g. an acetate or a phenylacetate) that is recognized by the biocatalyst and bound by an enzyme-labile linkage (an ester) to (2) a functional group (here a *p*-hydroxybenzylurethane) that undergoes a spontaneous fragmentation upon cleavage of the enzyme-sensitive bond resulting in (3) the liberation of a carbamic acid derivative which decarboxylates to give the desired peptide, e.g. **31**. The quinone methide **32** which is formed from the phenolate can be trapped by water or by an added nucleophile like $NaHSO_3$. These enzyme-labile urethanes turned out to be particularly useful for the construction of lipopeptides and phosphoglycopeptides (see below).

The initial attempts to achieve an enzyme-catalyzed deprotection of the carboxyl groups used endopeptidases such as chymotrypsin [32–34] or trypsin [33, 35, 36]. These transformations involve the danger of an undesired hydrolysis of the peptide bonds. This disadvantage can be overcome by the use of

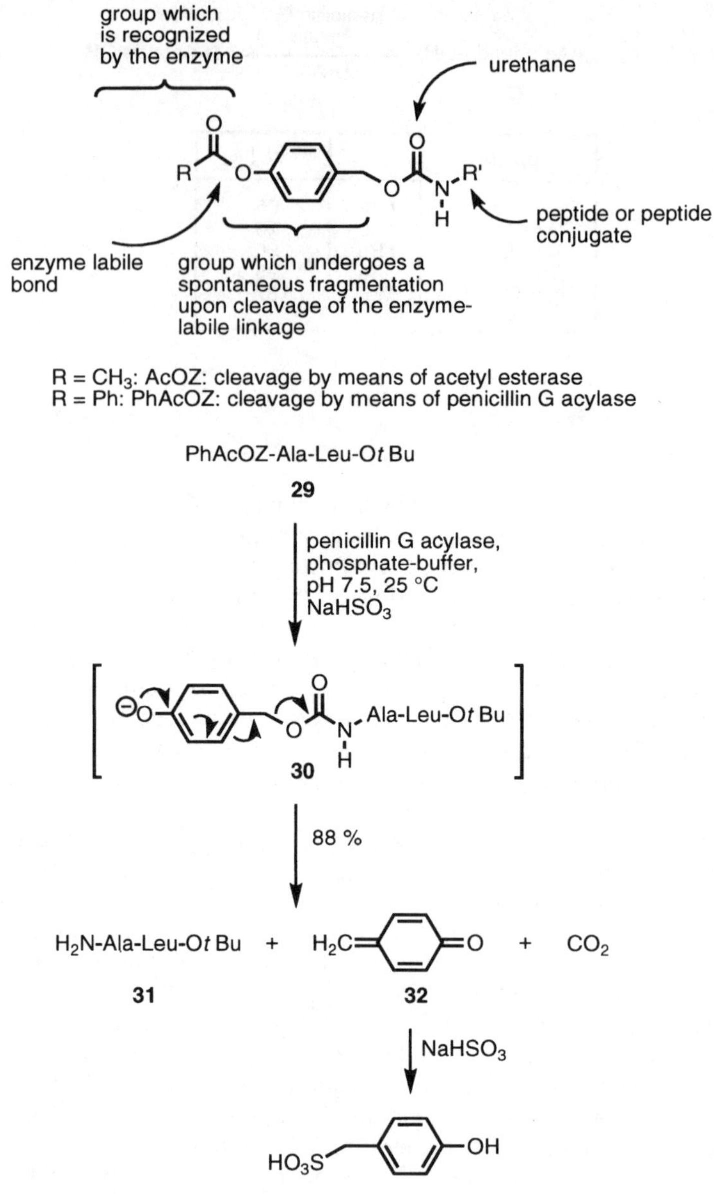

Fig. 9

carboxypeptidase Y from baker's yeast [37–39]. This serine-exopeptidase is characterized by different pH optima for the peptidase and the esterase activity (pH >8.5). By virtue of this property, it selectively removes the carboxyl-protecting groups from a variety of differently protected di- and oligopeptide methyl and ethyl esters [37, 39] without attacking the peptide bonds. An additional attractive feature is that the esterase activity is restricted to α-esters, the β- and the γ-esters of aspartic and glutamic acid not being attacked. The same regioselectivity is a feature of the thermostable extracellular serine protease thermitase from the thermophilic microorganism *Thermoactinomyces vulgaris*. Its esterase/protease ratio is >1000:1. The enzyme shows a broad tolerance for amino acids and cleaves methyl, ethyl, benzyl, ethoxybenzyl and even *tert*-butyl esters from a variety of Nps-, Boc-, Bpoc- and Z-protected di- and oligopeptides, delivering the peptides in high yields at pH 8 and 35–55 °C [25c,d, 40–42]. Even highly hydrophobic and water-insoluble peptides were accepted as substrates by alcalase from *Bacillus licheniformis*, whose major component is subtilisin A (subtilisin Carlsberg). This serine endopeptidase selectively saponifies peptide methyl and benzyl esters in a solvent system consisting of 90% *tert*-butanol and 10% buffer (pH 8.2) [43–45]. The hydrolysis of *C*-terminal amides is possible by means of a peptide amidase from the *flavedo* of oranges, which posesses a broad substrate tolerance and accepts Boc-, Trt-, Z- and Bz-protected and *N*-terminally unprotected peptide amides [46,47].

In these protease-catalyzed cleavages of the *C*-terminal protecting groups, it has to be taken into consideration that an undesired hydrolysis of peptide bonds can occur, especially if unnatural and poor substrates are subjected to enzyme-mediated transformations. The use of enzymes which cannot cleave amides at all enables this undesired side reaction to be overcome. This principle has been realized in the development of the heptyl (Hep) ester [13,14,48,49] as carboxyl protecting group, which can be enzymatically removed by means of lipases (Fig. 10).

The lipase from the fungus *Rhizopus niveus* accepts a variety of protected dipeptide heptyl esters (**33**) and hydrolyzes the ester function in high yields at pH 7 and 37 °C without damaging the urethane protecting groups and the peptide bonds. The *N*- and *C*-terminal amino acid can vary over a wide range, but, with increasing steric bulkiness and hydrophobicity of the amino acids, especially the *C*-terminal one, the reaction rates decrease.

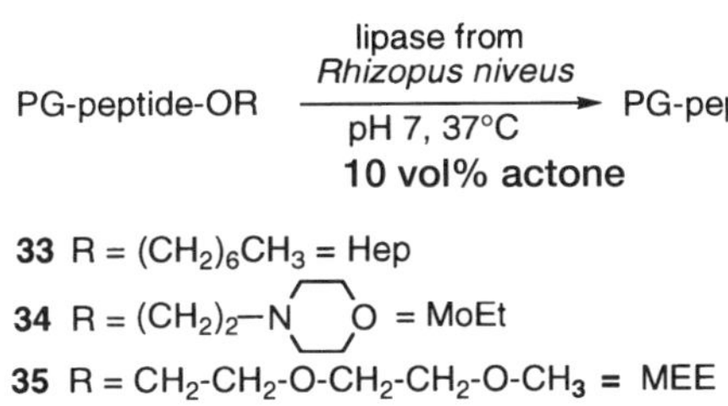

PG	peptide	R	yield [%]
Boc	Ser-Thr	Hep	95
Z	Thr-Ala	Hep	85
Aloc	Met-Gly	Hep	90
Boc	Val-Phe	MEE	97
Boc	Ser-Phe	MEE	94
Boc	Val-Phe	MoEt	91

Fig. 10

To overcome the problems associated with the use of hydrophobic Hep-esters, hydrophilic esters have been introduced which guarantee better accessibility of the substrate to the enzyme. Thus, the hydrophilic 2-(*N*-morpholino)ethyl ester (MoEt) (**34**) of a protected dipeptide was cleaved selectively and in high yield by means of lipase from *Rhizopus niveus* [50] (Fig. 10). The respective dipeptide heptyl ester was not attacked by the enzyme at all [48, 49]. The development of the diethyleneglycol- **35** and oligoethyleneglycol-derived esters [51] has to be seen as further progress. These can be hydrolyzed by lipases, although they are not particularly hydrophobic and so do not imitate the natural substrates of lipases. Even in cases in which the heptyl ester failed, the diethyleneglycol esters **35** were readily deblocked by lipase N.

A further advantageous solubility-enhancing protecting group was found by introducing a charged centre into the alcohol part of the *C*-terminal ester. The choline ester investigated for this purpose can be removed from peptides **36** under the mildest conditions (pH 6.5, room temperature) by using the enzyme butyrylcholine esterase from horse serum (Fig. 11) [52].

The solubilizing capacity of the choline residue is so pronounced that even substrates combining two hydrophobic amino acids are homogeneously soluble in aqueous buffer without any additional cosolvent. These favorable physical properties were also used in the enzymatic formation of peptide bonds. The amino acid choline ester **38** acts as the carboxyl component in kinetically controlled peptide syntheses with the amino acid amides **39** and **40** [52] (Fig. 11). The fully protected peptides **41** and **42** were built up by means of chymotrypsin in good yields. Other proteases like papain accept choline esters as substrates also, and even butyrylcholine esterase itself is able to generate peptides from these electrophiles.

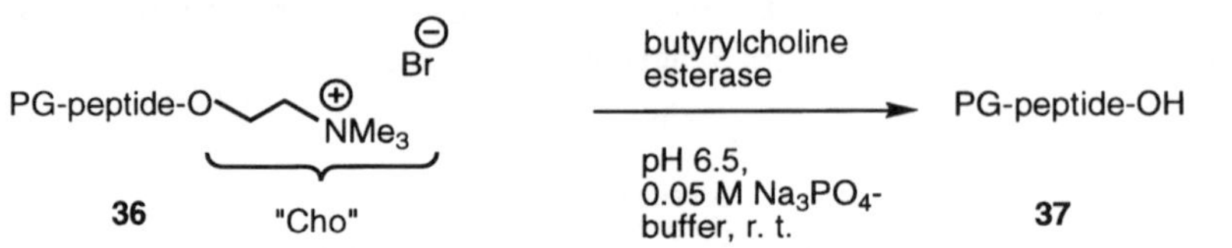

PG	peptide	yield
Boc	Phe-Phe	65
Boc	Ile-Ala	76
PhAc	Ile-Phe	91
Aloc	Leu-Pro	95

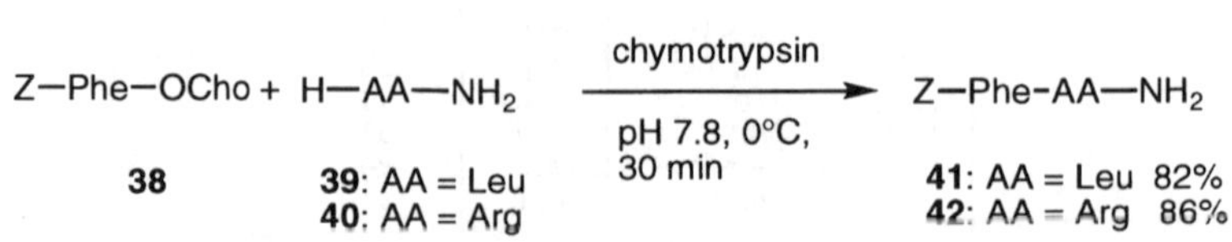

Fig. 11

5 Enzymatic Synthesis of Glycopeptides

The construction of peptide conjugates such as lipo-, phospho-, nucleo-, and glycopeptides requires various orthogonally stable protecting groups which have to be deprotected under mildest conditions because of the presence of structures sensitive to acids and bases (see above). Enzymatic removals of protecting groups offer the advantage that they often can be carried out close to neutral pH, so that the application of biocatalysts to effect the manipulation of suitable blocking groups can be the method of choice [5, 9–11, 53].

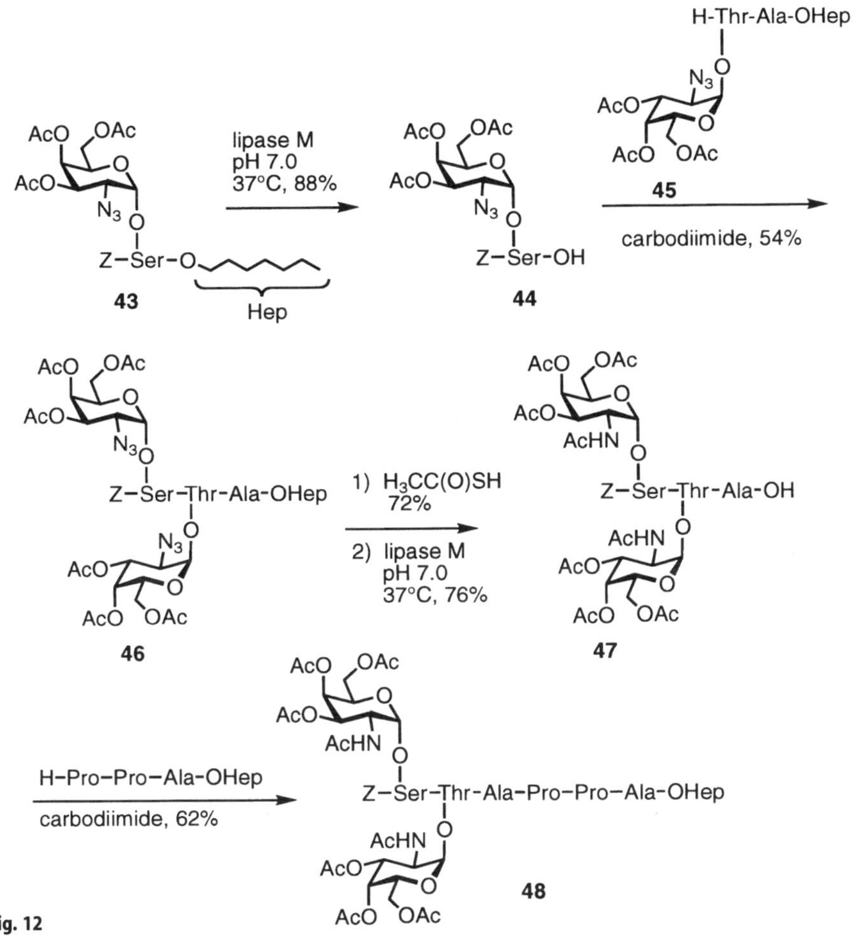

Fig. 12

The whole capacity of the lipase-mediated deprotection techniques described above was demonstrated in the synthesis of complex and sensitive glycopeptides carrying the characteristic linkage region present in biologically important *O*-glycoproteins [48, 54]. Thus, the serine glycoside heptyl ester **43** was selectively deprotected at the *C*-terminus by the lipase from the fungus *Mucor javanicus* (Lipase M from Amano) (Fig. 12). The carboxylic acid **44** liberated thereby was condensed with the *N*-terminally deprotected glycodipeptide **45** to yield the diglycotripeptide **46**. After conversion of its azido groups into acetamides, a second enzyme-mediated deprotection yielded the diglyco-tripeptide carboxylic acid **47**. The further elongation of the peptide chain with a tripeptide resulted in the formation of the complex diglycohexapeptide **48**, which carries the characteristic linkage region of a tumor-associated glycoprotein antigen found on the surface of human breast cancer cells. The enzymatic transformations left the *N*-terminal urethanes, the peptide bonds, the acid- and base-labile glycosidic linkages, and the acetyl protecting groups unattacked.

In addition to these examples, the lipase from *Mucor javanicus* accepts various further glycosylated amino acids and peptides as substrates and selectively deprotects them at the *C*-terminus (Fig. 13). The enzyme tolerates varia-

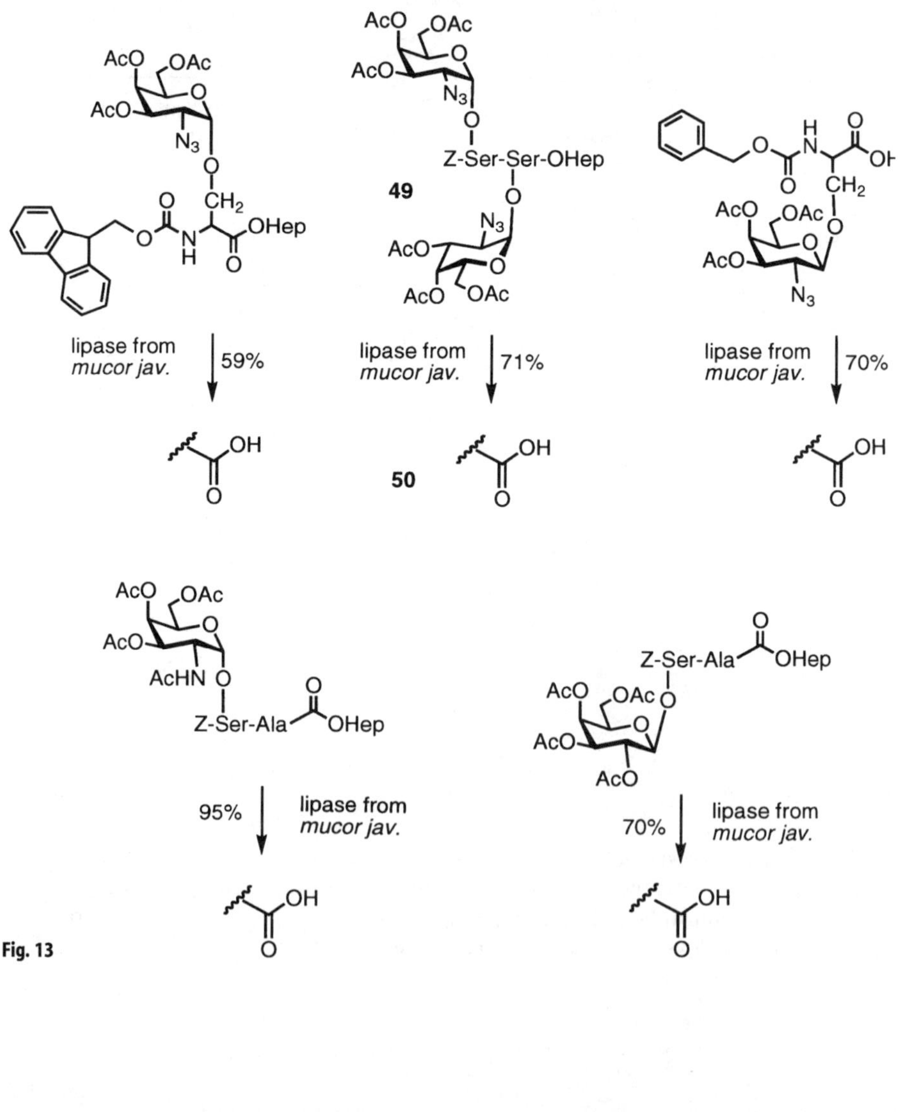

Fig. 13

tions of the *N*-terminal protecting group, of the configuration of the anomeric center, and finally of the structure of the entire carbohydrate. A particularly demanding example is the diglycodipeptide **49,** which carries two glycosylated amino acids linked to each other. Also in this case, the biocatalyst makes the desired selectively deprotected carboxylic acid **50** available in high yield.

In addition to the heptyl ester, the 2-(*N*-morpholino)ethyl (MoEt) protecting group can also be removed from *O*-glycopeptides by means of lipase N [50]. In the case of the diethyleneglycol esters [51], it has been shown that azidogalactosyl amino acid MEE esters are hydrolyzed by lipase M or papain. By means of the latter protease, the methyl ester could also be removed from other serine glycosides (e.g. **51**) [55, 56] and from an asparagine conjugate (**52**) [57] without disturbing side reactions (Fig. 14). Similarly, the liberation of the *C*-terminal carboxyl group of glycosylated dipeptides (e.g. **53** and **54**) was achieved by using subtilisin as biocatalyst [58] (Fig. 14). In these cases papain could not be employed, since the peptide bonds are cleaved preferably, highlighting the danger associated with the use of a protease for the removal of protecting groups from peptides.

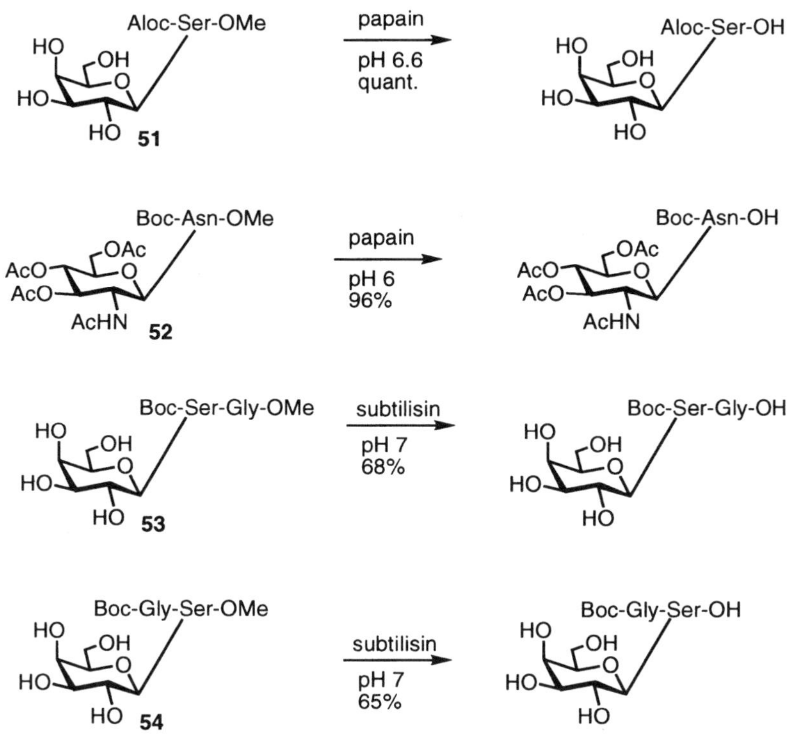

Fig. 14

The selective *N*-terminal deprotection of *O*-glycopeptides by means of an enzymatic technique was achieved using the *p*-phenylacetoxybenzyloxycarbonyl

(PhAcOZ) group (Fig. 15) [31]. The fully protected serine glycoside **55** could be deprotected under the mildest conditions by penicillin G acylase-catalyzed cleavage of the PhAcOZ-urethane. After condensation of the amine **56** obtained thereby with the glycosylated fragment **57**, once again the *N*-terminal blocking group could be removed from the diglycotripeptide **58** by enzyme-triggered fragmentation. The selectively deprotected diglycotripeptide **59** was further elongated with the PhAcOZ-masked dipeptide carboxylic acid **60**. From the resulting complex and sensitive diglycopentapeptide **61**, once more the PhAcOZ-group was

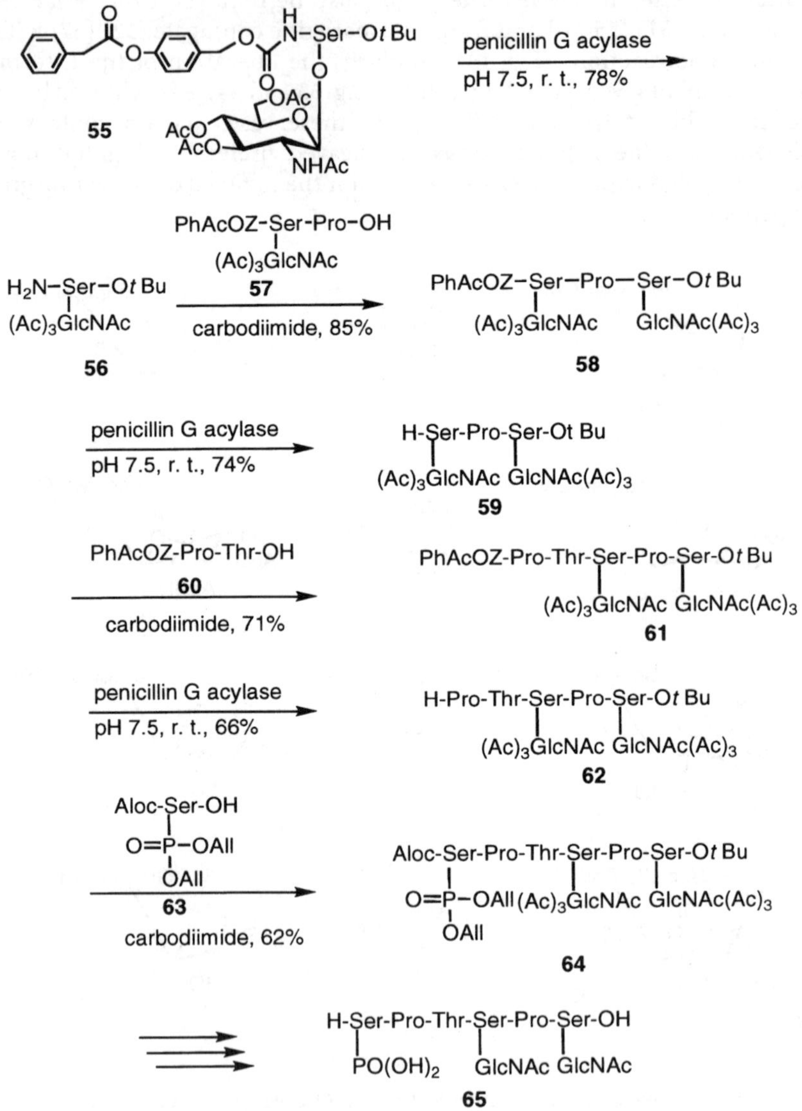

Fig. 15

cleaved off by means of penicillin G acylase to yield the selectively unmasked intermediate **62**. The coupling with the phosphorylated serine **63** resulted in the phosphoglycohexapeptide **64**, which carries one phosphorylated and two glycosylated hydroxyamino acids, and one unmodified. After the final deprotection steps, the complex multifunctional phosphodiglycohexapeptide **65** was obtained, which represents a characteristic partial structure of the *C*-terminal repeating domain of the large subunit of mammalian RNA polymerase II.

6
Enzymatic Synthesis of Lipopeptides

Another advanced application of enzymatic protecting group techniques for the construction of complex multifunctional and sensitive targets is the synthesis of the characteristic *S*-palmitoylated and *S*-farnesylated lipohexapeptide of the human N-*Ras* protein **69**. The presence of the base-labile thioester group and the acid-sensitive farnesyl group demands the use of mild protecting group techniques (see above). The synthesis of the lipopeptide **69** has been successfully carried out by two different routes, both employing enzyme-labile protecting groups. On the one hand, the choline esterase-mediated saponification of the choline ester was used as the key step in the construction of the lipohexapeptide **69** (Fig. 16). From the very base-labile palmitoylated tripeptide choline ester **66**, the biocatalyst removed exclusively the *C*-terminal blocking group at pH 6.5 and room temperature [52]. After *C*-terminal chain elongation with the farnesylated tripeptide **68** the desired lipopeptide **69** was obtained. It should be noted that the thioester present in **66** and **69** is so base-labile that even at pH 7–7.5 a spontaneous hydrolysis of this activated functional group occurs in aqueous solution. Thus, attempts to deblock a palmitoylated peptide ester by means of classical chemical techniques will result in a loss of the side chain. The enzymatic removal of the choline ester, however, permits the complete reversal of the chemoselectivity and thereby opens up new opportunities to construct biologically relevant peptide conjugatives like **69**.

In addition, the *p*-acetoxybenzyloxycarbonyl-(AcOZ) group, as an enzymatically cleavable urethane, served for an *N*-terminal coupling strategy [30]. To this end, the farnesylated tripeptide **74** was selectively deprotected by a lipase-initiated spontaneous fragmentation analogous to the cleavage of the PhAcOZ-group (see above). After elongation of the peptide chain with the AcOZ-protected dipeptide **72** and subsequent enzymatic deprotection, the lipopentapeptide **71** was obtained, which was condensed with the *S*-palmitoylated cysteine to give the acid- and base-sensitive lipohexapeptide **69** also.

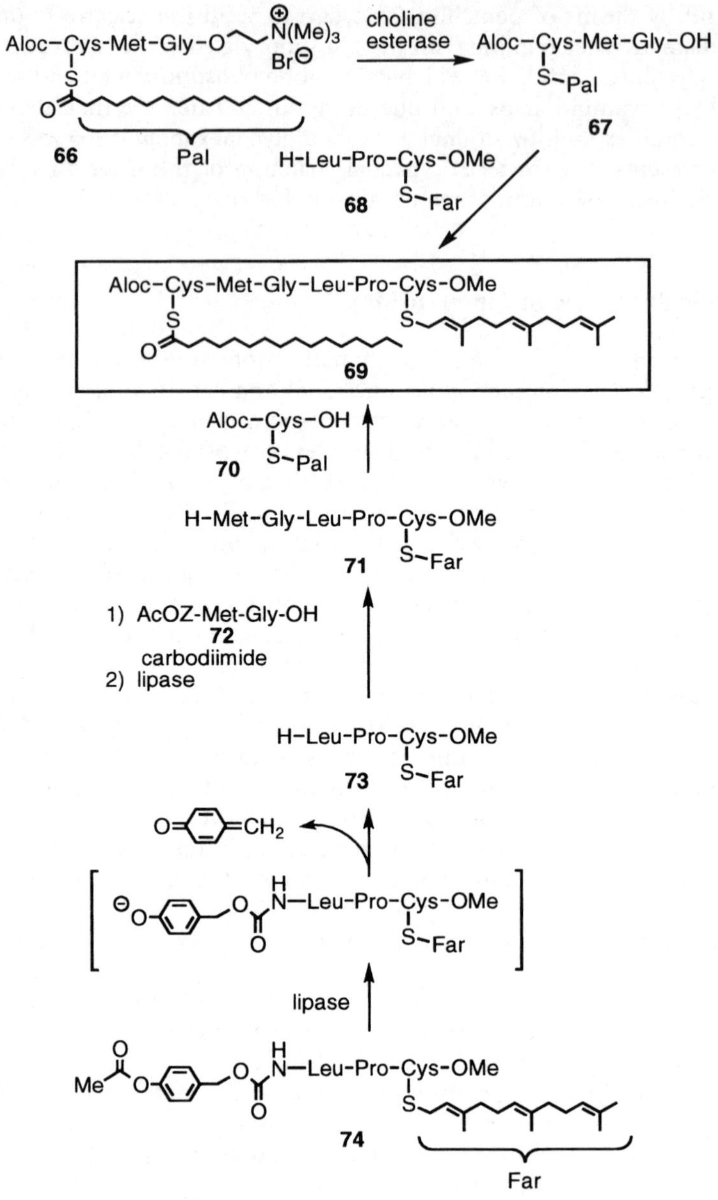
Aloc-Cys-Met-Gly-O N(Me)3 Br
S
O
66
Pal
choline esterase
Aloc-Cys-Met-Gly-OH
S-Pal
67
H-Leu-Pro-Cys-OMe
S-Far
68
Aloc-Cys-Met-Gly-Leu-Pro-Cys-OMe
S
S
O
69
Aloc-Cys-OH
S-Pal
70
H-Met-Gly-Leu-Pro-Cys-OMe
S-Far
71
1) AcOZ-Met-Gly-OH
72
carbodiimide
2) lipase
H-Leu-Pro-Cys-OMe
S-Far
73
O=CH2
H
O
N-Leu-Pro-Cys-OMe
S-Far
lipase
Me
O
N-Leu-Pro-Cys-OMe
S
74
Far

Fig. 16

7 Outlook

The ability of biocatalysts to perform chemical transformations under the mildest conditions and with high chemo- and regioselectivity can advantageously be exploited for efficient syntheses of sensitive polyfunctional peptide conjugates. Thus, compounds of high biological relevance, whose structures are unambiguously ascertained, can be made available, opening up new perspectives for advanced bioorganic investigations. For instance, from lipopeptides like **69**, so-called neo-lipoproteins can be built up, which carry characteristic structural elements of *Ras*-proteins as immunogenic determinants, making them promising probes for further immunological and biological studies. Peptide conjugates such as glyco-, phospho- and lipopeptides should be especially useful for studying the properties of proteins participating in signal transduction cascades and may thereby prove to be invaluable tools for the elucidation of the functions and interactions of these information-processing proteins. Relevant work at the borderline of biology and chemistry is in progress.

References

1. Rodbell M (1995) Angew Chem Int Ed Engl 34:1420
2. Egan SE, Weinberg RA (1993) Nature 365:781
3. Boguski MS, McCormick F (1993) Nature 366:643
4. Levitzki A (1994) Eur J Biochem 226:1
5. Schelhaas M, Waldmann H (1996) Angew Chem Int Ed Engl 35:2056
6. Synthesis of glycopeptides: (a) Kunz H (1987) Angew Chem Int Ed Eng 26:294 (b) Garg H, Jeanloz RW (1985) Adv Carbohydr Chem Biochem 43:135
7. Synthesis of phosphopeptides: (a) Perich JW (1990) In: Kemp BE (ed) Peptides and protein phosphorylation. CRC Press, Boca Raton, p 289 (b) Shapiro G, Buechler D (1994) Tetrahedron Lett 35:5421 (and refs. given therein)
8. Synthesis of nucleopeptides: (a) Dreef-Tromp CM, van den Elst H, van den Boogaart JE, van der Marel GA, van Boom JH (1992) Nucleic Acids Res 20:2435 (b) Dreef-Tromp CM, van der Maarel JCM, van den Elst H, van der Marel GA, van Boom JH (1992) Nucleic Acids Res 20:4015 (c) Kuyl-Yeheskiely E, Tromp CM, Lefeber AWM, van der Marel GA, van Boom JH (1988) Tetrahedron 44:6515
9. Waldmann H, Sebastian D (1994) Chem Rev 94:911
10. Waldmann H (1995) In: Drauz K, Waldmann H (eds) Enzyme catalysis in organic synthesis. VCH, Weinheim
11. Reidel A, Waldmann H (1993) J Prakt Chem 335:109
12. Waldmann H (1988) Liebigs Ann Chem 1175
13. Waldmann H, Braun P, Kunz H (1991) Biomed Biochim Acta 50:243
14. Waldmann H, Heuser A, Braun P, Schulz M, Kunz H (1992) In: Servi S (ed) Microbial reagents in organic synthesis. Kluwer, Dordrecht, p 113
15. Holla EW (1990) J Carbohydr Chem 9:113
16. Waldmann H, Heuser A, Reidel A (1994) Synlett 65
17. Waldmann H, Heuser A (1994) Bioorg Med Chem 2:477
18. Waldmann H, Reidel A, Angew Chem Int Ed Engl (in press)
19. Waldmann H, Pohl T (1995) Tetrahedron Lett 36:2963
20. Baldaro E, D'Arrigo P, Pedrocchi-Fantoni G, Rosell CM, Servi S, Casati P (1993) Tetrahedron Asymm 4:1031
21. Fuganti C, Grasselli P, Seneci PF, Servi S, Casati P (1986) Tetrahedron Lett 27:2061

22. Fuganti C, Grasselli P, Servi S, Lazzarini A, Casati P (1987) J Chem Soc Chem Commun 538
23. Fuganti C, Grasselli P, Servi S, Lazzarini A, Casati P (1988) Tetrahedron 44:2575
24. Waldmann H (1989) Tetrahedron Lett 30:3057
25. (a) Fuganti C, Grasselli P, Casati P (1986) Tetrahedron Lett 27:3191 (b) Waldmann H (1988) Tetrahedron Lett 29:1131 (c) Hermann P (1987) Wiss Z Univ Halle 36:17 (d) Hermann P (1991) Biomed Biochim Acta 50:19 (e) Greiner G, Hermann P (1991) In: Giralt E, Andreu D (eds) Peptides 1990. Escom, Amsterdam, p 277 (f) Didziapetris R, Drabnig B, Schellenberger V, Jakubke H-D, Svedas V (1991) FEBS Lett 287:31
26. Widmer F, Ohno M, Smith N, Nelson N, Anfinsen C (1983) In: Malon P (ed) Peptides 1982. De Gruyter, Berlin, p 375
27. Gerisch S, Jakubke H-D, Kreuzfeld H-J (1995) Tetrahedron Asymm 6:3039
28. Itoh K, Sekizaki H, Toyota E, Fujiwara N, Tanizawa K (1996) Bioorg Chem 24:59
29. Schuetz HJ, Wandrey C, Leuchtenberger W (1987) Abstracts of the Ninth Engineering Foundation Conf on Enzyme Engineering. New York, p 1
30. Waldmann H, Nägele E (1995) Angew Chem Int Ed Engl 34:2259
31. Pohl T, Waldmann H (1996) Angew Chem Int Ed Engl 35:1720
32. Walton E, Rodin JO, Stammer CH, Holly FW (1962) J Org Chem 27:2255
33. Kloss G, Schröder E (1964) Hoppe-Seyler's Z Physiol Chem 336:248
34. Xaus N, Clapes P, Bardaji E, Torres JL, Jorba X, Mata J, Valencia G (1989) Biotechnol Lett 11: 393
35. Hayward CF, Offord RE (1971) In: Scoffone E (ed) Peptides 1969. North-Holland, Amsterdam, p 116
36. Anantharamaiah GM, Roeske RW (1982) In: Rich DH, Gross E (eds) Peptides: synthesis, structure, function. Pierce Chemical Co, Rockford, p 45
37. Widmer F, Breddam K, Johansen JT (1981) In: Brunfeldt K (ed) Peptides 1980. Scriptor, Copenhagen, p 46
38. Royer G, Anantharamaiah GM (1979) J Am Chem Soc 101:3394
39. Royer GP, Hsiao HY, Anantharamaiah GM (1980) Biochimie 62:537
40. Hermann P, Salewski L (1983) In: Blaha K, Malon P (eds) Peptides 1982. De Gruyter, Berlin, p 399
41. Hermann P, Baumann H, Herrnstadt Ch, Glanz D (1992) Amino Acids 3:105
42. Dudek S, Friebe S, Hermann P (1990) J Chromatogr 520:333
43. Chen S T, Hsiao S-C, Chang C-H, Wang K-T (1992) Synth Commun 22:391
44. Chen S T, Chen S-Y, Hsiao S-C, Wang K-T (1991) Biomed Biochim Acta 50:181
45. Chen S-T, Wu S-H, Wang K-T (1991) Int J Pept Prot Res 37:347
46. Steinke D, Kula M-R (1990) Angew Chem Int Ed Engl 29:1139
47. Steinke D, Kula M-R (1991) Biomed Biochim Acta 50:143
48. Braun P, Waldmann H, Vogt W, Kunz H (1990) Synlett 105
49. Braun P, Waldmann H, Vogt W, Kunz H (1991) Liebigs Ann Chem 165
50. Braum G, Braun P, Kowalczyk D, Kunz H (1993) Tetrahedron Lett 34:3111
51. Kunz H, Kowalczyk D, Braun P, Braum G (1994) Angew Chem Int Ed Engl 33:336
52. Schelhaas M, Glomsda S, Hänsler M, Jakubke H-D, Waldmann H (1996) Angew Chem Int Ed Engl 35:106
53. Pohl T, Nägele E, Waldmann H (1994) Catalysis Today 22:407
54. Braun P, Waldmann H, Kunz H (1993) Bioorg Med Chem 1:197
55. Cantacuzene D, Attal S, Bay S (1991) Bioorg Med Chem Lett 197
56. Cantacuzene D, Attal S, Bay S (1991) Biomed Biochim Acta 50:231
57. Ishii H, Unabashi K, Mimura Y, Inoue Y (1990) Bull Chem Soc Jpn 63:3042
58. Attal S, Bay S, Cantacuzene D (1992) Tetrahedron 48:9251

Tandem Asymmetric C–C Bond Formations by Enzyme Catalysis

Michael Petersen · Maria Teresa Zannetti · Wolf-Dieter Fessner*

Institut für Organische Chemie der Rheinisch-Westfälischen Technischen Hochschule Aachen, Professor-Pirlet-Str. 1, D-52056 Aachen, Germany

Catalytic aldol reactions are among the most useful synthetic methods for highly stereocontrolled asymmetric synthesis. In this account we discuss the recent development of a novel synthetic technique which uses tandem enzyme catalysis for the bi-directional chain elongation of simple dialdehydes and related multi-step procedures. The scope and the limitations of multiple one-pot enzymatic C–C bond formations is evaluated for the synthesis of unique and structurally complex carbohydrate-related compounds that may be regarded as metabolically stable mimetics of oligosaccharides and that are thus of interest because of their potential bioactivity.

Table of Contents

Topics in Current Chemistry, Vol. 186

1 Introduction

A growing number of intercellular communication events are being identified to be encoded in oligosaccharide structures, which involve the specific binding of particular types of oligosaccharides on one cell surface to (glyco)protein receptors on the surface of another cell. Not the least by the medicinal importance of such cellular interactions through oligosaccharide structures in central biological recognition phenomena – such as cell adhesion, viral infection, or cell differentiation in organ development and tumor metastasis [1–7] – this has shaped the highly interdisciplinary new field of "glycobiology" and has underscored the increasing demand for novel, more efficient approaches to the synthesis and chemistry of complex carbohydrates and glycoconjugates. However, because natural oligosaccharides and glycoconjugates are chemically labile and orally inactive, a currently prominent area of research is dedicated to the development of potential mimetics of natural oligosaccharide effectors that will procure a therapeutic utility by reasonable metabolic stability as well as oral bioavailability. Of particular interest are C-disaccharides [8–10], a class of compounds in which the glycosidic oxygen is replaced by a methylene group. These compounds hold promise to be potentially active agonists or unreactive antimetabolites, i.e. inhibitors of glycoside processing enzymes, by virtue of their resistance to chemical and enzymatic hydrolysis of the "glycosidic" linkage and their ability to interact with protein receptors analogously to their *O*-linked counterparts [8, 10].

1.1 Nucleoside antibiotics

In this respect it is revealing to note that Nature itself, in fact, has evolved very similar strategies in the design of chemical weaponry. Especially illustrative for this principle are the various nucleoside antibiotics [11–13] that have been isolated from different microorganisms and that are aimed against competing organisms (Scheme 1). The core structure of these natural products typically consists of a conjugate of one of the common pyrimidine or purine bases (or analogs thereof) glycosidically linked to a larger carbohydrate backbone. In addition to the primary glycosidic ring moiety, the residual chain of the latter may extend to mimic further saccharides (e.g. hikizimycin) or amino acid conjugates (e.g. mildiomycin, sinefungin, miharamycin, amipurimycin), or alternatively may be conformationally locked into linked (e.g. tunicamycin) or joined (e.g. herbicidin, derivatives of octosyl acid) furanose/pyranose ring systems that rather seem to mimic disaccharide moieties.

Although so far most of these biologically active agents have proved not to be useful for human therapeutic purposes, the highly specific nature of these entities have made them to important tools in glycobiology research. The tunicamycins [14], for example, in prokaryotic systems block the exchange of UDP activated *N*-acetylmuramic acid pentapeptide with a phospholipid carrier, thus inhibiting cell wall biosynthesis [15]; in eukaryotic systems, they block the

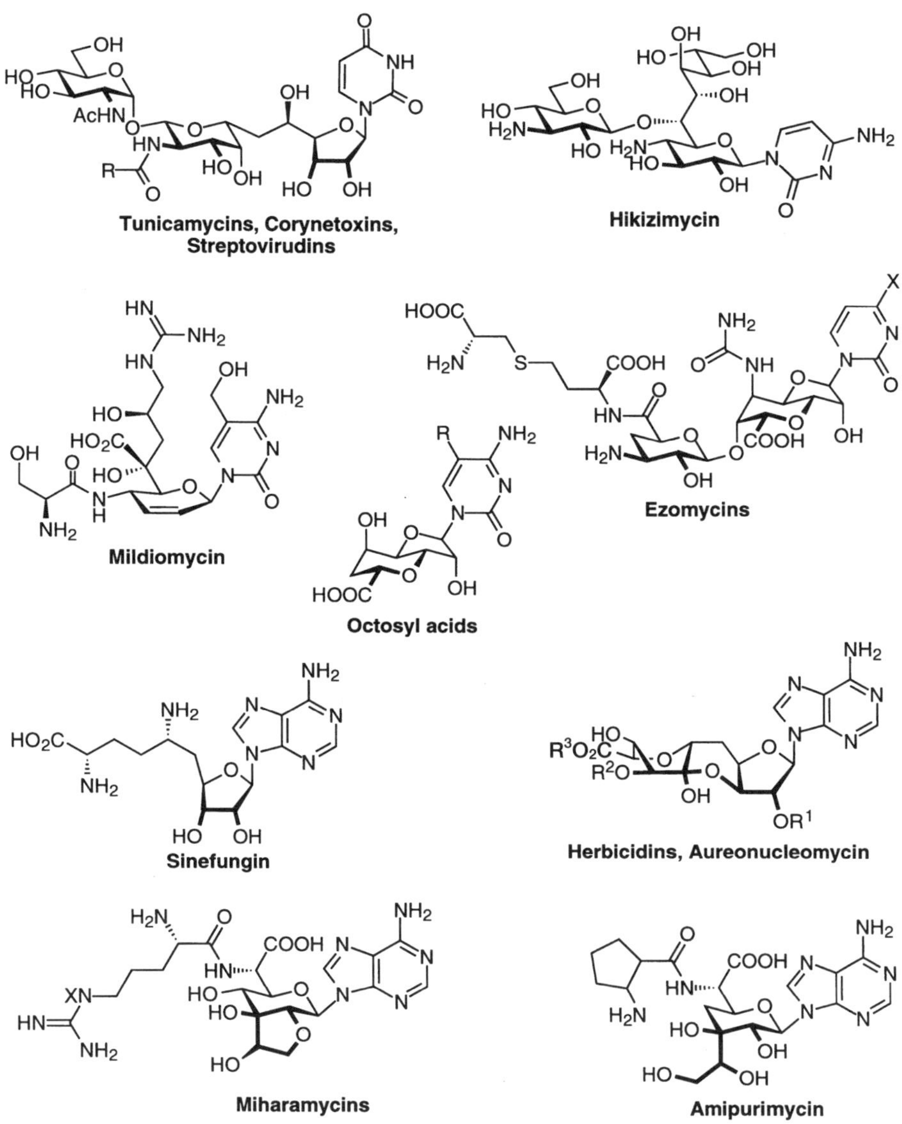

Scheme 1. Examples of naturally occurring nucleoside antibiotics containing an extended carbohydrate-type core structure

phosphoryltransferase reaction between nucleotide activated *N*-acetylglucosamine and the phospholipid anchor dolichol phosphate (Scheme 2), thereby inhibiting the first dedicated step of glycoprotein biosynthesis at the endoplasmatic reticulum [16]. Acting as a bisubstrate inhibitor, the transition state of the phosphoryl transfer is believed to be effectively mimicked by the spatially directed functionalities of the carbohydrate core of tunicamycin, the 11-carbon

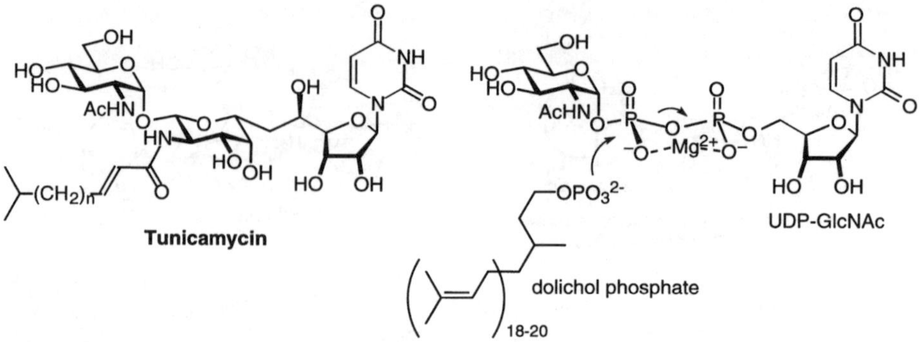

Scheme 2. Rationalization of inhibitory activity of the tunicamycins as a mimic of the transition state of the UDP-GlcNAc phosphotransferase reaction

aminodialdose tunicamine which is common to other families of nucleoside antibiotics such as the corynetoxins or streptovirudins [14]. Although by retrosynthetic operations D-ribose and *N*-acetyl-D-galactosamine are immediately identified as useful and commercially available building blocks for tunicamine, the currently best synthesis from such chiral-pool materials requires *more steps than the compound contains carbon atoms* [17, 18]. In contrast, Nature obviously succeeds in preparing such intricate entities essentially by a *one-pot synthetic operation*, and it does so without the need for those ingenious, but laborious schemes for functional group protection [19] that are the hallmark of classic carbohydrate chemistry!

Indeed, enzymatic catalysts are finding increasing acceptance in modern chemical research for the in vitro synthesis of valuable asymmetric compounds – as a supplement, or as an alternative, to classical chemical methodology – because of their high selectivity (substrate, chemo-, regio-, diastereo-) and their potential for high chiral induction and high catalytic efficiency [20–42]. In addition, biocatalytic conversions can usually be performed on underivatized substrates because of the usually very mild reaction conditions that are compatible with most functional groups, making tedious and costly protecting group manipulations superfluous [43]. Thus, particular advantages are evident for biocatalysis in the synthesis of biologically relevant carbohydrate conjugates which are typically polyfunctional and water soluble [22, 28, 30, 37, 38, 42, 44, 45].

1.2 Stereodivergent Enzymatic C–C Bond Formations

Current strategies for the synthesis of higher-carbon sugars, as exemplified by tunicamine – and particularly that of C-disaccharides –, fall largely into the two categories of chain elongation or of convergent block condensations on the basis of natural pentoses or hexoses: they either perform a (possibly iterative) extension of a given sugar from its carbonyl followed by diastereoselective placement of requisite substituents along the extended chain [46–48] or they use a coupling of two preformed sugar chains in a head to tail fashion after installation of an appropriate mutual functionalization [49]. In case of the generation of C-disaccharides, this translates into the tasks of extending a pyranose unit from the anomeric center, followed by subsequent pyranose cyclization or that of a suitable coupling of two preformed pyranose moieties [8, 9]. Obviously, limitations to these approaches are set by a need for the free, unmasked carbonyl group of the sugars and thus intensive protecting group manipulations, and by the limited availability or reactivity of suitably configurated precursors from the "chiral pool" [50, 51].

An alternative approach may also consist of the de novo synthesis of the two carbohydrate substructures that would have to be grown from the termini of much simpler precursors. For classical chemical syntheses, tremendous difficulties are immediately obvious by the simultaneous needs of high relative stereocontrol and terminus differentiation for each individual step. A somewhat better perspective may be seen for the use of stereoselective biocatalysts for asymmetric C–C bond coupling reactions. Indeed, more than 30 aldolases are known from which recently a larger number has been studied extensively for their utility in asymmetric synthesis [30, 37, 38, 41, 42].

For synthetic purposes, the most useful types are those lyases that depend on pyruvate or dihydroxyacetone phosphate (DHAP) as the aldol donor component [38, 42]. The latter enzymes catalyze the addition of DHAP to the carbonyl group of an aldehyde to form 3,4-dihydroxyketose 1-phosphates which contain two new vicinal chiral centers at the termini of the C–C bond created. Because the aldolization is doubly chirogenic this calls for four corresponding enzymes with stereoselectivities matching those of the four possible diastereomeric products,

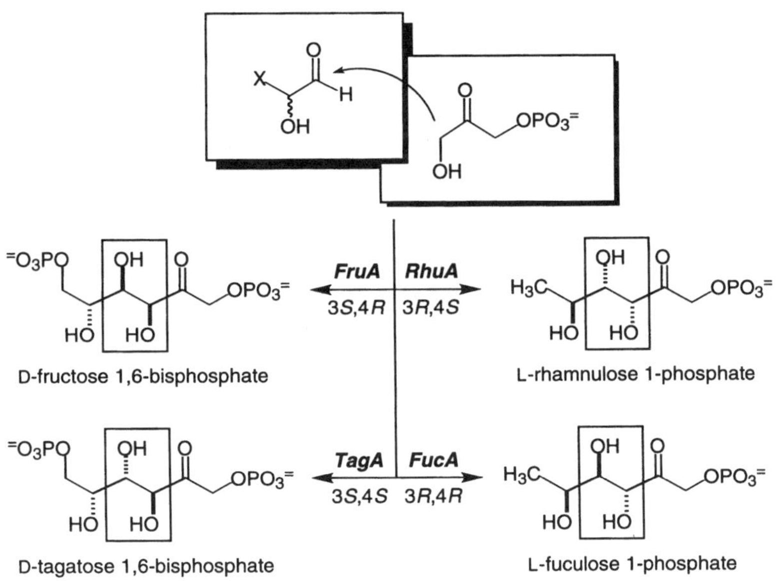

Scheme 3. Stereochemically complementary set of dihydroxyacetone phosphate dependent aldolases

ideally also with comparable synthetic capacities such as stability, substrate tolerance or kinetic features. Indeed, Nature has evolved a full set of stereochemically complementary aldolases [52] which act on four diastereoisomeric ketose 1-phosphates – D-fructose 1,6-bisphosphate aldolase=FruA, [EC 4.1.2.13]; D-tagatose 1,6-bisphosphate aldolase=TagA, [EC 4.1.2.n]; L-fuculose 1-phosphate aldolase=FucA, [EC 4.1.2.17]; L-rhamnulose 1-phosphate aldolase=RhuA, [EC 4.1.2.19] (Scheme 3). Recently, we have made all four types of these catalysts readily available [52–56], three of which can be overproduced from genetically engineered *E. coli* strains [52, 53, 56] and which are now commercial. Also, we and a number of other groups have established that the aldol acceptor component can be broadly varied, usually without affecting diastereoselectivity [22, 37, 42]. Formally, this has set the stage for a combinatorial biocatalytic approach to asymmetric synthesis [52, 57], for the deliberate preparation of chiral aldol adducts in a building block fashion by simply choosing the appropriate enzyme and starting materials to achieve full control over constitution and absolute configuration of the desired product.

1.3
Synthetic Utility of Aldolases

Mechanistically, the activation of the aldol donor substrates by stereospecific deprotonation is achieved in two different ways [58]: while class I aldolases (mostly found in animals and higher plants) bind their nucleophilic substrates covalently via imine/enamine formation to an active site lysine residue to initiate bond cleavage or formation [59], class II aldolases (commonly found in bacteria) utilize Zn^{2+} ions as a Lewis acid cofactor which by a bidentate ligand coordination facilitates deprotonation to give the enediolate nucleophile [60]. Particularly, the class I FruA isolated from rabbit muscle, which for long was the only aldolase commercially available, has been most extensively investigated for preparative purposes [61] and must be regarded as the prototypical reference catalyst. The stereochemistry of the C–C bond formation is controlled by the enzyme with high absolute and relative chiral induction, usually irrespective of the constitution or configuration of the substrate. Particularly, polar substituted aldehydes are good substrates, with highest conversion rates, diastereoselectivities, and yields generally being achieved with 2- or 3-hydroxyaldehydes, where the ketose products readily cyclize in aqueous solution to form stable furanoid or pyranoid rings. Individual enzymes even have the capacity for a powerful kinetic resolution of racemic 2-hydroxyaldehyde substrates (FucA, RhuA) [62, 63] to simultaneously determine additional contiguous chiral centers. Because of the reversible nature of the C–C bond forming processes, the thermodynamically most stable product configuration can be obtained selectively upon prolonged equilibration when starting from a given stereochemically related set of acceptor aldehydes (enantiomers or diastereomers) [61, 64]. Such techniques are facilitated by the fact that most of the aldolases are sufficiently robust to allow their use in solution for an extended period of time, often for several days [53, 65]. As a consequence, the absolute and relative configurations in the products can be controlled by the choice of the aldolase (C-3, -4) and by adopt-

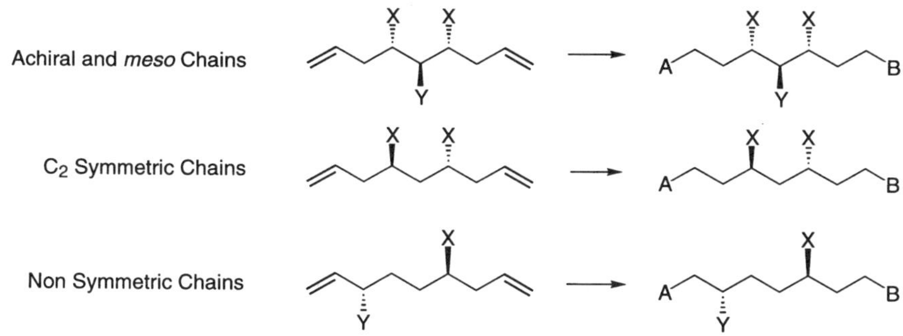

Scheme 4. Stereochemical consequences of bi-directional chain synthesis [70] on appended groups (A, B) as imposed by symmetry

ing a kinetically or thermodynamically selective reaction mode (C-5, -6) (*vide infra*).

In principle, such processes should also be applicable to bifunctional aldehydes for a two-directional chain elongation in which two equivalents of DHAP nucleophiles would be added sequentially to both the acceptor carbonyls in a fashion that can be classified as a "tandem" reaction [66, 67], without the need for isolation of any intermediates. Depending on the specificity of the enzyme used and on the number and position of hydroxyl functions in the starting material, the isomeric constitution, as well as the absolute and relative stereochemistry should be deliberately addressable. Thus, in a preparatively simple manner, such tandem aldolizations [68] should permit to rapidly construct larger carbohydrate molecules that would rival the carbohydrate core of tunicamine and related nucleoside antibiotics in structural complexity.

Within the scheme of a bi-directional chain synthesis [69–71] (Scheme 4) because of the lack of any protecting groups on the functional groups present in starting material, intermediate(s), and product, the positioning of OH groups relative to each other and relative to the aldehydic function will strongly influence the nature and prevalence of intermediate structures and thus, the likelihood of potentially alternative reaction pathways. Consequently, an inherent feature of reactions of this type would be the intricacy of the sequence of enzymatic transformations and potential chemical interconversion of the various open-chain or cyclic intermediates or products. On the other hand, careful selection of individual biocatalysts may offer also effective simultaneous solutions to the problem of terminus differentiation [69–71] for a bi-directionally elongated chain. Principally, this may be effected by a kinetically or thermodynamically based distinction of those dialdehyde termini that are stereochemically unequal, such as in the case of *meso* or racemic precursors.

For preparative scale applications DHAP may be produced by chemical [72–75] or enzymatic syntheses [76, 77]. Problematic, however, is the fact that DHAP is relatively unstable in solution and, particularly under alkaline conditions, readily decomposes into methylglyoxal and inorganic phosphate [78–80]. Hence, DHAP is preferably generated and consumed in situ to avoid the buildup of high stationary concentrations. This can be achieved conveniently either by the cleavage of fructose 1,6-bisphosphate (FBP) by a combination of FruA and triose phosphate isomerase (EC 5.3.1.1) [61, 81, 82] to give two equivalents of DHAP or, alternatively, DHAP can be formed by enzymatic oxidation of glycerol *via* L-glycerol phosphate (G3P) [83, 84].

2 Open-Chain Precursors

2.1 Simple Aliphatic Dialdehydes

Several attempts of FruA-catalyzed DHAP additions to simple aliphatic dialdehydes like glyoxal or glutaric dialdehyde have been reported in the literature, but in no case had a product been isolated and characterized [72, 85]. Malonic dialdehyde cannot be used because it tends to enolize under protic conditions and engages in polycondensations. Our own extensive studies corroborate that enzymatic assays indicate a consumption of DHAP, but no defined products result according to t.l.c. or NMR analysis. Problematic also is the fact that aliphatic dialdehydes irreversibly destroy enzymatic activity by protein cross-linking [86, 87].

On the assumption that intermediary mono adduct may decompose because of their lower stability of cyclic forms and thus resulting higher reactivity, 3-hydroxylated glutaric dialdehydes (*e.g.*, **1**) were tested in an attempt to allow the formation of stabile pyranosides, however so far with no avail [88].

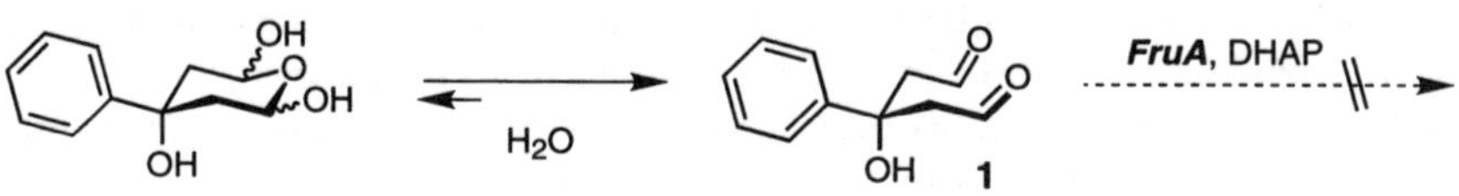

2.2 Dihydroxy-α,ω-Dienes

In reactions catalyzed by DHAP-aldolases, hydroxylated aldehydes are generally superior to unsubstituted aldehydes presumably because of their higher reactivity (electrophilicity), higher affinity to the enzyme active site (lower K_m values), and the fact that the products are stabilized by the formation of cyclic isomers [42]. Accordingly, substrates with dual 2- or 3-hydroxyaldehyde termini seemed to be a logical choice for potential tandem aldolizations.

The smallest member of such homologous series are the 2,3-dihydroxy succinic (tartaric) dialdehydes. Three stereoisomeric forms are possible, i. e. one *meso* (*erythro*) and two enantiomeric (D- and L-*threo*) compounds (i. e. **3**). Because of the enantiotopic nature of the termini of a meso chain, any twofold aldol addition under reagent control imposed by the same chiral biocatalyst would eliminate the element of σ-symmetry and thus effect a terminus differentiation (cf. Scheme 4). For the same reason, enzymes that cannot easily differentiate the two enantiotopic aldehyde groups would lead to the formation of two different,

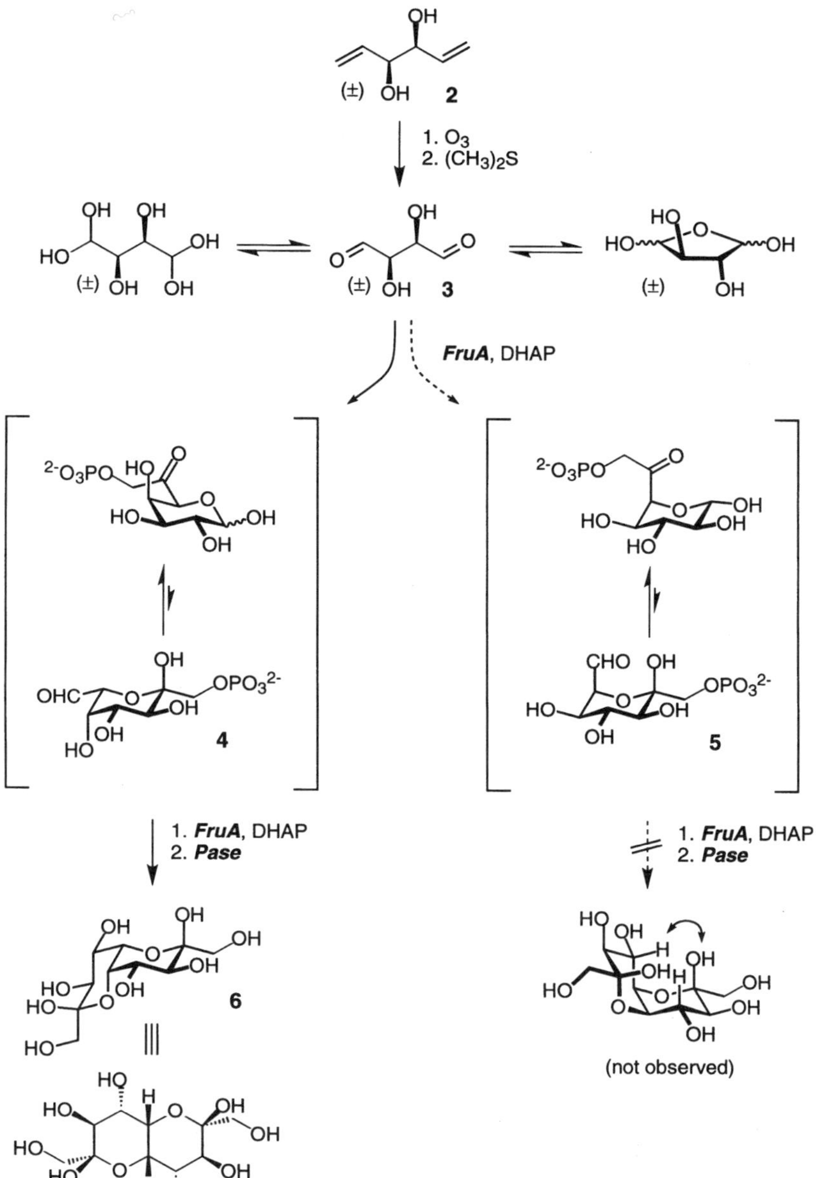

Scheme 5. Divergent pathways for tandem aldol additions of dihydroxyacetone phosphate (DHAP) to *rac*-tartaric dialdehyde catalyzed by fructose 1,6-bisphosphate aldolase (FruA)

diastereomeric mono adducts. In contrast, the *threo* isomers **3** are C_2-symmetrical and contain homotopic groups at the termini, so that the bi-directional homologation procedure would maintain the overall chain symmetry. Also, for each of the enantiomers only a single mono adduct (**4** and **5**) would be formed.

For the sake of analytical simplicity, an enantiomerically pure staring material would therefore be ideal. On the other hand, the employment of the racemate **3**, which is accessible by much simpler means [89], seemed to be similarly feasible because of the distinct properties that have to be expected for the pair of diastereoisomeric bisadducts. The stereospecific addition of one equivalent of DHAP to each of the dialdehyde enantiomers should produce two diastereomeric 6-heptosulose 7-phosphates with D-*galacto* (**4**; from L-*threo*) and D-*ido* (**5**; from D-*threo*) configuration (Scheme 5). Because of the presence of both aldehydic and keto carbonyl groups at the respective ends of the sugar backbones which can be utilized for the generation of cyclic hemiacetals, in aqueous solution compounds **4** and **5** must be expected to exist as a mixture containing two furanoses and pyranoses each in both anomeric configurations. If present at a sufficient concentration to be accepted by the aldolase, an isomeric form having a free (hydrated) aldehyde function could be converted in a consecutive addition step. Although the presence of a sugar ring next to the electrophilic site will likely reduce acceptance for steric reasons, the negative charge introduced with the first DHAP equivalent in fact might even enhance the substrate quality for a FruA catalyzed conversion, because it resembles the charge in its natural acceptor substrate glyceraldehyde 3-phosphate. Indeed, some aldose phosphates have been shown to be substrates for FruA [90].

The racemic *threo*-precursor **3** is readily accessible by pinacol dimerization of acrolein [89], followed by separation of stereoisomers [91] (**2**) and ozonolysis. In aqueous solution it exists as a complex equilibrium mixture, in which according to NMR analysis furanoid structures and open-chain hydrates predominate, and only a minute fraction of free aldehyde forms are detectable which are required for an enzymatic conversion. In the event, DHAP addition catalyzed by the FruA from rabbit muscle proceeded smoothly to the desired bisadduct stage [92]. As expected, transitory formation of a mono adduct could be detected by t.l.c., if only at low concentration, which indicates that the rate for the second addition step is of the same order of magnitude as that for the initial step. Interestingly, after isolation of the bisphosphate fraction by ion exchange chromatography followed by dephosphorylation using alkaline phosphatase, NMR analysis indicated that only a single diastereomer **6** derived from the L-*threo* dialdehyde was present. Its constitution as the C_2-symmetrical bicyclic D-*threo*-L-*galacto*-deco-2,9-diulo-2,6;9,5-dipyranose was readily established by the number of NMR resonances and magnitude of H,H coupling constants. The corresponding bisadduct derived from the D-*threo* antipode could not be detected. In fact, in a separate experiment this product could also not be generated under more forcing conditions using an excess of DHAP and larger catalyst quantities. Upon inspection of molecular models it becomes obvious that this elusive product would be highly destabilized in an anulated bipyranose form because of unfavorable twofold 1,3-diaxial interactions and other steric crowding within its concave interior face.

In summary, the unprotected ten-carbon diketose **6** could be obtained in an enantio- and diastereomerically pure state and 63% overall yield through quite simple laboratory manipulations (3 steps from racemic diene **2**) [92]. Interesting features are the C_2-symmetrical hemispherical ring arrangement related to *cis*-decalin which induces a chiral hydrophobic cavity, as well as the complete and relatively rigidly presented oxygen functionalization around its outer rim. During the one-pot, tandem enzymatic operation a remarkable number of six asymmetric centers are determined: 2 × 2 stereogenic centers are newly created upon C–C bond formations and two more are selected from the racemic precursor due to thermodynamic preference. When compared to the unfruitful attempts for twofold homologation of generic dialdehydes, the success of the above experiments must clearly be attributed to the presence of hydroxyl functions which enable intermediates and products to form stable hemiacetals.

Considerations as to the pathways open for conversion of the bishomologous *erythro*-3,4-dihydroxyhexanedial **8** are illustrative as an example for *meso* dialdehydes (Scheme 6). Its two enantiotopic termini most likely cannot be easily distinguished by an aldolase because of their more remote position from asymmetric centers. Therefore, the generation of two diastereomeric mono adducts (**9, 10**) in comparable quantities is to be expected. These mono adducts can exist as an equilibrium mixture of phosphorylated aldofuranoses and ketopyranoses. The latter species quite likely are the prevalent components, at least for the D-*gulo*-diastereomer **9** (because of the greater stability of an equatorially substituted pyranose [93] and the higher electrophilicity of the keto carbonyl due to electron withdrawing substitution), which should favor a consecutive tandem addition to take place. In any case should the negative charge introduced by the phosphate group enhance the substrate quality. By such a second addition of DHAP to the respective intermediates both pathways would converge with the formation of a single, non-symmetrical bisadduct **11**.

The substrate **8** can be conveniently prepared from the corresponding α,ω-diene **7** by ozonolysis (for cyclic precursors, see Sect. 3.2) [94]. In aqueous solution, this dialdehyde also forms several cyclic hemiacetal structures as was evident by its complex NMR spectra. Because of its lower electrophilicity, a certain proportion exists as free aldehyde which suggested improved substrate qualities. Indeed, from the enzymatic reaction with FruA catalysis relatively rapidly a bisadduct **11** accumulated directly with no intermediate mono adduct being detectable, so that a potential kinetic preference of the aldolase for either of the competing enantiotopic hydroxyaldehyde moieties within the **8** could not be uncovered. Elucidation of product structure was facilitated by dephosphorylation and conversion of the single, enantiomerically and diastereomerically pure bisadduct **12** (70% overall yield) into a dimethyl glycoside derivative [68]. This revealed that the product exists as a bipyranose in which the rings are directly connected by a C–C bond. Remarkably, the pyranose subunit stemming from the (*S*)-configurated part of the dialdehyde has to tolerate two axial oxygen substituents at C-3/4 in order to obviate an unfavorable equatorial to axial ring connection.

Extending on these observations, a series of related homologous dihydroxy-α,ω-dienes **13** has been investigated (Scheme 7) to study the influence of the

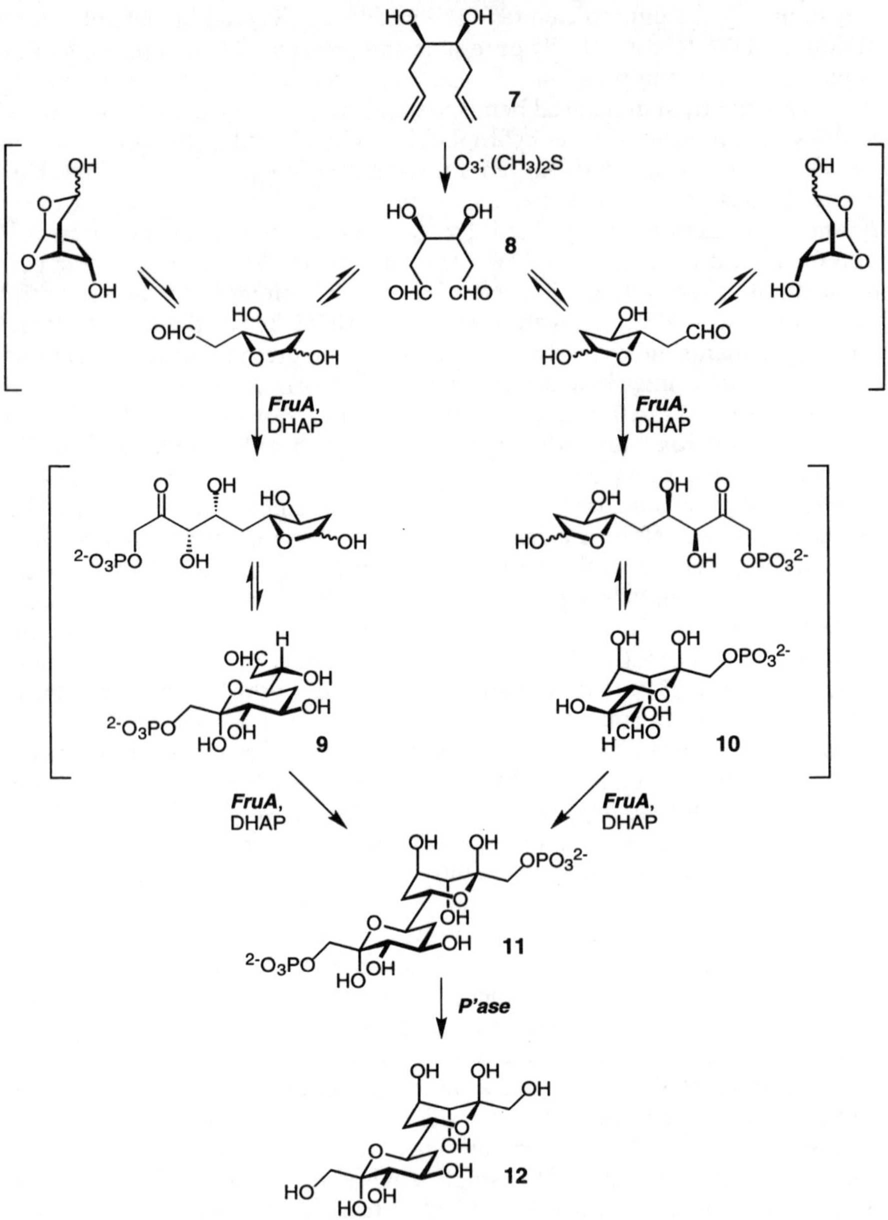

Scheme 6. Convergent pathways for tandem aldol additions of dihydroxyacetone phosphate (DHAP) to *meso*-3,4-dihydroxyhexane dialdehyde catalyzed by fructose 1,6-bisphosphate aldolase (FruA)

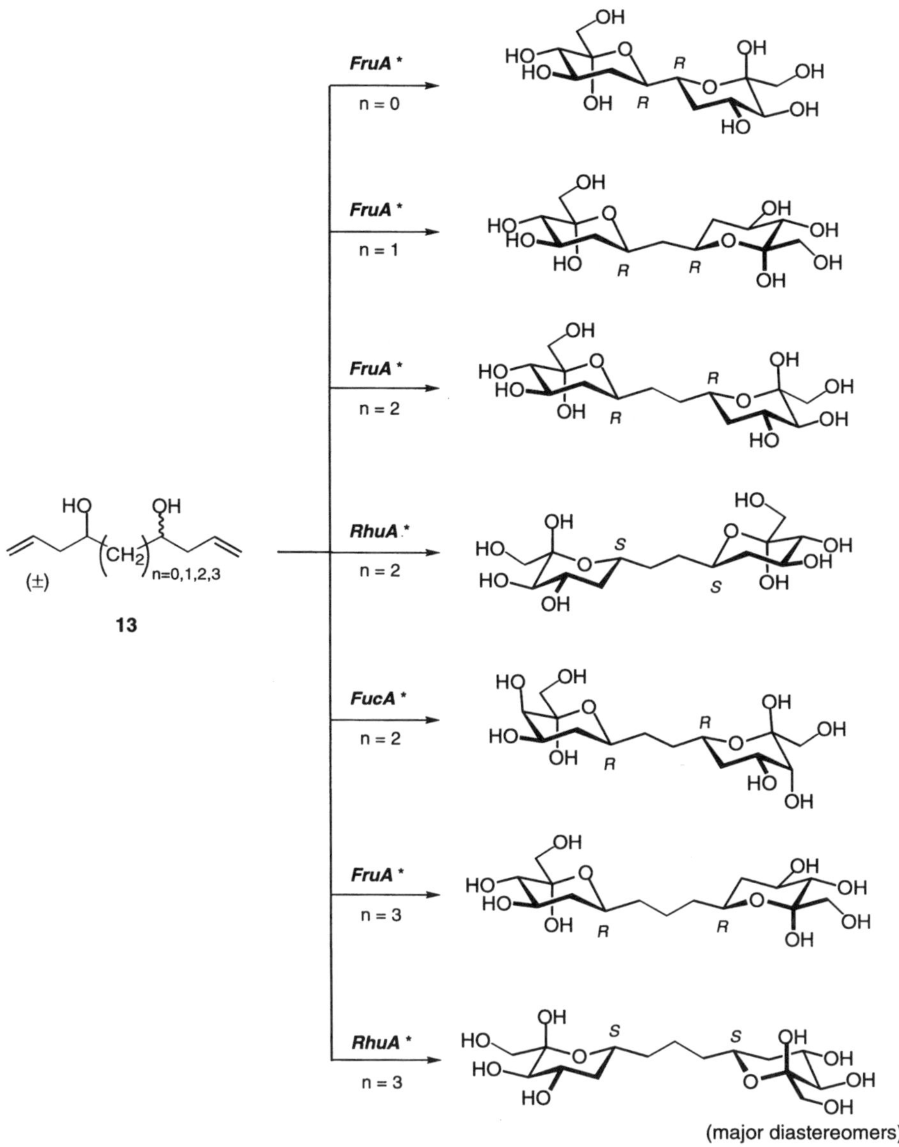

Scheme 7. Formation of a series of hydrolytically stable dipyranoid disaccharide mimetics from homologous dihydroxy-α,ω-dienes by tandem enzymatic aldolizations. Abbreviations for reaction conditions: FruA*/RhuA*/FucA* = 1.) O_3, then $(CH_3)_2S$; 2.) aldolase, DHAP; 3.) phosphatase. This notation is used throughout subsequent diagrams (except when indicated otherwise)

interlinking tether length on the effectiveness of the synthesis protocol, which proved indeed that this scheme for the preparation of disaccharide mimetics is rather flexible [94]. Starting materials were made by bi-directional allylation of simple dialdehydes (glyoxal to glutaric dialdehyde) in protic solvents followed by ozonolysis to provide the respective γ,(ω-2)-dihydroxydialdehydes. Due to the use of unresolved mixtures of meso/DL stereoisomers, in each case the enzymatic addition of DHAP entailed considerable analytical difficulties because of the several optional reaction pathways, and overall yields of bisadducts were less satisfactory so far. Lower than anticipated yields are likely also a reflection of inferior substrate qualities of aldehydes masked as hemiacetals and likely having high apparent K_m-values, and of the limited stability of DHAP over extended reaction periods (often several days). However, the study revealed that among the mixture of diastereomers obtained for each individual reaction, generally the C_2-symmetrical diastereomers featuring an equatorial attachment of the tether to the sugar ring predominated by far, obviously as a consequence of the thermodynamic advantage, so that these can be obtained highly selectively. This preference is related to the thermodynamic preference previously observed for 3-hydroxybutanal (94% de) [61, 64] and related compounds [88]. Under less selective conditions, this technique allows an easy construction of small libraries of C-disaccharide mimetics using aldolase methodology [94].

In contrast to FruA catalysts, the RhuA and FucA enzymes will typically not accept small anionically charged aldehydes (carboxylates, phosphates) [42, 83]. In this respect it is remarkable that with a tether length of equal to or larger than two methylene units all the different DHAP aldolases could be employed to catalyze tandem-type additions (Scheme 7) [94]. Obviously this is due to the increasing distance of the negatively charged phosphate group in the mono adducts from the remaining aldehyde carbonyl group which will thus not be repelled from the enzyme active site.

There is only a single literature example of a bi-directional skeleton elongation performed with an enzyme different from the DHAP dependent aldolases. The 2-deoxy-D-ribose 5-phosphate aldolase (RibA; EC 4.1.2.4), a bacterial class I enzyme which requires acetaldehyde as the nucleophilic substrate [42], has been applied to the tandem aldolization of a thioether dialdehyde [95]. The substrate **14** was synthesized from optically homogenous (*R*)-glycidaldehyde and, because of its C_2-symmetry, diastereoselective twofold addition of acetaldehyde led to a symmetrical 5,5'-sulfide-linked dipentofuranose **15**.

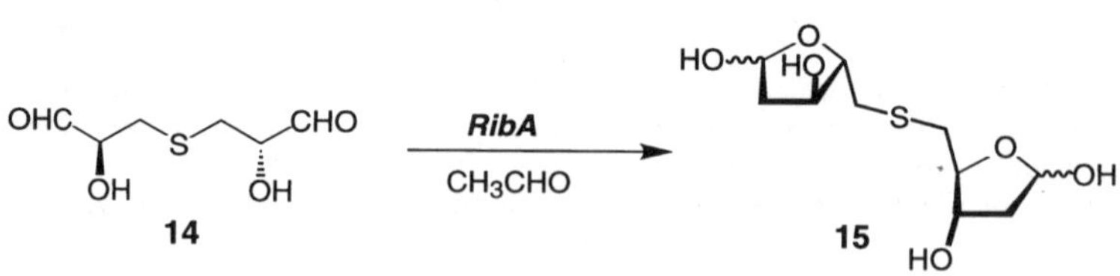

14 **15**

3
Cyclic Precursors

As was evident from the earlier studies, the application of mixtures of stereoisomers during method development was unpractical for several reasons. Control of relative (if not absolute) stereochemistry is facilitated for cycloolefinic precursors, which have the added advantage not to lead to concomitant formation of formaldehyde, itself being a good aldolase substrate [53, 61], upon ozonolytic generation of dialdehyde substrates. Following simple considerations, proper selection of candidate substrates would likely lead to a useful prediction of product structure: total chain length is determined by precursor ring size plus twice the nucleophile length (*i.e.*, 2 × C_3 for DHAP), while relative positioning of hydroxyl functions determines type of sugar ring, tether length, and overall symmetry. Additionally, connectivity of the sugar rings could be possibly varied by placing hydroxyl functions at or external to the cycloolefin. For racemic precursors, relative thermodynamic stability of the two possible products for a given choice of enzymatic catalyst can be used to estimate diastereoselectivity.

3.1
Cyclopentene Derivatives

Several five-membered probes can be derived from cyclopentadiene by employing different modes of oxygenation [96–98]. A racemic dihydroxydialdehyde was obtained from *trans*-3,5-dihydroxycyclopentene **16** [98] upon ozonolysis. Because of the electron withdrawing α-substituent, cyclic hemiacetal forms predominate with the pendent aldehyde function being nearly completely hydrated. The FruA catalyzed aldol addition of DHAP proceeded smoothly to furnish a single bisadduct in 18% overall yield which, according to NMR analysis, had the C_2-symmetrical bisfuranoid structure **17** with the methylene bridge clearly connected in a *trans* fashion [56]. Taking into account that the aldehyde was used

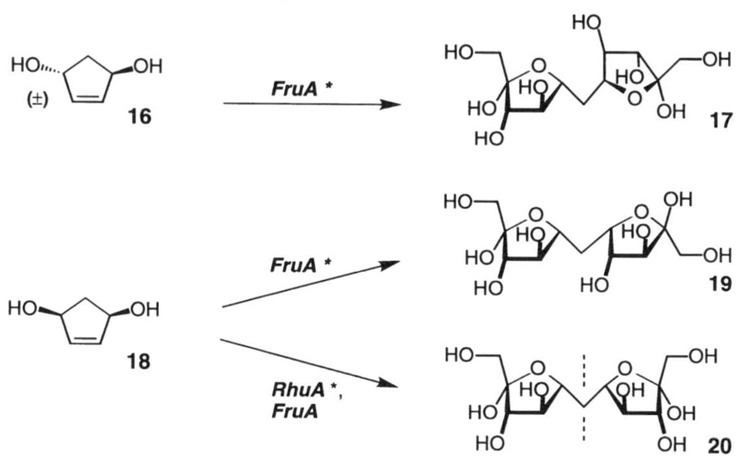

in excess, the high diastereoselectivity can be rationalized by the considerable thermodynamic preference for a bis-*trans* as compared to a bis-*cis* connected linkage.

Ozonolysis of the corresponding *meso*-precursor **18** [97] gave the dialdehyde also as a complex mixture of isomeric forms, from which tandem aldolization with FruA expectedly delivered a non symmetrical, bisfuranoid undecodiulose **19** as the sole product which was isolated in 25% yield [56]. No intermediary mono adduct could be detected by t.l.c. from which follows that, no matter which of the enantiotopic aldehyde groups was attacked first, the second addition step must be kinetically faster, most likely due to steric reasons and the presence of anionic charge in the intermediates.

When the same *meso*-dihydroxydialdehyde was treated with DHAP in the presence of the "enantiocomplementary" RhuA enzyme, however, the addition proceeded only to the stage of monophosphate product(s) with no indication of any further addition taking place upon prolonged reaction time [56]. This result is not surprising given the known refusal of charged substrates by this catalyst [42, 83]. A second, rather rapid aldolization could only be brought about by addition of FruA to the reaction mixture, from which the σ-symmetrical bisfuranoid bisadduct **20** was isolated in 15% yield as the only tandem product from this one-pot conversion. In retrospect, this result requires that the intermediate formed by the RhuA in the first step already had arisen with a preference for a *trans*-linked tether, although it remains open so far if this result is thermodynamic or kinetic in origin. Concerning the rules of bi-directional synthesis [69–71], these examples nicely underscore the potential effectiveness of reagent-based stereoselectivity for homotopic or enantiotopic terminus manipulations when making use of suitable biocatalysts.

To illustrate the opportunity to pre-determine different types of ring structures by the deliberate choice of substitution patterns within an cycloolefinic precursor, racemic *cis*-3,4-dihydroxycyclopentene **21** [96, 98] was subjected to the routine reaction sequence (ozonolysis, aldolization, dephosphorylation). A bisadduct fraction was obtained in 25% for which NMR analysis showed the presence of two components in a 4:1 ratio [56]. After chromatographic separation, the main component was established to have the mixed furanoid-pyranoid structure **22** with *trans*-all equatorial ring junction that indeed was to be expected for the thermodynamically more favorable product derived from the (2*R*,3*R*) enantiomer of the dialdehyde. The minor product **23** arising from the (2*S*,3*S*) enantiomer, however, did not contain the presumed *trans*-diaxial substitution at C-3/4 of the pyranose substructure which would be forced upon by

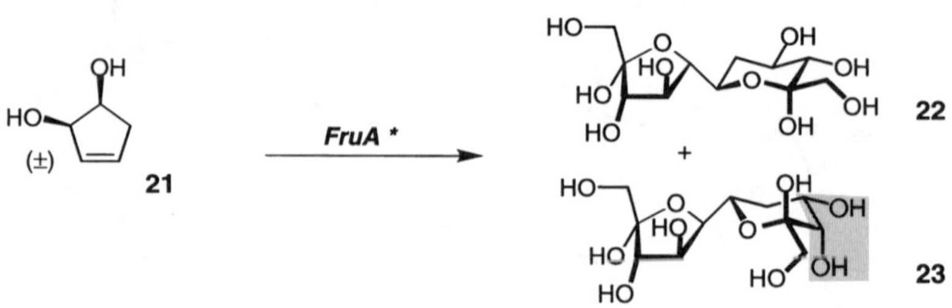

the higher priority of equatorial accommodation of the sterically demanding second ring. Instead, the actual *cis*-relationship points to a stereo error made by the enzyme which allowed the more stable diastereomer to accumulate during a long equilibration.

In order to test a substrate with external hydroxy groups for a possible transposition of the location of ring linkages, the *meso*-cyclopentene-3,5-dimethanol **24** which is easily accessible from norbornadiene [99] was transformed into the corresponding dialdehyde. Although the relatively large proportion of free aldehyde present in aqueous solution looked promising at first sight, in fact this led to the rapid deactivation of the FruA from rabbit muscle, with copious formation of a precipitate within minutes which most likely consisted of the extensively cross-linked catalyst. No trace of product was detectable from this attempt. Since it is well established that the rabbit enzyme is a less robust catalyst [61], which as a homotetramer may be more sensitive to cross-linking agents, another attempt was started with the monomeric FruA from *Staphylococcus carnosus* (FruA$_{sca}$). This enzyme, which we have made abundantly available very recently by recombinant expression in *E. coli* [56], had been reported to confer a much higher stability [85, 100]. The supposition that the monomeric biocatalyst would be more resistant against a denaturing agent indeed proved correct, because under the same reaction conditions no precipitate formed. T.l.c. monitoring indicated the formation of monophosphate(s), but no trace of a bisadduct. Following dephosphorylation, the material isolated in 26% overall yield was shown to consist of a 3:1 mixture (major component **25**) of the two diastereomers predicted for DHAP attack at the respective enantiotopic aldehyde groups [56]. As these structures can each form a rather stable arrangement of anulated dipyranoses, a second addition in tandem mode is at a considerable kinetic disadvantage. Gratifyingly, when incubated under the same conditions but employing the RhuA enzyme instead, also no enzyme denaturation took place. Spectral analysis of the mono adducts (32% yield) corroborated that the same diastereomers were present, only in enantiomeric form. Interestingly, the 1:3 ratio (major component **26**) was opposite to that registered for the FruA$_{sca}$ experiment which indicates that both catalysts had preferentially attacked the carbonyl group with the same enantiotopicity.

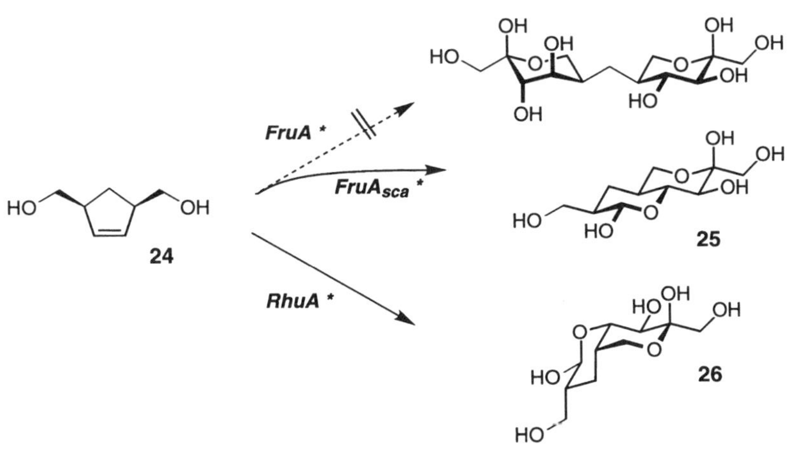

3.2
Cyclohexene Derivatives

With the larger ring size of cyclohexene, a full formal transposition of both hydroxy functions between allylic and homoallylic positions is possible. A plausible candidate for initial tests was racemic cyclohex-2-ene-1,4-diol **27** [101] which produced 2,5-dihydroxyhexanedial by ozonolysis. Its coupling with DHAP under FruA catalysis proceeded at a satisfactory rate with well detectable, transient formation of a monoadduct which points to a certain kinetic disadvantage for the secondary addition step. The single bisadduct was isolated in 37% overall yield as the free sugar which according to its spectra unequivocally had the constitution of the C_2-symmetrical, all-*trans*-configurated bisfuranose **28** – a 6,6'-linked dimer of D-fructose [68]. Not surprisingly, the diastereomer stemming from the antipode, demanding a twofold *cis*-vicinal tether connection, was not detectable.

The isomeric *threo*-3,4-dihydroxyhexanedial was generated by oxidative ring cleavage of racemic bis(homoallylic) diol **29** [102] and characterized by NMR in aqueous solution to form stable bicyclic hemiacetals. Accordingly, the primary aldolization step proceeded rather sluggishly. With no intermediate being detectable, a singular bisadduct was isolated in 26% total yield. As anticipated, the crystalline bipyranoid diketose was conclusively shown to be the thermodynamically favored C_2-symmetrical all-equatorially substituted diastereomer **30** also obtained from open-chain precursors (Sect. 2.2) [68]. Conversion of the same dialdehyde using the RhuA ceased with formation of a monoadduct, from which a bisadduct **31** (46% overall yield) could only be induced upon addition

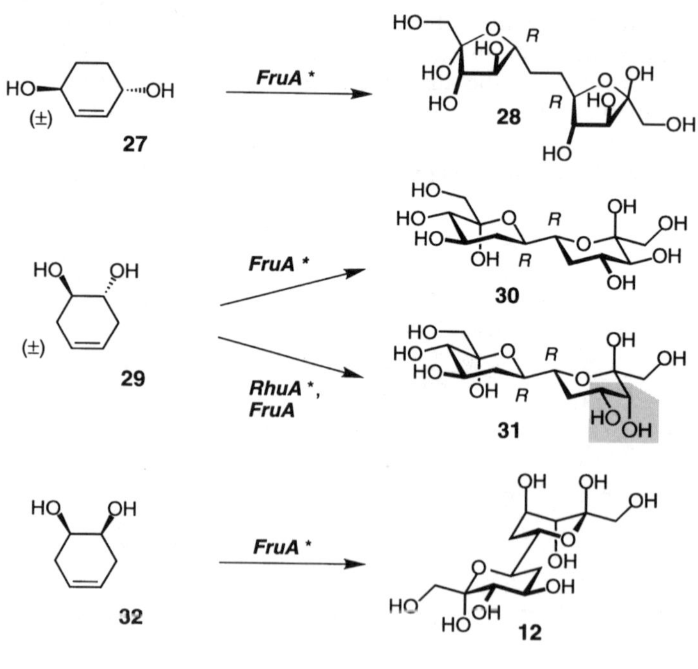

of FruA. NMR analysis unambiguously established a *cis*-vicinal substitution pattern in one of the two pyranose rings, rather than the expected two *trans* forms. With respect to related observations, it became clear that the incorrect stereochemistry was due to RhuA catalysis, where equilibration at longer reaction times probably allowed the erroneously produced *cis*-adduct to accumulate [42]. Of cause, from the corresponding *meso*-precursor **32** the tandem product **12** was obtained identical to that described above but derived from an open-chain precursor [68].

3.3 Azido Substituted Substrates

As a prominent class of carbohydrate mimetics, the so-called "aza sugars" have commanded the interest in recent years as potent glycoprocessing inhibitors [103–105]. Because of the potential value of aza sugars as therapeutic agents for the treatment of diabetes and other metabolic disorders as well as for the blocking of viral or microbial infection and metastasis [106], an enormous variety of these compounds has been generated for biological testing. A highly popular avenue into the preparation of stereoisomeric monocyclic aza sugars has been the submission of azido substituted aldehydes to enzymatic aldol reactions [104].

By application of the tandem chain extension strategy, a novel route to aza C-disaccharides seemed plausible. A very simple precursor **33** can be easily obtained through ring opening of the monoepoxide of 1,4-cyclohexadiene by azide [107]. Although the dialdehyde generated by ozonolysis proved rather stable in aqueous solution, a rather rapid decomposition was noted after incubation with DHAP in the presence of an aldolase. A bisadduct **36** could be secured only after extensive experimentation in a very poor overall yield of 2% (Scheme 8) [108]. Because the azido substituent cannot engage in ring cyclization at neither the mono or bisadduct stage, only one carbonyl group can be locked by cyclization, and an elimination of HN_3 from the intermediary adduct **34** to form an α,β-unsaturated aldehyde **35** is therefore facilitated. Because the aldolase establishes an equilibrium between mono and bisadduct, a decomposition of the intermediate must also deplete the tandem material.

Consequently, a dihydroxylated azidodialdehyde was examined next which indeed behaved as anticipated. When generated from the racemic allylic azide **37** [109], FruA catalysis effected a smooth tandem addition to the dialdehyde to provide a diastereoisomerically pure bipyranoid azido C-disaccharide **38**, from which the pyrrolidine type aza sugar **39** was highly stereoselectively produced by standard reductive amination [108]. Model considerations suggest a close resemblance of the protonated aza C-disaccharide to transition states of saccharase or maltase. Indeed, several of the glycosidases tested were inhibited by **39** at concentrations below 1 mM.

For an interesting C_2-symmetrical diaza C-disaccharide of pyrrolidine character, the requisite precursor diazidocyclohexenediol **40** could be effectively obtained from racemic *anti*-benzene bisepoxide [110]. Unfortunately, the

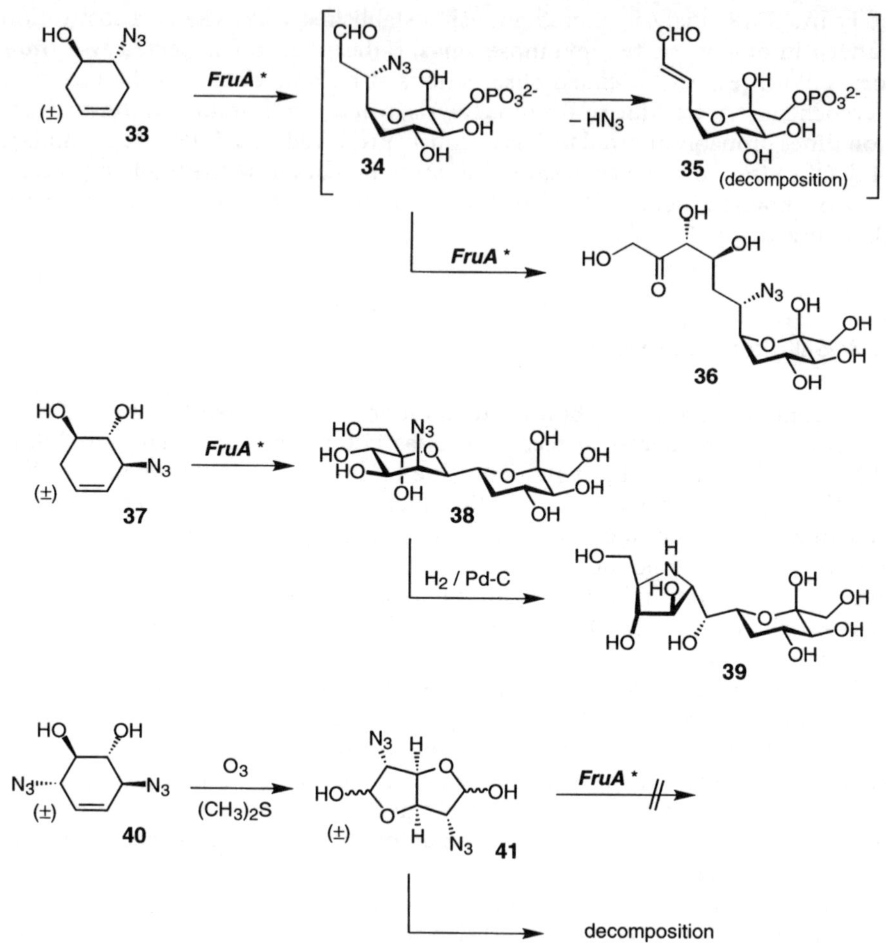

Scheme 8. Synthetic entry to aza C-disaccharides based on tandem enzymatic aldolizations

derived dialdehyde **41** proved to be too labile in aqueous solution to be useful for tandem enzymatic aldol reactions [56].

3.4 Bicycloalkene Derivatives

When hydroxy and carbaldehyde groups are positioned as vicinal substituents on a cyclic scaffold, enzymatic DHAP addition will effect a ring anulation by a pyranose [111]. In this respect, the dihydroxydialdehyde **43** generated by ozonolysis of *exo*-norbornenediol **42** [99] seemed to be a promising platform for a tandem sequence towards twofold pyranose anulation (Scheme 9). FruA catalysis indeed induced the desired aldolization to furnish a unique bisadduct in 20%

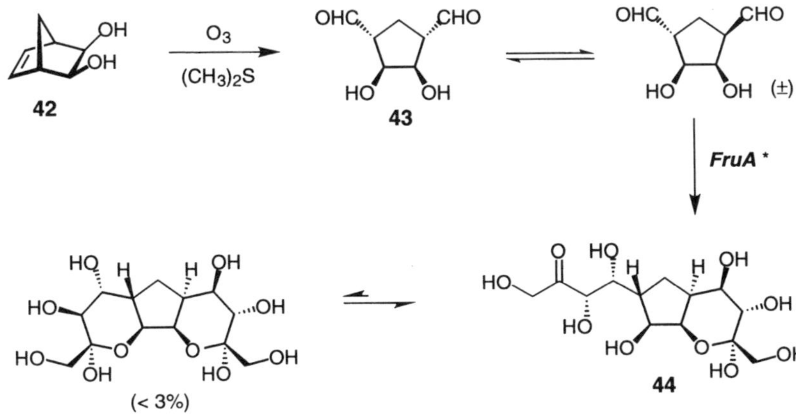

Scheme 9. Sugar ring anulation by tandem enzymatic aldol additions to cyclic dihydroxydialdehydes

overall yield [56]. Surprisingly, careful spectroscopic analysis unveiled that the product did not adopt the expected tricyclic but only a bicyclic structure **44**, and that the rings were fused in a *cis* rather than the *trans* fashion predicted from the norbornene precursor stereochemistry. This stereochemistry requires epimerization at the aldehyde group at some intermediary state, which is corroborated by the presence of a considerable fraction of free aldehyde in aqueous solution, and a previously unknown, strong preference of the rabbit muscle FruA towards one enantiomer of the aldehyde. Clearly, the twist-boat pyranose conformation enforced by the substitution pattern rules out a thermodynamic contribution.

4 Spirosugars

The occurrence of natural *spiro*-cyclic sugars is very limited, although the rigidity of such a type of scaffold would present hydroxyl substituents in a precisely defined manner in three-dimensional space for potentially highly specific receptor interactions. Compounds of this type can be constructed by enzymatic aldol reactions, when hydroxy and aldehyde groups are positioned as geminal substituents on an existing cyclic support [82]. Direct placement of functionalities preselects for furanose attachment, while homologation (of either substituent) will bring about a pyranoid cyclization mode. Transliteration to the tandem aldolization concept, this requires suitable placement of paired functionalities, with due consideration of the consequences of inherent symmetry elements to the relative topicity of reactive groups.

Out of several conceivable probes, a cyclohexane derivative was selected initially because twofold allylation of cyclohexane-1,4-dione offered a ready entry and led to an easily separable mixture of the corresponding *cis*-(**47**) and *trans*- (**45**) tertiary alcohols. In each case ozonolysis resulted in the formation of the respective C_{2h}- and C_{2v}-symmetrical dialdehydes. Both individual dialdehyde reacted under FruA catalysis to give single enantio- and diastereomerically pure bisadducts **46/48** each [92]. The moderate yields of 38% and 14% are partially accounted for by the capacity of the dialdehyde precursors to induce enzyme denaturation due to the fact that none of them is able to form hemiacetal structures. The C_2-symmetry element remaining in each of the bis-*spiro* systems was clearly manifested in the very simple NMR spectra. From inspection of molecular models it is evident that the two *spiro*-pyranoses synthesized de novo are held at a distance quite similar to a 1,6-linked disaccharide with conformation and relative orientation of the sugar rings to one another, however, rigidly confined by the cyclohexane core in a non-hydrolyzable manner.

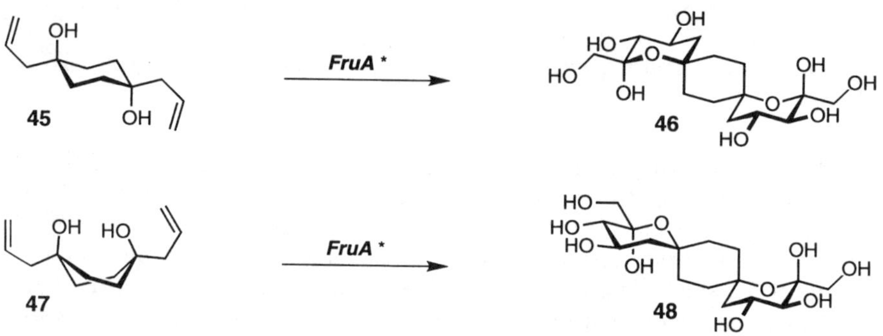

As a more complex target derived from a carbohydrate core structure, our attention was attracted by the high symmetry of the diketone obtained by oxidation of 1,4:3,6-dianhydro-D-mannitol (isomannitol) [112, 113]. Allylation using allyl bromide/Zn in aqueous medium [114] proceeded with complete stereocontrol from the convex outer face to give the enantiomerically pure C_2-symmetrical diol **49** which was ozonized to the corresponding dialdehyde. In spite of its steric hindrance, the substrate was smoothly converted by FruA catalysis in the desired tandem fashion. Because of the identical topicity of the carbonyl groups in the optically homogenous starting material, only a single symmetrical bisadduct **50** was isolated in 25% overall yield [56]. From a comparison reaction performed with the complementary RhuA enzyme, a bisadduct **51** could also be secured in a 20% yield. It is worth noting that, due to the rigid-skeleton, the anionic charge within the intermediate mono adduct is placed at a very remote position relative to the remaining second reactive group which greatly facilitates acceptance by the RhuA for a tandem activity.

Both tandem *spiro*-frameworks **50/51** bear a certain resemblance to trisaccharides. Generally, in the aldol adducts from 3-hydroxyaldehydes the respective pyranose conformations are induced by the preference of the hydroxy functions generated during C–C coupling for an equatorial position. Therefore, the opposite absolute configurations induced by the aldolases consequently force a change of relative orientation of the sugar chairs with respect to the bicyclic

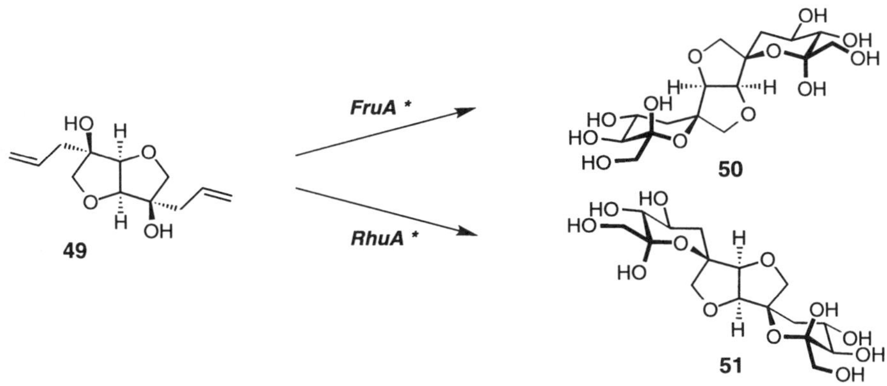

fused system which grossly influences the overall appearance, as well as the location of polar functionalities projecting into three-dimensional space.

5
Miscellaneous Systems

5.1
2-Deoxy-D-ribose 5-Phosphate Aldolase

A small number of reactions has been developed that follow the concept of enzymatic tandem C–C bond formations, however, which do not proceed in a bi-directional but in a repetitive manner from a single functional group, which relates such reactions to polymerization. However, only controlled, stereoselective examples of oligomerizations are considered to be relevant for the scope of this account.

During investigations with RibA it had been realized by serendipity that the initial product **52** from catalyzed self-aldolization of acetaldehyde again serves as a suitable acceptor for the sequential addition of a second donor molecule to

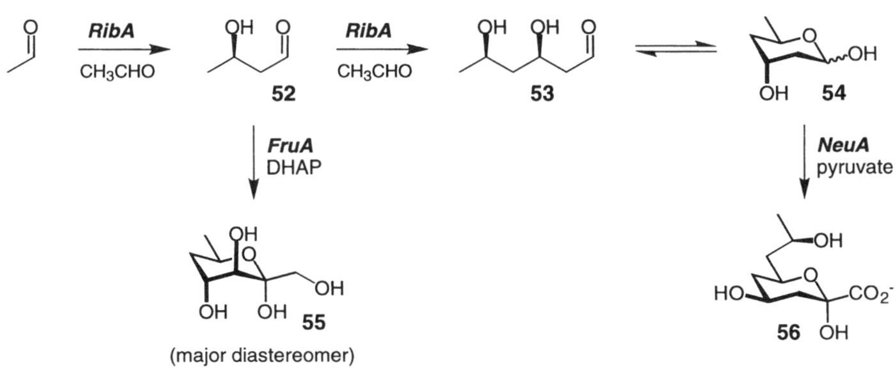

Scheme 10. Stereoselective oligomerization of acetaldehyde, and heterocatalytic tandem enzymatic aldolizations by using combinations of different aldolases

give the (3*R*,5*R*)-2,4,6-trideoxyhexose **53** as the major stereoisomer in 20% yield (Scheme 10) [115]. Formation of higher order adducts, *i.e.* polymerization, is effectively precluded after the second monomer addition by rearrangement into a stable pyranose hemiacetal **54** which masks the requisite free aldehyde form.

Under carefully chosen reaction conditions by using a combination of RibA and another aldolase of different specificity, the intermediate **52** from the first RibA reaction can formally be intercepted by the second aldolase through the addition of a different nucleophile. In the presence of DHAP and FruA indeed predominant formation of the dideoxyketose **55** (17% yield) could be induced [116]. Due to the reversible nature of a thermodynamically unfavorable product, and because of the competition of both enzymes for the same substrate and the long reaction times required, a number of different side products and several stereoisomers have to be accounted for.

In a related protocol, the acetaldehyde trimer **54** from the generic RibA oligomerization was found to be a substrate for the *N*-acetylneuraminic acid aldolase (NeuA; EC 4.1.3.3) which catalyzed the addition of pyruvate. By this means, a tetradeoxy-L-*arabino*-2-nonulosonic acid **56** was obtained in 55% yield [116]. A one-pot, tandem operation was complicated by the fact that temperature requirements for optimum activity and stability of the two catalysts were not compatible.

5.2 Transketolase

The transketolase (TK; EC 2.2.1.1) catalyzes the reversible transfer of a hydroxyacetyl fragment from a ketose to an aldehyde [42]. A notable feature for applications in asymmetric synthesis is that it only accepts the D-enantiomer of 2-hydroxyaldehydes with effective kinetic resolution [117, 118] and adds the nucleophile stereospecifically to the *re*-face of the acceptor. In effect, this allows to control the stereochemistry of two adjacent stereogenic centers in the generation of (3*S*,4*R*)-configurated ketoses by starting from racemic aldehydes; thus this provides products stereochemically equivalent to those obtained by FruA catalysis. The natural donor component can be replaced by hydroxypyruvate from which the reactive intermediate is formed by a spontaneous decarboxylation, which for preparative purposes renders the overall addition to aldehydic substrates essentially irreversible [42].

Because the ketose products formed by transketolase reactions are not acceptors for a consecutive transformation by the same enzyme, we have investigated the effect of adding a xylose isomerase (XylI; EC 5.3.1.5), which has similar stereochemical specificity, for ketose to aldose equilibration (Scheme 11). Starting from racemic lactaldehyde, the transketolase forms 5-deoxy-D-xylulose **57** which indeed was accepted in situ by the XylI for diastereospecific conversion into 5-deoxy-D-xylose **58** [119]. The latter again proved to be a substrate of transketolase which completed a tandem operation to furnish 7-deoxy-D-*ido*-2-heptulose **59** as the sole bisadduct in 24% overall yield and in enantio- and diastereomerically pure quality [119]. This sequence for mono-directional carbohydrate chain elongation, however, is different from

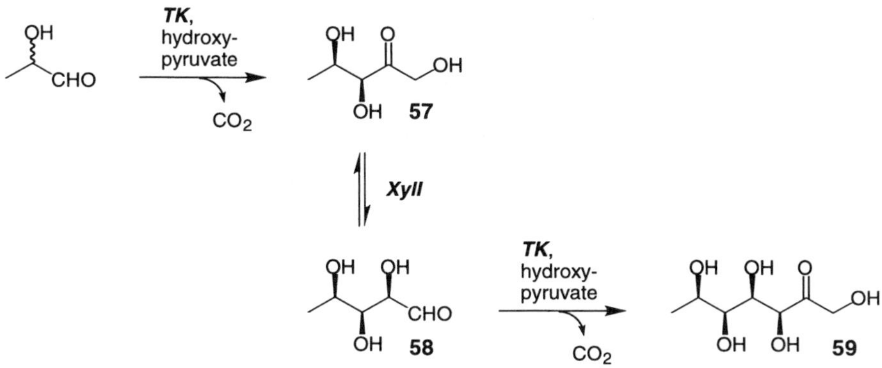

Scheme 11. Tandem enzymatic chain extension catalyzed by transketolase, aided by the combination with a suitable ketose-aldose isomerase

the previous example in that the C–C bond forming steps are not immediately consecutive but require an auxiliary step. Nevertheless, all four stereogenic centers of the resulting heptulose **59** are completely enzymatically controlled during this one-pot operation.

5.3 In vitro Biosynthesis

The reactions discussed in the foregoing sections are examples of operationally rather simple laboratory manipulations which allow, by application of readily available (commercial) enzymes, to construct rather intricate products with several asymmetric centers in stereochemically homogenous form from rather simple, mostly achiral or racemic precursors. It is pertinent to emphasize, as eluded to in the introduction, that the construction of even more complex natural products is achieved by living organisms essentially in one-pot reactions. The recent wave of developments in molecular biology has placed the

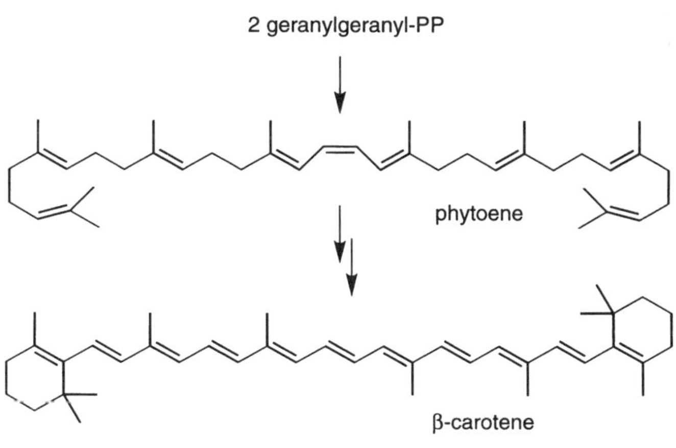

very same variety of enzymic catalysts that Nature uses within reach of the synthetic chemist [120, 121], and it seems appropriate in this context to highlight the most promising recent developments that will allow to widen the scope of the rational use of biocatalysts towards tandem asymmetric C–C bond formations in the synthesis of large, multifunctionalized structures of high complexity.

Biosynthetic pathways very rarely seem to take advantage of the concept of bi-directional synthesis – plausibly for the same problems of terminus differentiation that synthetic chemists have to face in the case of (typically) non-symmetric targets –, although there are related examples such as in the elaboration of geranylgeranyl pyrophosphate on the path to carotenes.

More typical among biosynthetic schemes are repetitive C–C bond formations of an oligomerization type for which the synthesis of polyketides is a characteristic example. Polyketides are a large, structurally diverse family of natural compounds possessing a broad range of biological activities, including pharmaceutically useful properties. Their biosynthesis is catalyzed by multifunctional polyketide synthases (related to fatty acid synthases) and proceeds by tandem poly-Claisen-type condensations up to a certain degree of oligomerization. Structural variability is introduced by proper control of the chain lengths and by degree and position of ketoreduction, dehydration, and enoylreduction, and by the mode of cyclization involving a specific cyclase enzyme. The latter folds the nascent chain appropriately to allow the induction of specific ring cyclizations through subsequent multiple C–C bond formations (followed by further alterations) [122, 123].

Newer developments in this area draw from the fact that the biosynthetic machinery for different polycyclic aromatic polyketides in different species follows the same modular principle of organization in assembling a PKS from specific subunits that are dedicated to analogous tasks. Facilitated by a certain tolerance of such systems to genetic reprogramming, this allows the swapping of corresponding gene fragments between distinct synthetic pathways to delibe-

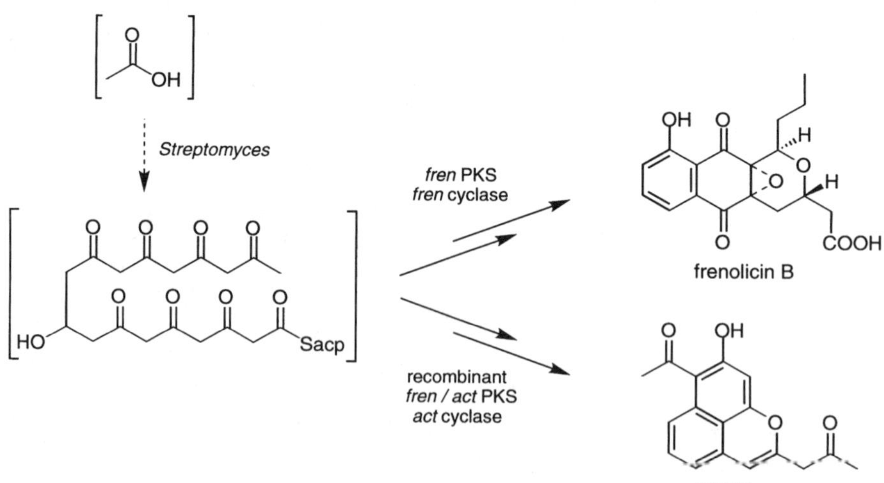

rately manipulate the assembly of recombinant PKS enzymes or their pairing with heterologous enzymes downstream of the biosynthetic pathways [124, 125]. When co-expressed in a suitable host organism, such non-natural catalyst combinations can generate hybrid polyketides that are not found in Nature ("combinatorial biosynthesis" [126, 127]) and that are inaccessible by applying conventional mutagenesis techniques to production strains. As an early example using this strategy, genes of *Streptomyces roseofulvus* from the *fren* genetic cluster, which normally code for enzymes responsible for the synthesis of the nonaketide frenolicin B, have been interchanged in an empirical format with those from the *act* genetic cluster of S. coelicolor, which normally code for octaketide actinorhodin synthesis. Indeed, in certain cases new products with different carbon backbones such as RM18 could be isolated [128].

Most spectacular achievements regarding in vitro biosynthesis stems from the work on elucidating the biosynthetic pathway of vitamin B_{12}, one of the tetrapyrrol "pigments of life" [129, 130]. Building on data accumulated by a combination of genetics, molecular biology, enzymology, chemistry, and spectroscopy the sequence of the tandem C–C bond forming steps required for the biogenetic build-up of uroporphyrinogen III from 5-aminolevulinic acid and the multitude of methylation reactions, as well as the mode and timing of the ring contraction could finally be unraveled. With all the genes and details now ready at hand, it seemed a rather straightforward endeavor to separately express all of the 12 enzymes involved in the aerobic synthesis of corrins and reconstitute the entire pathway in vitro [131, 132].

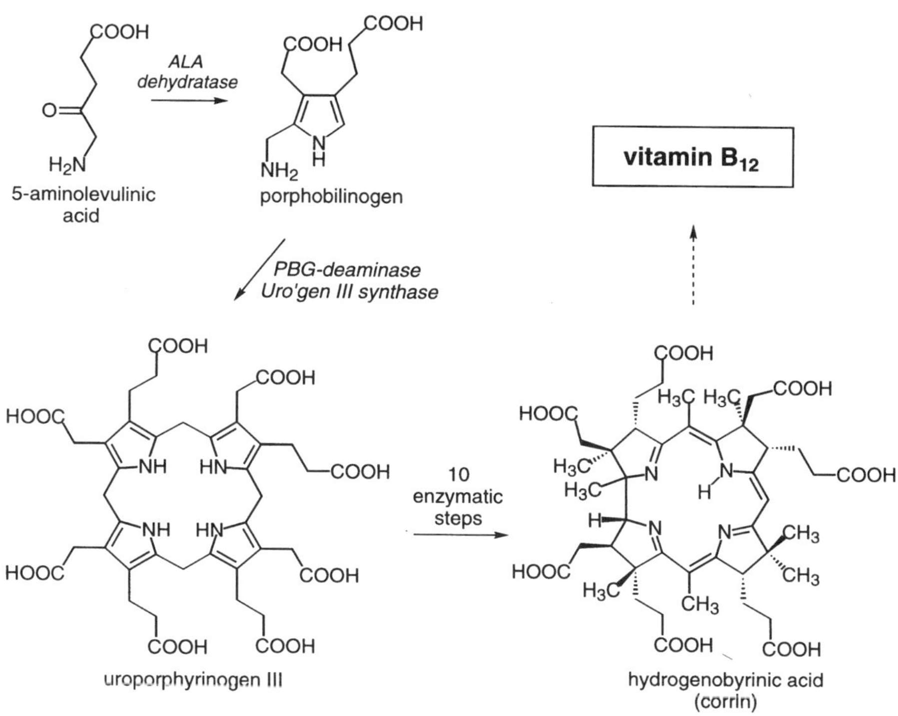

6
Summary and Outlook

This account has attempted to demonstrate that contemporary tactical schemes in modern organic synthesis such as bi-directional chain synthesis [69–71] and tandem transformations [66, 67] can indeed be transported to the arena of biocatalysis in asymmetric synthesis, often with similar benefits when applied to C–C bond formations by aldol additions and related reactions. The power and simplicity of this methodology is certainly best appreciated if one considers the advantages of one-pot reactions that can be performed on readily available and inexpensive starting materials which do not require protection of functional groups or optical purity but still provide a rapid entry, with few exceptions, to enantiomerically and diastereomerically pure products of high structural variability [68]. In a single synthetic operation, depending on the type of enzyme(s) used, thus up to four new asymmetric centers can be created and one or or more dependent ones may be differentiated with high selectivity and efficiency to generate novel carbohydrate-type structures which are not found in Nature but, by their resemblance to natural effectors, may prove bioactive.

By nature of the catalysts employed for the exploratory work which has been summarized here, applications were directed primarily towards complex sugar skeletons. However, by taking advantage of enzymes from other than carbohydrate metabolic pathways, this principle should be applicable to other classes of compounds as well [37, 42]. With genetic knowledge on whole organisms advancing rapidly [133–136], such prospects will certainly foster further exciting developments in this fascinating research area in the years ahead.

Acknowledgement. This work has been generously supported by the *Deutsche Forschungsgemeinschaft* (Fe 244/6-2), the *Sonderforschungsbereich: Asymmetrische Synthesen mit chemischen und biologischen Methoden* (SFB380/B25), the *Fonds der Chemischen Industrie*, and *Boehringer Mannheim GmbH*. M. T. Z. is grateful to the *Alexander-von-Humboldt Foundation* for a postdoctoral fellowship.

References

1. Feizi T (1991) Curr Opin Struct Biol 1:766
2. Hynes RO, Lander AD (1992) Cell 68:303
3. Furukawa K, Kobata A (1992) Cell Surface Carbohydrates – Their Involvement in Cell Adhesion. In: Ogura H, Hasegawa A, Suami T (eds) Carbohydrates. Kodansha, Tokyo, p 369
4. Varki A (1994) Proc Natl Acad Sci USA 91:7390
5. Carlos TM, Harlan JM (1994) Blood 84:2068
6. Wegner CD (1994) Adhesion Molecules. Academic Press, London
7. Hakomori S, Igarashi Y (1995) J Biochem 118:1091
8. Postema MHD (1995) *C*-Glycoside Synthesis. CRC Press, London
9. Levy DE, Tang C (1995) The Chemistry of *C*-Glycosides. Pergamon, Oxford
10. Bertozzi C, Bednarski M (1996) Synthesis of *C*-Glycosides; Stable Mimics of *O*-Glycosidic Linkages. In: Khan SH, O'Neill RA (eds) Modern Methods in Carbohydrate Synthesis. Harwood Academic, Amsterdam, p 316
11. Isono K (1988) J Antibiot 41:1711
12. Isono K (1991) Pharmacol Ther 52:269

13. Lerner LM (1991) Synthesis and Properties of Various Disaccharide Nucleosides. In: Townsend LB (ed) Chemistry of Nucleosides and Nucleotides. Plenum Press, New York, vol. 2, p 27
14. Tamura G (1982) Tunicamycin. Japan Scientific Press, Tokyo
15. Tamura G, Sasaki T, Matsuhashi M, Takatsuki A, Yamasaki M (1976) Agric Biol Chem 40:447
16. Takatsuki A, Kohno K, Tamura G (1975) Agric Biol Chem 39:2089
17. Garner P (1988) Synthetic Approaches to Complex Nucleoside Antibiotics. In: Atta-ur-Rahman (ed) Studies in Natural Product Chemistry. Elsevier, Amsterdam, vol. 1, p 397
18. Knapp S (1995) Chem Rev 95:1859
19. Grindley TB (1996) Protecting Groups in Oligosaccharide Synthesis. In: Khan SH, O'Neill RA (eds) Modern Methods in Carbohydrate Synthesis. Harwood Academic Publishers, Amsterdam, p 225
20. Whitesides GM, Wong C-H (1985) Angew Chem Int Ed Engl 24:617
21. Jones JB (1986) Tetrahedron 42:3351
22. Toone EJ, Simon ES, Bednarski MD, Whitesides GM (1989) Tetrahedron 45:5365
23. Davies HG, Green RH, Kelly DR, Roberts SM (1989) Biotransformations in Preparative Organic Chemistry. Academic Press, London
24. Crout DHG, Christen M (1989) Biotransformations in Organic Synthesis. In: Scheffold R (ed) Modern Synthetic Methods. Springer, Berlin, vol. 5, p 1
25. Gerhartz W (1991) Enzymes in Industry: Production and Applications. VCH, Weinheim
26. Boland W, Froessl C, Lorenz M (1991) Synthesis 1049
27. Schellenberger V, Jakubke HD (1991) Angew Chem Int Ed Engl 30:1437
28. David S, Augé C, Gautheron C (1991) Adv Carbohydr Chem Biochem 49:175
29. Blanch HW, Clark DS (eds) (1991) Applied Biocatalysis, vol. 1. Marcel Dekker, New York
30. Fessner W-D (1992) Kontakte (Darmstadt) (3): 3; ibid (1993) (1):23
31. Holland HL (1992) Organic Synthesis with Oxidative Enzymes. VCH, Weinheim
32. Rozzell D, Wagner F (1992) Biocatalytic Production of Amino Acids & Derivatives. Hanser, Munich
33. Santaniello E, Ferraboschi P, Grisenti P, Manzocchi A (1992) Chem Rev 92:1071
34. Servi S (1992) Microbial Reagents in Organic Synthesis. Kluwer Academic, Dordrecht
35. Halgas J (1992) Biocatalysts in Organic Synthesis. Elsevier, Amsterdam
36. Poppe L, Novák L (1992) Selective Biocatalysis. A Synthetic Approach. VCH, Weinheim
37. Wong C-H, Whitesides GM (1994) Enzymes in Synthetic Organic Chemistry. Pergamon, Oxford
38. Fessner W-D (1995) Enzyme Catalyzed Aldol Additions. In: Helmchen G, Hoffmann RW, Mulzer J, Schaumann E (eds) Houben-Weyl, Methods of Organic Chemistry. Stereoselective Synthesis. Thieme, Stuttgart, vol. E 21b, p 1736
39. Faber K (1995) Biotransformations in Organic Chemistry, 2nd edn. Springer, Berlin
40. Azerad R (1995) Bull Soc Chim Fr 132:17
41. Drauz K, Waldmann H (1995) Enzyme Catalysis in Organic Synthesis. A Comprehensive Handbook. VCH, Weinheim
42. Fessner W-D, Walter C (1996) Top Curr Chem 184:97
43. Waldmann H, Sebastian D (1994) Chem Rev 94:911
44. Drueckhammer DG, Hennen WJ, Pederson RL, Barbas CF, Gautheron CM, Krach T, Wong C-H (1991) Synthesis 499
45. Thiem J (1995) FEMS Microbiol Rev 16:193
46. Brimacombe JS, Hanna R, Kabir AKMS, Bennett F, Taylor ID (1986) J Chem Soc Perkin Trans I 815
47. Dondoni A, Fantin G, Fogagnolo M, Medici A, Pedrini P (1989) J Org Chem 54:693
48. Ko SY, Lee AWM, Masamune S, Reed LA, Sharpless KB, Walker FJ (1990) Tetrahedron 46:245
49. Jarosz S (1993) J Carbohydr Chem 12:1149
50. Scott JW (1984) Readily Available Chiral Carbon Fragments and Their Use in Synthesis. In: Morrison JD, Scott JW (eds) Asymmetric Synthesis. Academic Press, Orlando, vol. 4, p 1

51. Bols M (1996) Carbohydrate Building Blocks. Wiley-Interscience, New York
52. Fessner W-D (1992) A Building Block Strategy for Asymmetric Synthesis: The DHAP Aldolases. In: Servi S (ed) Microbial Reagents in Organic Synthesis. Kluwer Academic Publishers, Dordrecht, p 43
53. Fessner W-D, Sinerius G, Schneider A, Dreyer M, Schulz GE, Badia J, Aguilar J (1991) Angew Chem Int Ed Engl 30:555
54. Fessner W-D, Eyrisch O (1992) Angew Chem Int Ed Engl 31:56
55. Eyrisch O, Sinerius G, Fessner W-D (1993) Carbohydr Res 238:287
56. Zannetti MT, Fessner W-D, unpublished results
57. Fessner W-D (1993) GIT Fachz Lab 37:951
58. Horecker BL, Tsolas O, Lai CY (1972) Aldolases. In: Boyer PD (ed) The Enzymes, 3 edn. Academic Press, New York, vol. VII, p 213
59. Gefflaut T, Blonski C, Perie J, Willson M (1995) Prog Biophys Mol Biol 63:301
60. Fessner W-D, Schneider A, Held H, Sinerius G, Walter C, Hixon M, Schloss JV (1996) Angew Chem Int Ed Engl 35:2219
61. Bednarski MD, Simon ES, Bischofberger N, Fessner W-D, Kim MJ, Lees W, Saito T, Waldmann H, Whitesides GM (1989) J Am Chem Soc 111:627
62. Fessner W-D, Badia J, Eyrisch O, Schneider A, Sinerius G (1992) Tetrahedron Lett 33:5231
63. Fessner W-D, Schneider A, Eyrisch O, Sinerius G, Badia J (1993) Tetrahedron Asymm 4:1183
64. Durrwachter JR, Wong C-H (1988) J Org Chem 53:4175
65. von der Osten CH, Sinskey AJ, Barbas CF, Pederson RL, Wang YF, Wong C-H (1989) J Am Chem Soc 111:3924
66. Ho T-L (1992) Tandem Organic Reactions. Wiley-Interscience, New York
67. Tietze LF, Beifuss U (1993) Angew Chem Int Ed Engl 32:131
68. Eyrisch O, Fessner WD (1995) Angew Chem Int Ed Engl 34:1639
69. Schreiber SL (1987) Chem Scripta 27:563
70. Poss CS, Schreiber SL (1994) Acc Chem Res 27:9
71. Magnuson SR (1995) Tetrahedron 51:2167
72. Effenberger F, Straub A (1987) Tetrahedron Lett 28:1641
73. Pederson RL, Esker J, Wong C-H (1991) Tetrahedron 47:2643
74. Colbran RL, Jones JKN, Matheson NK, Rozema I (1967) Carbohydr Res 4:355
75. Jung SH, Jeong JH, Miller P, Wong C-H (1994) J Org Chem 59:7182
76. Crans DC, Kazlauskas RJ, Hirschbein BL, Wong C-H, Abril O, Whitesides GM (1987) Methods Enzymol 136:263
77. Yanase H, Okuda M, Kita K, Shibata K, Sakai Y, Kato Y (1995) Appl Microbiol Biotechnol 43:228
78. Richard JP (1984) J Am Chem Soc 106:4926
79. Richard JP (1993) Biochem Soc Trans 21:549
80. Phillips SA, Thornalley PJ (1993) Eur J Biochem 212:101
81. Meyerhof O, Lohmann K, Schuster P (1936) Biochem Zeitschr 286:301
82. Fessner W-D, Walter C (1992) Angew Chem Int Ed Engl 31:614
83. Fessner W-D, Sinerius G (1994) Bioorg Med Chem 2:639
84. Fessner W-D, Sinerius G (1994) Angew Chem Int Ed Engl 33:209
85. Brockamp HP, Kula MR (1990) Tetrahedron Lett 31:7123
86. Chibata I (1978) Immobilized Enzymes. Research and Development. Wiley, New York
87. Trevan MD (1980) Immobilized Enzymes. An Introduction and Applications in Biotechnology. Wiley, New York
88. Fessner W-D, unpublished results
89. Scharf H-D, Plum H, Fleischhauer J, Schleker W (1979) Chem Ber 112:862
90. Bednarski MD, Waldmann HJ, Whitesides GM (1986) Tetrahedron Lett 27:5807
91. Chuche J, Dana G, Monot M-R (1967) Bull Soc Chim Fr 9:3300
92. Petersen M, Fessner W-D, unpublished results
93. Angyal SJ (1969) Angew Chem Int Ed Engl 8:157
94. Walter C (1996) Dissertation, RWTH Aachen

95. Wong C-H, Garcia-Junceda E, Chen LR, Blanco O, Gijsen HJM, Steensma DH (1995) J Am Chem Soc 117:3333
96. Brutcher FV, Vara FJ (1956) J Am Chem Soc 78:5695
97. Kaneko C, Sugimoto A, Tanaka S (1974) Synthesis 876
98. Korach M, Nielsen DR, Rideout WH (1974) Org Synth, Coll Vol V 414
99. Shealy YF, Clayton JD (1969) J Am Chem Soc 91:3075
100. Brockamp HP, Kula MR (1990) Appl Microbiol Biotechnol 34:287
101. Bäckvall J-E, Byström SE, Nordberg RE (1984) J Org Chem 49:4619
102. Suemune H, Hizuka M (1989) Chem Pharm Bull 37:1379
103. Winchester B, Fleet GWJ (1992) Glycobiology 2:199
104. Look GC, Fotsch CH, Wong C-H (1993) Acc Chem Res 26:182
105. Hughes AB, Rudge AJ (1994) Nat Prod Rep 11:135
106. Karlsson GB, Butters TD, Dwek RA, Platt FM (1993) J Biol Chem 268:570
107. Zipperer B, Müller K-H, Gallenkamp P, Hildebrand R, Fletschinger M, Burger D, Pillat M, Hunkler D, Knothe L, Fritz H, Prinzbach H (1988) Chem Ber 121:757
108. Eyrisch O (1994) Dissertation, Universität Freiburg
109. Kühlmeyer R, Keller R, Schwesinger R, Netscher T, Fritz H, Prinzbach H (1984) Chem Ber 117:1765
110. Vogel E, Altenbach H-J, Schmidbauer E (1973) Angew Chem Int Ed Engl 12:838
111. Eyrisch O, Keller M, Fessner W-D (1994) Tetrahedron Lett 35:9013
112. Heyns K, Trautwein W-P, Paulsen H (1963) Chem Ber 96:3195
113. Limberg G, Thiem J (1994) Synthesis 317
114. Pétrier C, Luche J-L (1985) J Org Chem 50:912
115. Gijsen HJM, Wong C-H (1994) J Am Chem Soc 116:8422
116. Gijsen HJM, Wong CH (1995) J Am Chem Soc 117:7585
117. Effenberger F, Null V, Ziegler T (1992) Tetrahedron Lett 33:5157
118. Kobori Y, Myles DC, Whitesides GM (1992) J Org Chem 57:5899
119. Zimmermann F, Fessner W-D, unpublished results
120. Schreiber SL, Verdine GL (1991) Tetrahedron 47:2543
121. Mullis KB (1994) Angew Chem Int Ed Engl 33:1209
122. O'Hagan D (1991) The Polyketide Metabolites. Ellis Horwood, Chichester
123. Katz L, Donadio S (1993) Annu Rev Microbiol 47:875
124. McDaniel R, Ebertkhosla S, Hopwood DA, Khosla C (1995) Nature 375:549
125. Pieper R, Luo GL, Cane DE, Khosla C (1995) Nature 378:263
126. Tsoi CJ, Khosla C (1995) Chem Biol 2:355
127. Rohr J (1995) Angew Chem Int Ed Engl 34:881
128. McDaniel R, Ebert-Khosla S, Hopwood DA, Khosla C (1994) J Am Chem Soc 116:10855
129. Battersby AR (1994) Science 264:1551
130. Scott AI (1994) Tetrahedron 50:13315
131. Scott AI (1994) Synlett 871
132. Roessner CA, Spencer JB, Ozaki S, Min CH, Atshaves BP, Nayar P, Anousis N, Stolowich NJ, Holderman MT, Scott AI (1995) Protein Express Purif 6:155
133. Fleischmann RD, Adams MD, White O, Clayton RA, Kirkness EF, Kerlavage AR, Bult CJ, Tomb J-F, Dougherty BA, Merrick JM, McKenney K, Sutton G, FitzHugh W, Fields CA, Gocayne JD, Scott JD, Shirley R, Liu L-I, Glodek A, Kelley JM, Weidman JF, Phillips CA, Spriggs T, Hedblom E, Cotton MD, Utterback TR, Hanna MC, Nguyen DT, Saudek DM, Brandon RC, Fine LD, Fritchman JL, Fuhrmann JL, Geoghagen NSM, Gnehm CL, McDonald LA, Small KV, Fraser CM, Smith HO, Venter JC (1995) Science 269:496
134. Burland V, Plunkett G, Sofia HJ, Daniels DL, Blattner FR (1995) Nucleic Acids Res 23:2105
135. Fraser CM, Gocayne JD, White O, Adams MD, Clayton RA, Fleischmann RD, Bult CJ, Kerlavage AR, Sutton G, Kelley JM, Fritchman JL, Weidman JF, Small KV, Sandusky M, Fuhrmann J, Nguyen D, Utterback TR, Saudek DM, Phillips CA, Merrick JM, Tomb JF, Dougherty BA, Bott KF, Hu PC, Lucier TS, Peterson SN, Smith HO, Hutchison CA, Venter JC (1995) Science 270:397
136. Hodgkin J, Plasterk RHA, Waterston RH (1995) Science 270:410

The Synthesis of Novel Sialic Acids as Biological Probes

Mark von Itzstein · Robin J. Thomson

Department of Medicinal Chemistry, Monash University (Parkville Campus), 381 Royal Parade, Parkville, Victoria, 3052, Australia

Sialic acids are involved in a number of biological processes including cell-to-cell, cell-to-microorganism, -toxin, and -antibody binding. Their importance in these processes, especially those with relevance to human disease states, has led to interest in the synthesis of both natural and modified sialic acids. This review examines the most recent methods used for the synthesis and modification of sialic acids and for the preparation of sialyl glycosides as biological probes of sialic acid-recognising proteins.

Table of Contents

List of Symbols and Abbreviations

Symbol	*Description*
Ac	$CH_3C(O)$-
Bn	$PhCH_2$-
Boc	$(CH_3)_3COC(O)$-
CBz	$PhCH_2OC(O)$-
CMP	cytidine monophosphate
ds	diastereoselectivity
Gal	galactose
GlcNAc	*N*-acetyl-D-glucosamine
KDN	2-keto-3-deoxy-D-*glycero*-D-*galacto*-nononic acid (3-deoxy-D-*glycero*-D-*galacto*-nonulosonic acid)
KDO	2-keto-3-deoxy-D-*manno*-octonic acid (3-deoxy-D-*manno*-octulosonic acid)
Lys	lysine
ManNAc	*N*-acetyl-D-mannosamine
Neu5Ac	*N*-acetylneuraminic acid (5-acetamido-3,5-dideoxy-D-*glycero*-D-*galacto*-nonulosonic acid)
Neu5Ac1Me	methyl *N*-acetylneuraminate
TBDMS	*tert*-butyldimethylsilyl

1 Introduction

Sialic acids are a family of 3-deoxy-2-ulosonic acids found most frequently as α-glycosidically linked terminal residues of glycoproteins and glycolipids. The most abundant sialic acid is *N*-acetylneuraminic acid (5-acetamido-3,5-dideoxy-D-*glycero*-D-*galacto*-nonulosonic acid, Neu5Ac, **1**), which was first isolated in the 1930s. To date, 36 sialic acids have been isolated, many of which are *O*-acetylated derivatives of *N*-acylated neuraminic acid [1].

The biological functions of sialic acids derive both from their size and overall negative charge and from their natural position as the terminal residue on cell surface glycoconjugates. Several accounts describe the range of biological properties which are endowed by sialic acids on natural glycoconjugate structures [1–3]. As terminal residues on cell surface glycoconjugates, sialic acids are directly involved in cell-to-cell, and cell-to-microorganism, -toxin, and -antibody binding [1]. Pathogenic organisms often express sialic acid-recognising proteins on their cell surface, and use these to attach themselves to host cell surface sialoglycoconjugates, usually a crucial part of the infection process. If one assumes that these adhesion proteins play a key role in the organism's infection cycle, then they provide useful drug design targets. More complex sialylated oligosaccharide structures are known to be involved in the initial stages of leucocyte recruitment to activated endothelial cells in damaged or infected tissues.

The importance of sialic acids in biological processes has led to an increased interest in the synthesis of both natural and modified sialic acids. Modification of the functional groups on the sialic acid skeleton may lead to changes in steric and/or electronic properties of the molecule which can be used to probe space and charge requirements of sialic acid-recognising proteins. This review will examine the methods used for the synthesis of Neu5Ac and related compounds from hexose sugars, and give an overview of the transformations which have been carried out on the neuraminic acid skeleton resulting in modification at every carbon of Neu5Ac. Most, if not all, of the modified sialic acids have been used to probe the requirements of sialic acid-recognising proteins, in particular the enzymes involved in sialic acid metabolism [4].

2 Synthesis of *N*-Acetylneuraminic Acid and its Analogues

For synthetic purposes, Neu5Ac has been isolated from a number of natural sources, including edible bird's nest (2–11% sialic acid by weight) [5–7] and submaxillary gland mucins which contain as much as 15–25% sialic acid on a dry weight basis [8, 9]. However the need for a more convenient source of Neu5Ac has led to the investigation of both enzymic and chemical methods of preparation.

2.1 Enzymic Synthesis of Sialic Acids

The biosynthesis of *N*-acetylneuraminic acid (Neu5Ac, **1**) in mammalian systems is carried out by *N*-acetyl-D-neuraminyl-9-phosphate synthase [10]. However it is *N*-acylneuraminate pyruvate-lyase (Neu5Ac aldolase; EC 4.1.3.3) which is responsible for the catabolism of sialic acids in vivo [10–12] that has been used to synthesise a wide range of naturally occurring sialic acids [13, 14] and other ulosonic acids [15–23]. The use of aldolase enzymes in preparative organic chemistry has been reviewed [24–26]. Neu5Ac aldolase catalyses the reversible aldol condensation between *N*-acetyl-D-mannosamine (ManNAc, **2**) and pyruvate (Scheme 1). Although the natural role of Neu5Ac aldolase is the cleavage of Neu5Ac, the equilibrium of the reaction can be pushed to favour the formation of the aldol product by using excess pyruvate [26, 27]. If convenient retrieval of the protein is important, the enzyme can be used either encased in dialysis tubing [28] or bound to an agarose gel [13, 16]. ManNAc can be obtain-

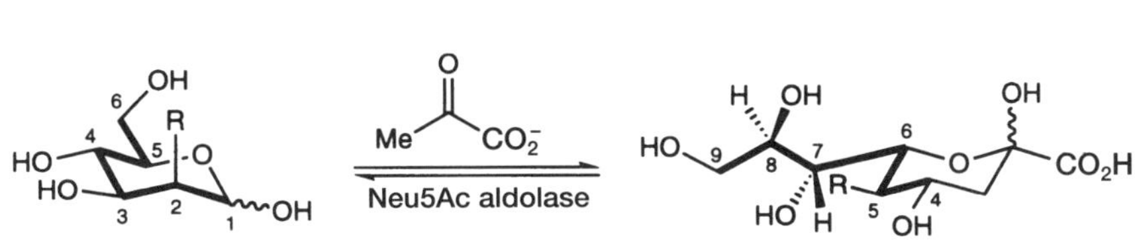

(**2**) R = NHAc ManNAc
(**4**) R = OH Mannose

(**1**) R = NHAc Neu5Ac
(**5**) R = OH KDN

Scheme 1

ed by base-catalysed epimerisation of the corresponding *gluco*-configurated sugar (GlcNAc, **3**) [13, 29, 30], which gives an equilibrium mixture of approximately 20:80 ManNAc:GlcNAc [29]. Kragl and coworkers have developed an "enzyme membrane reactor" in which the enzyme *N*-acetyl-D-glucosamine 2-epimerase (EC 5.1.3.8) is included with Neu5Ac aldolase [27]. This procedure allowed Neu5Ac to be prepared directly from relatively inexpensive GlcNAc.

Neu5Ac aldolase has been isolated from a number of bacterial and mammalian sources [10–12, 31, 32]. Generally the enzyme from either *Clostridium perfringens* [11, 13, 31, 33–35] or from *Escherichia coli* [11, 19, 22, 32] has been used in studies of the synthesis and catabolism of Neu5Ac and its analogues. A mechanism for the cleavage of Neu5Ac with aldolase from *C. perfringens* has been proposed by Schauer and coworkers [2, 31, 36]. The retro-aldol reaction is thought to proceed through a Schiff's base formed between the keto group of the open-chain form of Neu5Ac and a lysine residue of the protein. The imidazole ring of a histidine residue is also believed to be involved in the catalytic process, possibly in a proton transfer process [36]. Baumann et al. [35] have suggested that the pyranose forms of the reactant and product are important in interaction with the protein. This was based on the finding that, for reaction in both forward and reverse directions, the α-anomers of Neu5Ac and ManNAc are the primary substrates in reaction with the aldolase enzyme [35, 37]. The crystal structure of the enzyme from *E. coli*, determined to 2.2 Å resolution, was published by Colman and coworkers in 1994 [38]. The active site was tentatively identified and shows a suitable lysine residue (Lys-165), but there does not appear to be an appropriately positioned histidine residue that would satisfy the above mechanisms. Our preliminary investigations of the X-ray structure using molecular modelling and computer graphic techniques suggest that an active site tyrosine residue may be important in the reaction [39]. Although there is still some uncertainty about the mechanism of the reaction, both proposals have been used to develop an understanding of the substrate specificity of the aldolase enzyme [20, 23, 33].

The substrate specificity of Neu5Ac aldolase has been thoroughly investigated, particularly in the synthetic direction. Only pyruvate is accepted as the 3-carbon synthon [11, 15, 32]. One of the 'naturally occurring' substrates for Neu5Ac aldolase is ManNAc (**2**) [11], while GlcNAc (**3**) is not accepted by the enzyme [11, 15]. This has allowed epimeric mixtures of ManNAc and GlcNAc to be used, the unreacted GlcNAc being recovered after the reaction [13]. D-Mannose (**4**) is also accepted as a substrate, leading to 3-deoxy-D-*glycero*-

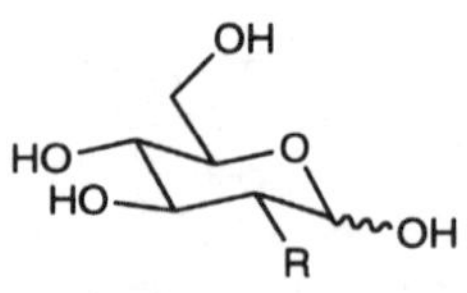

(**3**) R = NHAc GlcNAc
(**6**) R = OH Glucose

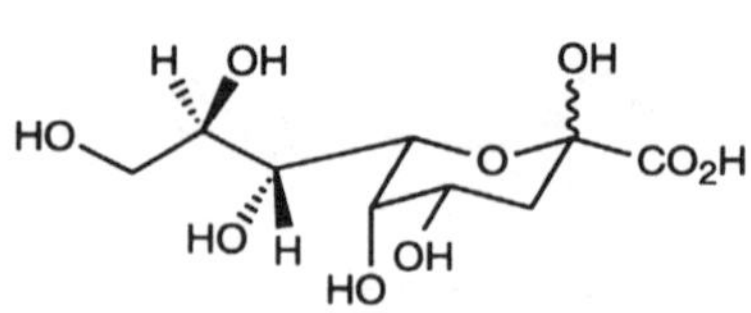

(**7**) 5-*epi* KDN

D-*galacto*-nonulosonic acid (KDN, **5**) [14, 15], a sialic acid of increasing biological importance. Interestingly, D-glucose (**6**) is a weak substrate for the enzyme, giving 5-*epi*-KDN (**7**) [15, 16]. The enzyme has been found to accept a range of substrates with the *manno* configuration and is quite tolerant of substitution at C-6, C-4, and particularly at C-2 of the *manno*-configurated sugars. An overview of the *manno*-configurated substrates that have been investigated is presented in Table 1. The yield of the sialic acid obtained depends on how well the substrate is accepted by the enzyme, but is generally in the range of 60–80% for a good substrate [24].

A large number of substitutions at C-2 of *manno*-configurated sugars are tolerated, allowing the synthesis of variously C-5 substituted sialic acids. *Manno*-configurated sugars substituted at C-3, however, failed to undergo enzymic aldol condensation [23], except for 3-deoxy-mannose, which gave a mixture of 4 acids [18]. The aldol condensation of C-4- derivatised mannose-based sugars leads to C-7-substituted sialic acids, which are generally more difficult to obtain through derivatisation of Neu5Ac itself. 7-Azido-7-deoxy-Neu5Ac (**8**) obtained by this method is the first reported example of Neu5Ac with a nitrogen substituent at C-7 [22, 40]. Functionalisation at C-5 of ManNAc or mannose should lead to the C-8 sialic acid derivatives. While 5-deoxy ManNAc or mannose, sugars in the furanose form, have been found to be good substrates for the enzyme [18], 5-methoxy-ManNAc gave only traces of 8-methoxy-Neu5Ac [13] (possibly due to steric constraints). Substitution at C-6 of ManNAc, even with bulky or extended groups, is accepted by Neu5Ac aldolase [13, 41, 42]. When ManNAc was replaced by the 5-carbon sugar 2-acetamido-2-deoxy-D-lyxose (**9**), effectively removing the sixth carbon atom, the 8-carbon Neu5Ac derivative **10** was produced [43]. Similarly, D-lyxose gave the truncated KDN analogue [15–17].

Apart from sugars in the *manno* configuration, a variety of other sugars, including 4- and 5-carbon sugars, are accepted as substrates by Neu5Ac aldolase [15–17, 19, 20, 23], giving entry into a range of 3-deoxy-2-ulosonic acids. Intensive examination of the substrate specificity of the enzyme has led the groups of Augé [20] and Wong [19, 23] to propose a difference in the facial selectivity of

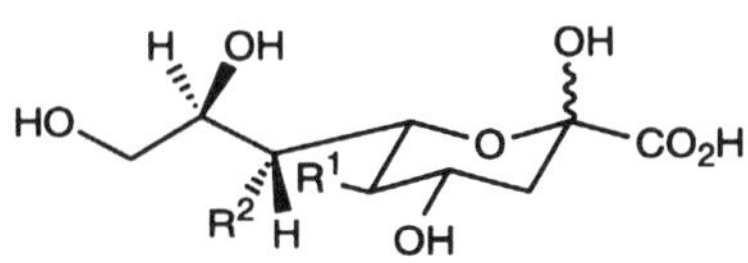

(**8**) R^1 = NHAc, R^2 = N_3

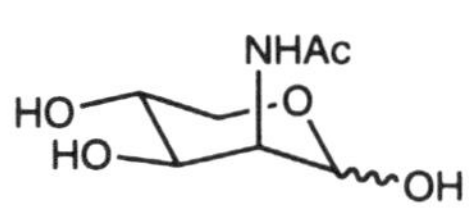

(**9**)

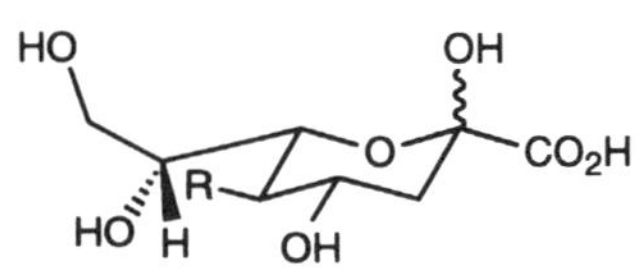

(**10**) R = NHAc

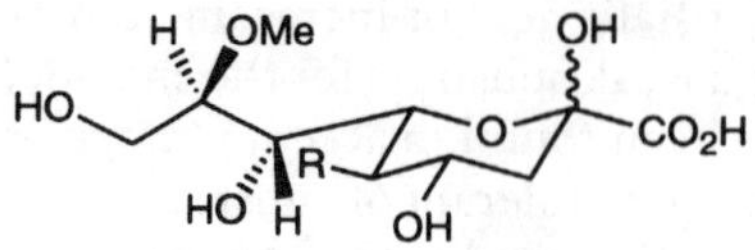

(11) R = NHC(O)CH$_2$OH

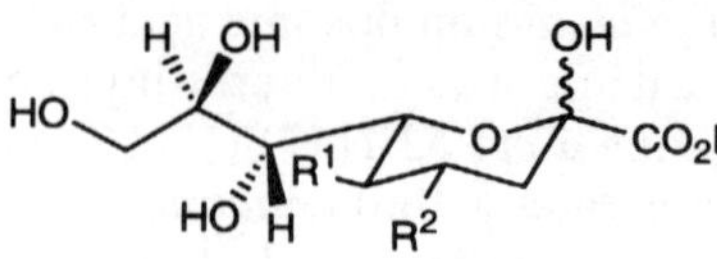

(12) R^1 = NHAc, R^2 = H
(15) R^1 = NHAc, R^2 = OMe
(16) R^1 = NHAc, R^2 = NHAc

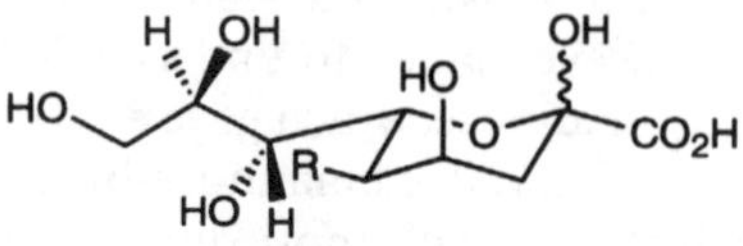

(13) R = NHAc

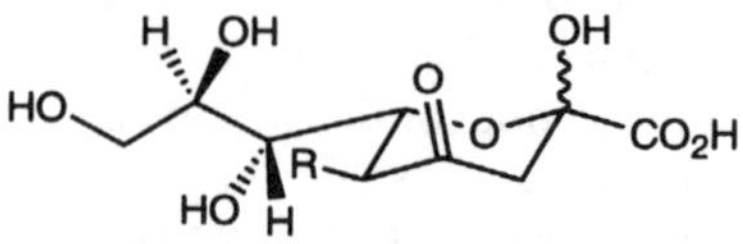

(14) R = NHAc

the attack by the pyruvate synthon on the carbohydrate aldehyde to explain the products obtained.

The substrate specificity of the enzyme in the reverse aldol, lyase, or catabolic, direction has also been studied [4, 33, 34, 44, 45]. The enzyme cleaves C-9 acetylated Neu5Ac at one fourth of the speed of Neu5Ac [44], which is in line with a slower aldolase reaction for C-6 acetylated ManNAc [15]. C-8 *O*-methylated *N*-glycolylneuraminic acid (**11**) is not a substrate for the enzyme [46], and 5-methoxy-ManNAc appears to be almost inert to the action of aldolase [13]. In some cases, however, the behaviour of Neu5Ac analogues towards the enzyme does not mirror the behaviour of the corresponding substrates in the aldol condensation reaction. For example, C-7 and C-8 deoxy-Neu5Ac were not susceptible to enzymic cleavage [33], whereas the corresponding C-4 [21] and C-5 [18] deoxy-ManNAc derivatives were utilised by the enzyme isolated from the same source. Of interest was the lack of reaction with 4-deoxy- (**12**) [33, 47], 4-*epi*- (**13**) [33, 48], and 4-oxo-Neu5Ac (**14**) [49], all of which have been reported as inhibitors of the enzyme, while 4-methoxy- (**15**) [50] and 4-acetamido-4-deoxy-Neu5Ac (**16**) [34] were not recognised at all. This is consistent with the proposed mechanism of the enzyme where the hydroxyl group at C-4 of Neu5Ac and the aldehyde of ManNAc are important functional groups in the enzyme reaction.

Neu5Ac aldolase has also been used for the synthesis of 3-deoxy-D-*manno*-octulosonic acid (KDO, **17**) [16, 17, 20]. Condensation of D-arabinose (**18**) with pyruvate gave a mixture of KDO and 4-*epi*-KDO (**19**). Wong and coworkers have since reported the isolation of a KDO aldolase which produces KDO with complete stereospecificity at C-4 and also accepts a wide variety of carbohydrate substrates [51]. *N*-Acetylneuraminate synthase, found in *Neisseria meningitidis*, has been used to catalyse the condensation of 6-azido-6-deoxy-*N*-acetylmannosamine with phosphoenolpyruvate to give 9-azido-9-deoxy-Neu5Ac [52].

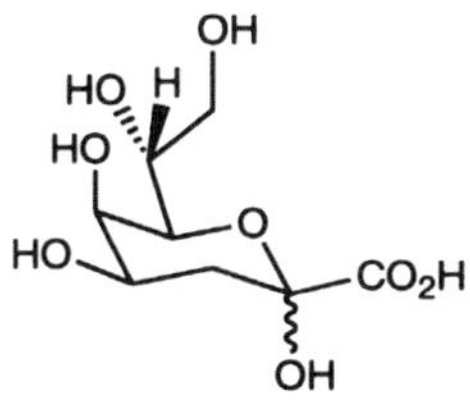

(17) KDO

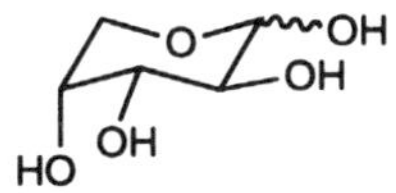

(18)

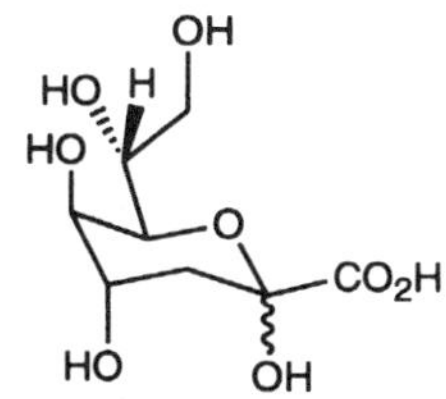

(19)

Table 1. A sample of *manno*-configurated sugars tested as substrates for Neu5Ac aldolase*

Carbon	Mannose	ManNAc
C-2	OH[a] *epi*-OH (glucose)[b, c]	NHC(O)CH_3[i] ***epi*-NHAc (GlcNAc)**[b, i] NHC(O)CH_2OAc[g] NHC(S)R[j] N_3[c, k, l] NHCBz[l] NHBoc[h] CH_2NHAc[m] CH=CH_2[n] OAc[n] SR[n] SAr[n] H[a, b] F[n] Br[n] Cl[n]
C-3	(H)[d] **N_3**[e] **NH_2**[e] **Br**[e]	**SH**[o]
C-4	H[c] *epi*-OH (talose)[f]	OMe[g, p] H[p] N_3[q] (OAc → 6-OAc)[g]
C-5	H[d]	**OMe**[g] H[d]
C-6	OAc[b, g] Br[h] N_3[h] –C-6 (lyxose)[b, c]	OAc[g] OMe[g] OP$(Me)_2$[r] F[r] N_3[r] OC(O)CH(OH)CH_3[g] OC(O)CH_2NHBoc[s] N_3-(2-NHCBz)[t] –C-6[u]

* Entries in bold were not substrates for Neu5Ac aldolase.
a [14]; *b* [15]; *c* [16]; *d* [18]; *e* [23]; *f* [19]; *g* [13]; *h* [53]; *i* [11]; *j* [54]; *k* [55]; *l* [56]; *m* [57]; *n* [58]; *o* [59]; *p* [21]; *q* [22]; *r* [42]; *s* [41]; *t* [60]; *u* [43].

2.2 Chemical Synthesis of Sialic Acids

The methods used to chemically synthesise sialic acids can be broadly grouped into two types (1) base-catalysed aldol condensation of an aldohexose with oxaloacetic acid (oxobutanedioic acid) as the pyruvate equivalent, and (2) homologation of the hexose sugar from the reducing end in a number of steps or in a single three-carbon extension. In each case, the crucial point is the diastereoselectivity of addition to C-1 of the aldohexose, establishing the configuration of the substituent at C-4 of the sialic acid. To date, only one total synthesis of Neu5Ac from non-carbohydrate precursors has been reported [61, 62]. The advantage of a chemical synthesis is that it can use and produce compounds which may not be accepted by the aldolase enzyme. The chemical synthesis of sialic acids has been reviewed by Tuppy and Gottschalk [63] and more recently by DeNinno [64].

The first chemical synthesis of Neu5Ac was reported by Cornforth, Gottschalk and coworkers in 1957 [65, 66]. They condensed *N*-acetyl-D-glucosamine (**3**) and oxaloacetic acid at pH 9–11, to give, after decarboxylation, Neu5Ac (**1**) in approximately 2% yield (Scheme 2). Under the basic conditions, the GlcNAc had partially epimerised to the *manno*-configurated sugar which then underwent the condensation. By using ManNAc (**2**) the yield of the reaction was raised to 8–11% [67]. Addition of borate ion to the reaction mixture increased the overall yield to 22% by suppressing epimerisation of ManNAc [68]. Further modification of the reaction conditions and the use of the potassium salt of di-*tert*-butyl oxobutanedioate led to an increase in yield to 34% [69]. The use of the 4,6-*O*-benzylidene derivative of GlcNAc (**20**) was also found to give yields in the region of 30% and allowed the use of the less expensive *gluco*-configurated sugar [69]. Although attempts to undertake the condensation with pyruvate itself were unsuccessful [64], 3-fluoropyruvate gave small yields of a compound considered to be 3-fluoro-Neu5Ac [70], while bromopyruvate and hydroxypyruvate have been used to synthesise 3-hydroxy-Neu5Ac [71]. GLC analysis of the products from the reaction with bromopyruvate showed that all of the eight possible isomeric products were formed. However, there was one major product, tentatively identified as the *β*-anomer with equatorial hydroxyl groups on C-3 and C-4 (**21**) [71].

The condensation with oxaloacetic acid, or equivalent reagents, has been used for the preparation of C-5 derivatives of Neu5Ac [72, 73], 8-methoxy-Neu5Ac [74], and the side-chain truncated, 8-carbon analogue **10** [75, 76].

The syntheses of KDN (**5**) and KDO (**17**) have also been carried out, by condensation of oxaloacetic acid with D-mannose (**4**) and D-arabinose (**18**)

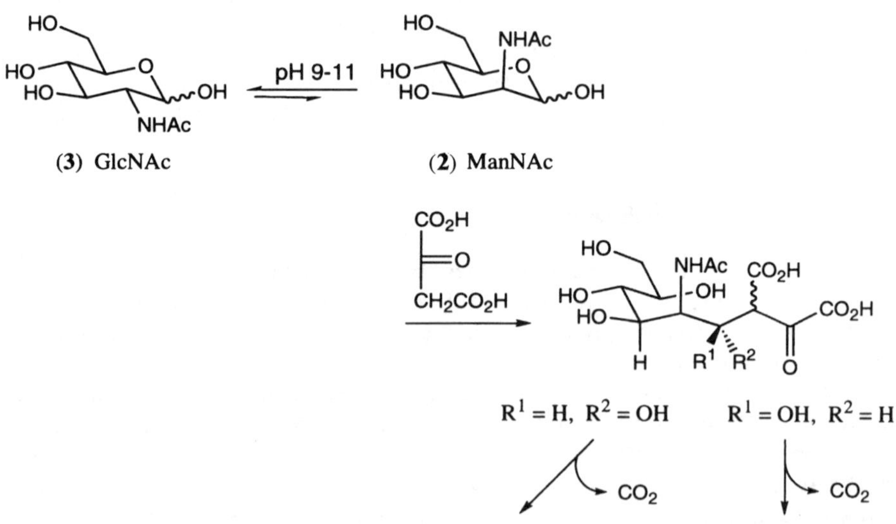

(**3**) GlcNAc

(**2**) ManNAc

R^1 = H, R^2 = OH

R^1 = OH, R^2 = H

Neu5Ac (**1**)

4-*epi*-Neu5Ac (**13**)

Scheme 2

respectively [77]. In a synthesis of KDN which utilised $NiCl_2$ in the decarboxylation step, yields of up to 70% were obtained with a selectivity of approximately 4.2:1 in favour of the equatorial substituent at C-4 [77]. The thioisostere of Neu5Ac has also been prepared using an aldol condensation with nickel(II)/oxaloacetic acid [78]. The effect of metal ions on both the condensation and decarboxylation processes has been discussed by Hagedorn et al. [79].

Homologation of the hexose sugar from the reducing end has been carried out using a number of methods (Fig. 1). A stepwise extension of ManNAc (**2**) by stereoselective addition of nitromethane (twice) and cyanide produced Neu5Ac (**1**) in 17% yield after 13 steps [80]. The 1-nitro-heptitol **22** has also been manipulated to give the 2-deoxy-heptose **23** [81]. A Wittig-Wadsworth-Emmons reaction, with the ylid derived from **24**, was then utilised and ultimately led to 4-deoxy-Neu5Ac (**12**) [81]. This reaction has also been reported using the methyl ester of same ylid [82]. Another approach employed a single carbon extension of ManNAc by stereoselective addition of hydrocyanic acid to form the cyanohydrin **25**, conversion to the aldehyde **26**, and then a Wittig reaction between the aldehyde and the phosphorane ylid **27** [83]. A modification of the latter method was used in the preparation of 4-methoxy-Neu5Ac (**15**) [50, 84, 85]. In that case, the Wittig reaction with the protected 2-methoxy-heptose **28** was accompanied by a partial inversion of configuration at the α-carbon, with the result that the two C-4 stereoisomers of the Neu5Ac derivative were produced in approximately equal amounts [85].

Dondoni et al. have reported the syntheses of KDN (**5**) and 4-*epi*-KDN (**29**) from protected D-mannose (**30a, b**) (Scheme 3) [86]. The ylid **31**, with the thiazole substituent as a masked formyl group, was added to the protected sugars to give the (E)-α,β-enones **32a, b** in good yield. Conjugate addition of benzyl oxide anion to the enones gave mixtures of *syn* and *anti* diastereomers

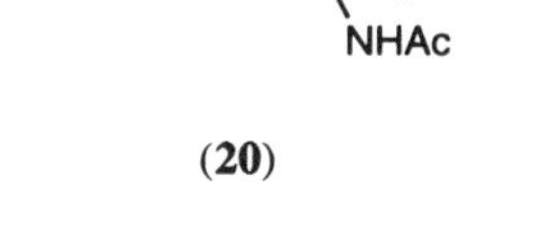

(**20**)

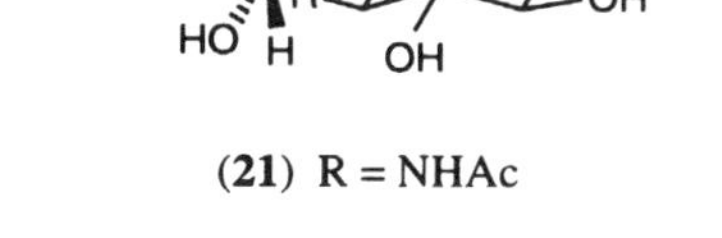

(**21**) R = NHAc

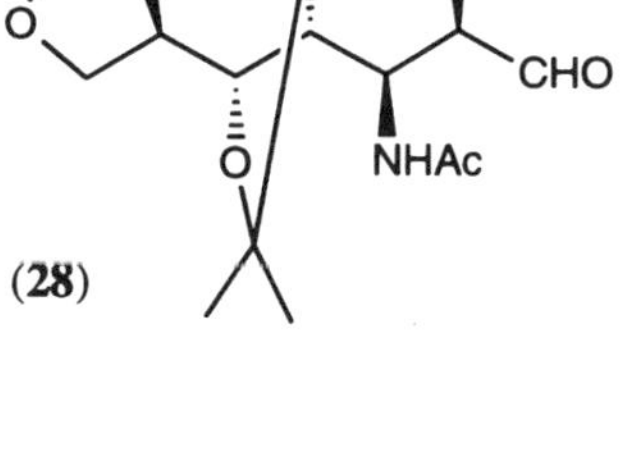

(**28**)

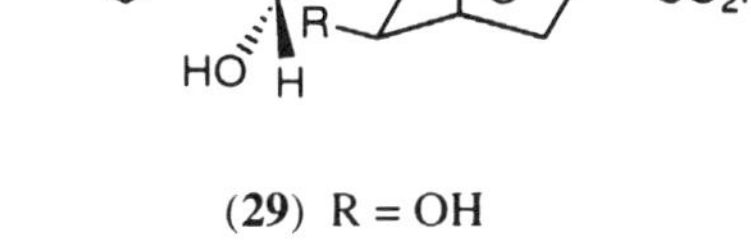

(**29**) R = OH

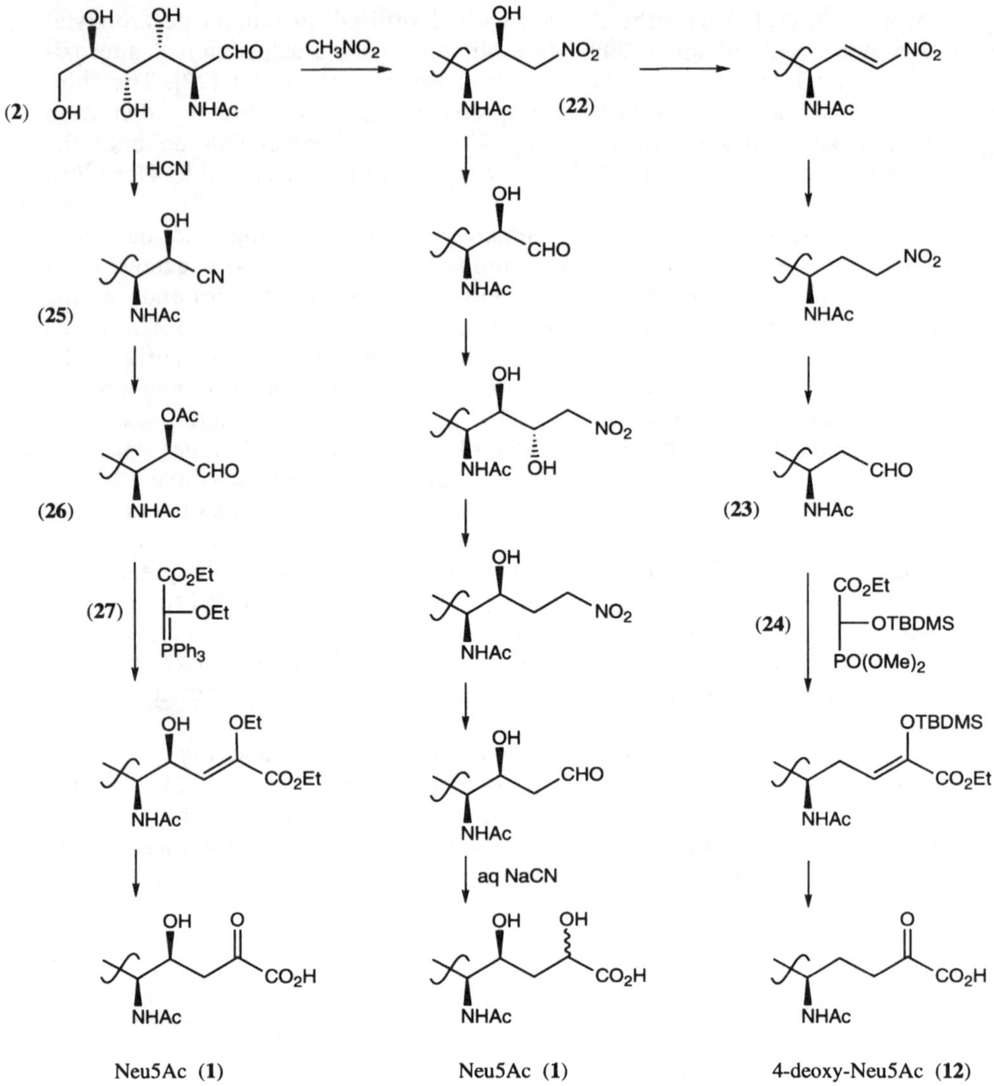

Fig. 1

(**33a, b**), which were separable by chromatography. Interestingly, the diastereoselectivity was mainly *syn* (70%) to an adjacent hydroxyl protected as an acetonide (**33a**), but *anti* (85%) to an adjacent benzyloxy group (**33b**). This allowed the preparation of both C-4 epimers of KDN. Conjugate addition of trimethylsilylazide to the enone **32a** proceeded to give predominantly the *syn* product (*syn*:*anti* 3:1). Cyclisation, and manipulation of the major product, the 4-azido-4-deoxy-KDN derivative **34**, ultimately led to 4-acetamido-4-deoxy-KDN (**35**) [87].

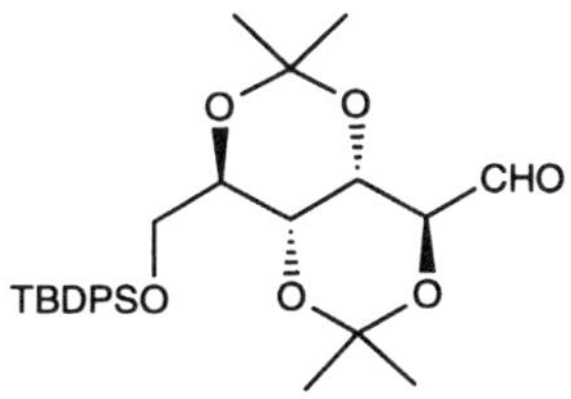

(**30a**)

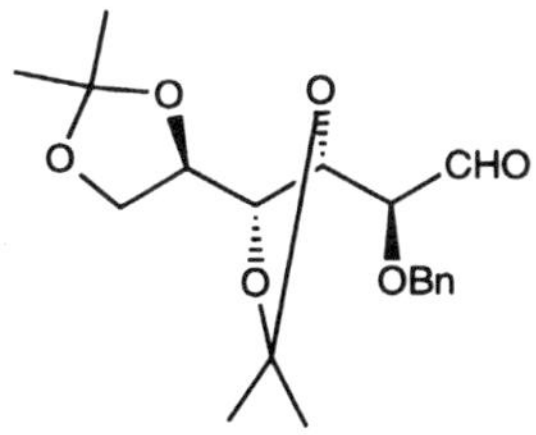

(**30b**)

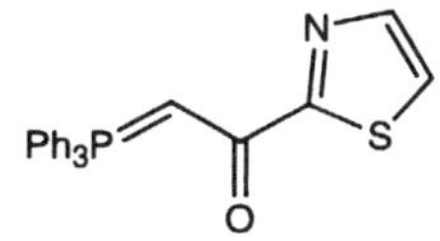

(**31**)

TBDPS = *t*-butyldiphenylsilyl

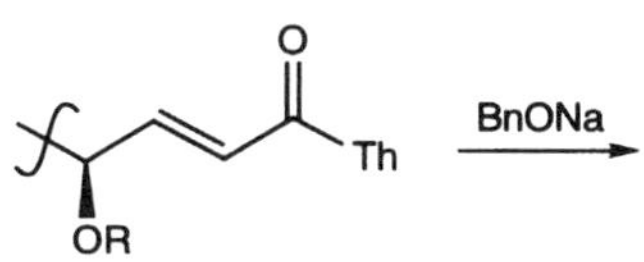

Th = 2-thiazole
(**32a**) R = acetonide
(**32b**) R = benzyl

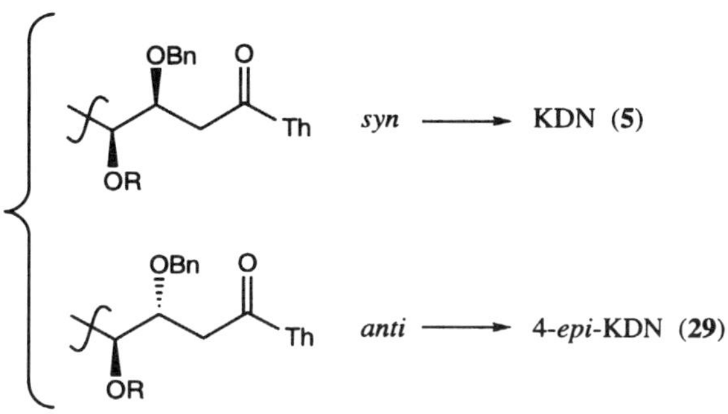

(**33a**) R = acetonide; *syn:anti* 70:30
(**33b**) R = benzyl; *syn:anti* 15:85

Scheme 3

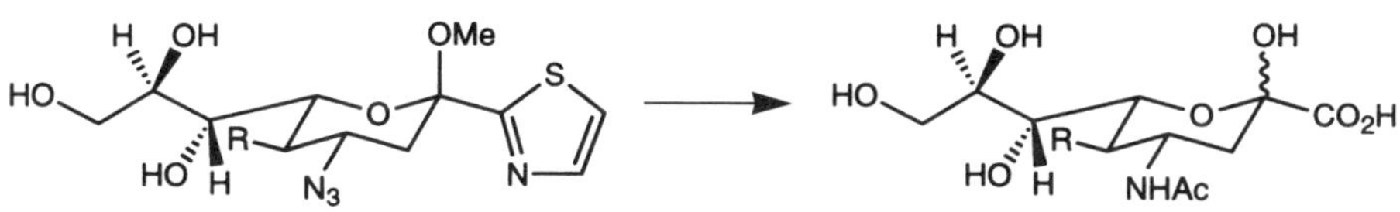

(**34**) R = OH (**35**) R = OH

Vasella and coworkers have utilised 2-bromomethyl acrylate to give the three-carbon extension required, in a base-catalysed addition to the 1-nitro-*N*-acetyl-D-mannosamine derivative **36** [88] (Scheme 4). The ketone of the chain-extended compound **37** could be stereoselectively reduced (~94% ds) with $NaBH_4$ in the presence of acetic acid. Ozonolysis of the alkene in **38a** produced the protected Neu5Ac analogue **39** (approx. 46% yield from the nitro compound). The diastereoselectivity of the ketone reduction was completely reversed in the absence of acetic acid to give **38b**, leading to the synthesis of 4-*epi*-Neu5Ac (**13**). This method was also used to prepare 4-deoxy-Neu5Ac (**12**) [89] and 6-amino-2,6-dideoxy Neu5Ac derivatives [90] which will be discussed in a later section.

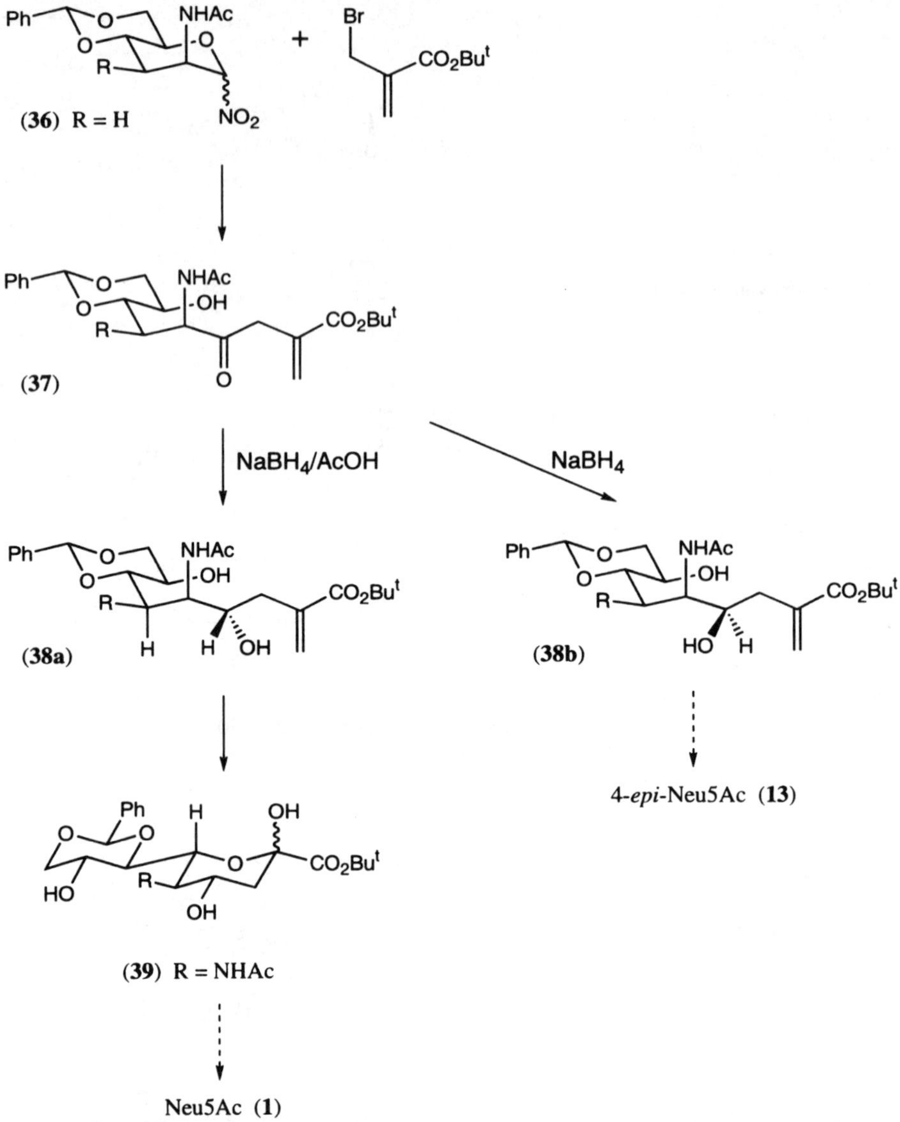

Scheme 4

Recently, the use of indium or tin to catalyse the coupling of 2-bromomethyl acrylic acid or its ester to unprotected ManNAc (**2**) [91, 92], or D-mannose (**4**) [92, 93] in aqueous media has been reported. The coupling leads directly to the C-4-hydroxylated nine carbon chain **40a, b** (Scheme 5) (in contrast to the C-4 ketone **37** obtained from the base-catalysed addition of 2-bromomethyl acrylate to **36** [88]). The reactions proceeded in good yield, and gave products with a predominantly *threo* relationship (*threo*:*erythro* 3 to 5:1) between the newly gene-

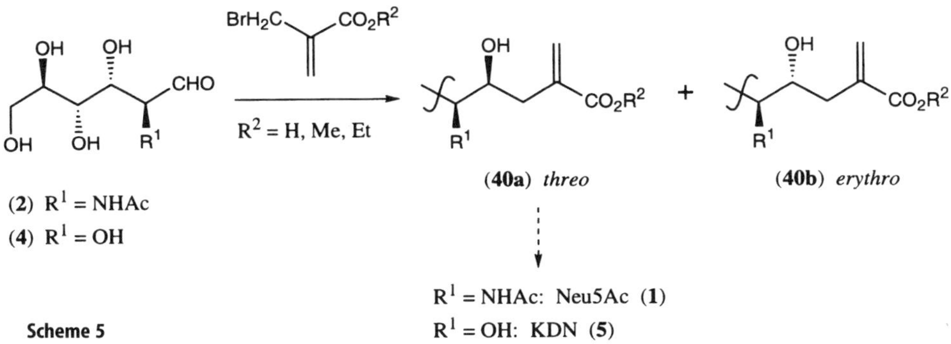

Scheme 5

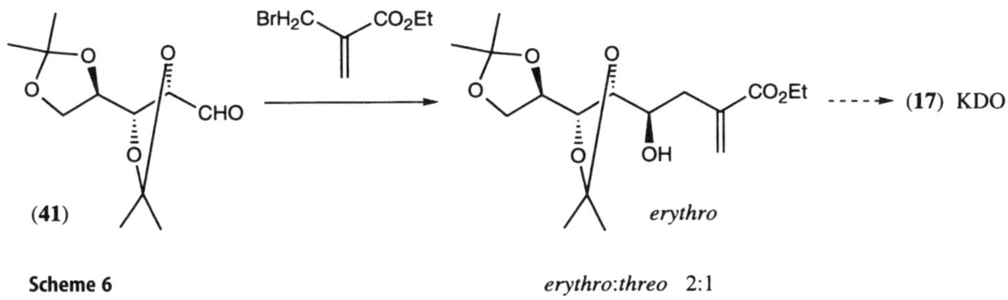

Scheme 6

rated hydroxyl group and the substituent at C-2 of the starting carbohydrate [91, 92, 94]. This leads to the correct, equatorial orientation for the C-4 substituent of Neu5Ac and KDN. In the syntheses of KDN [92, 93], the enoate **40a** (R^1=OH) or its acetate could be separated from the minor product. For Neu5Ac, however, the mixture of enoates **40a, b** (R^1=NHAc) was inseparable [91, 92]. The enoates were ozonolysed, and the ozonides decomposed to give the corresponding cyclised sialic acids. Interestingly, KDO (**17**), which has the reverse stereochemistry at C-4 to Neu5Ac, could be prepared by an indium-mediated coupling of ethyl 2-bromomethyl acrylate with acetonide-protected D-arabinose (**41**) (Scheme 6). By using the protected sugar, a majority of the *erythro* product was obtained [95].

Vasella and coworkers have also developed a sequence in which a 3-carbon extension to the reducing end of a 1-nitro derivative of *N*-acetyl-D-glucosamine (**42**) ultimately forms the glycerol side-chain (C-7, C-8, and C-9) of Neu5Ac [96] (Scheme 7). The addition reaction between **42** and cyclohexylidene-D-glyceraldehyde produced a single nitroalcohol, **43**. Acetylation, followed by reductive denitration with tri-*n*-butyltin hydride gave **44**. Debenzylidenation freed the C-6 hydroxyl group, which was oxidised to give what would eventually become the C-1 carboxyl group of Neu5Ac. The product was isolated as the peracetylated methyl ester **45** which was treated with base to produce the 2-deoxy-2,3-didehydro-Neu5Ac derivative **46**. This was converted to Neu5Ac via compound **47**, resulting from addition of *N*-bromosuccinimide in methanol. Reductive denitration of **43** with tri-*n*-butyltin [^{2}H]-hydride led to the formation of [6-^{2}H]-Neu5Ac

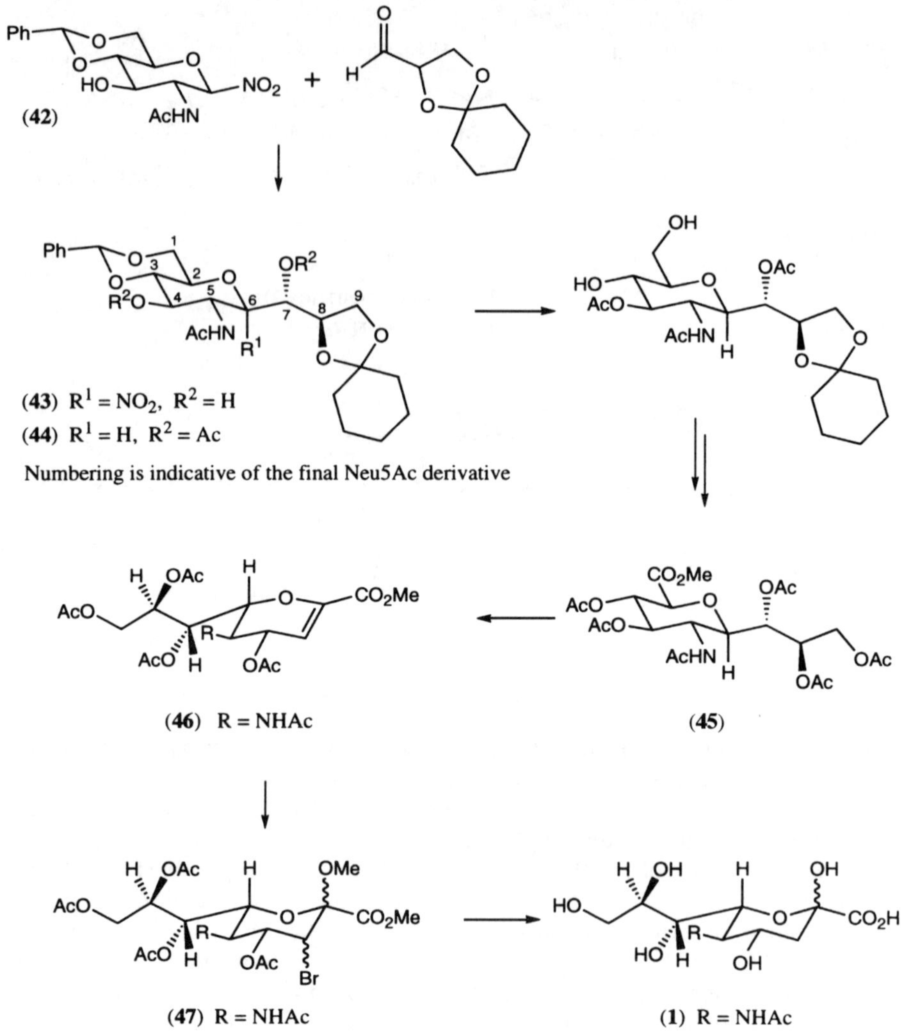

Scheme 7

(48) [96]. Manipulation of the nitro group of **43** allowed the introduction of methyl and hydroxymethyl substituents [97], which, by following a pathway similar to that shown in Scheme 7, led to the corresponding C-6-substituted Neu5Ac derivatives **49** and **50**. A 2-deoxy-2,3-didehydro-Neu5Ac analogue completely lacking the glycerol side-chain at C-6 **51** has been synthesised solely from GlcNAc, using similar chemistry for creation of the C-1 carboxyl group and the 2,3-double bond [98].

Carbocyclic analogues of Neu5Ac and KDO [99], and of side-chain truncated analogues of 4-substituted 2-deoxy-2,3-didehydro-Neu5Ac [100], have been synthesised using Diels-Alder chemistry.

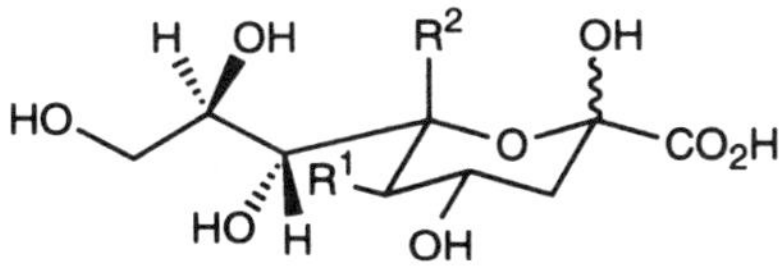

(**48**) R^1 = NHAc, R^2 = ^2H
(**49**) R^1 = NHAc, R^2 = CH_3
(**50**) R^1 = NHAc, R^2 = CH_2OH

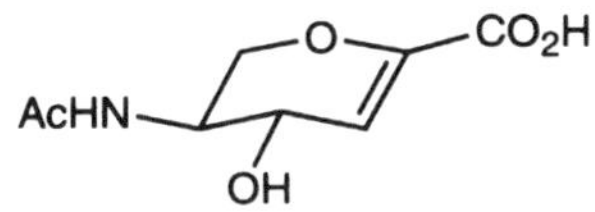

(**51**)

3 Derivatisation of Sialic Acids

The extensive manipulations which have been carried out on sialic acids themselves have been reviewed by Tuppy and Gottschalk [63], Holmquist [101], Vliegenthart and Kammerling [102], in detail by Zbiral [4], and recently by von Itzstein and Kiefel [103]. This account will therefore give an overview of the transformations which have been carried out, resulting in modification at every carbon of Neu5Ac. The derivatives prepared have been used to probe the requirements of sialic acid-recognising proteins, in particular the enzymes involved in sialic acid metabolism such as CMP-sialate synthase, sialidases, and Neu5Ac aldolase.

In general, manipulations on Neu5Ac have been carried out with the C-1 carboxyl group esterified, and the C-2 anomeric position protected as a glycoside, a thioglycoside, or an ester. The choice of an α- or β-glycoside for protection of the anomeric position during manipulations rests with the ease of preparation of the glycoside [104], the effect of the glycoside on subsequent reactions, and the ease of cleavage of the glycoside to obtain the unprotected sialic acid. The β-methyl glycoside **52** has been the most widely used for manipulations of Neu5Ac ([4] and references therein). It is readily prepared in high yield by refluxing a methanolic solution of methyl *N*-acetylneuraminate (Neu5Ac1Me, **53**) in the presence of an acid catalyst for 24–48 hours [105, 106] (Scheme 8). The α-methyl glycoside **54** can be prepared by reaction of peracetylated β-glycosyl chloride (2-β-chloro-2-deoxy-Neu5Ac1Me, **55**) [7, 105, 107] with sodium methoxide [107], or with methanol in the presence of a silver promoter [104, 105, 108]. A recent report of the preparation of the α-methyl glycoside in 96% yield by reaction of (**55**) with methanol without a promoter [109] makes this glycoside an attractive alternative to the normally employed β-methyl glycoside. The α-glycosides of Neu5Ac are reputedly more easily hydrolysed by mild acid than the corresponding β-anomers [110], and have the advantage that they can be cleaved enzymically by sialidase (provided, of course, that the sialic acid analogue is not a sialidase inhibitor) [47, 54, 111].

With the C-1 and C-2 positions protected, the most reactive hydroxyl group of those remaining is the primary hydroxyl group at C-9. Silylation studies with Neu5Ac have shown that the most reactive of the secondary hydroxyl groups is that at C-4 [112] (Scheme 9). For KDN, selective functionalisation may be

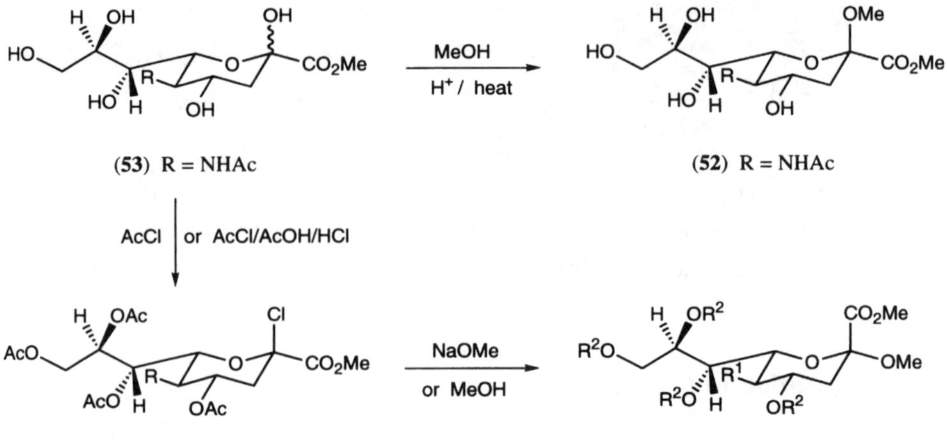

(**53**) R = NHAc

(**52**) R = NHAc

(**55**) R = NHAc

(**54**) R^1 = NHAc
with NaOMe: R^2 = H
with MeOH: R^2 = Ac

Scheme 8

complicated by the additional secondary hydroxyl group at the C-5 position. Methodology has been reported for the peracetylation of both the methyl ester and the free acid of Neu5Ac, and for preparation of 4,7,8,9-tetra-*O*-acetyl-Neu5Ac1Me (**56**) [113]. The protecting groups normally employed during synthesis are acetate, benzoate, silyl, benzyl, and the isopropylidene group across the C-8 and C-9 hydroxyls. The intricate protecting group manipulations sometimes required in the functionalisation of sialic acids are exemplified in the syntheses of the side-chain deoxy analogues reported by Zbiral and his co-workers [4, 114, 115].

A number of the naturally occurring sialic acids, in particular the *O*-acetylated Neu5Ac derivatives, have been prepared by manipulation of Neu5Ac. Deri-

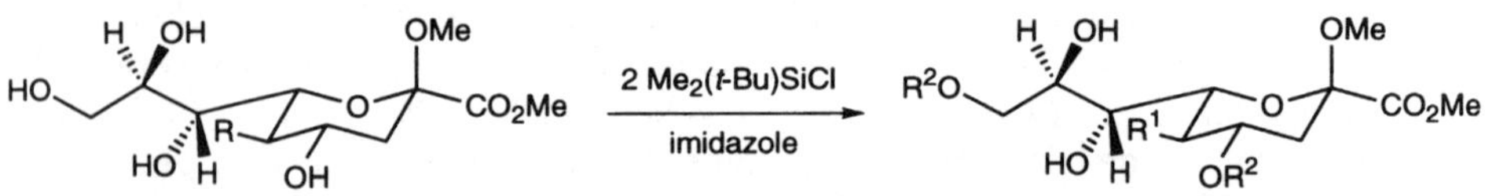

(**52**) R = NHAc

R^1 = NHAc
R^2 = Si(*t*-Bu)Me_2

Scheme 9

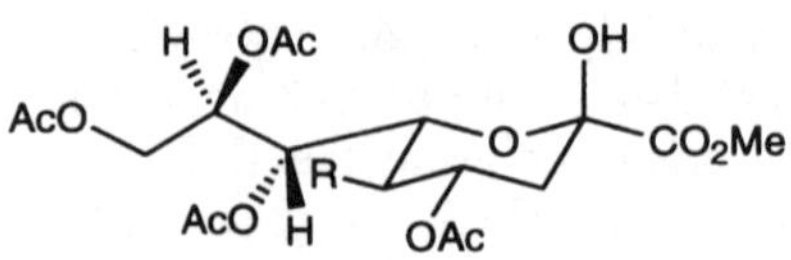

(**56**) R = NHAc

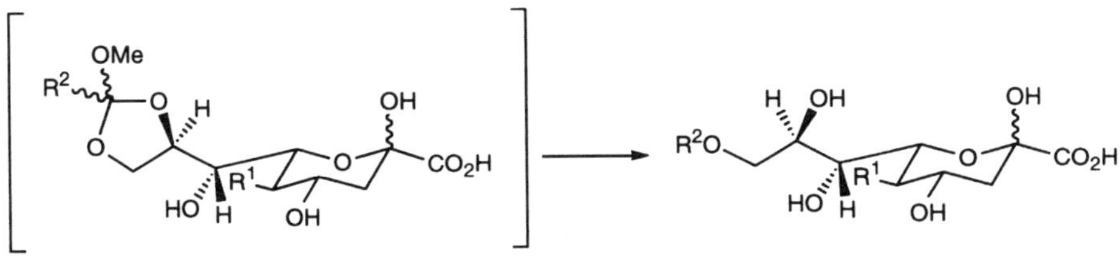

R^1 = NHAc, NHC(O)CH_2OH
R^2 = H, Me, Ph etc

(**57**) R^1 = NHAc, NHC(O)CH_2OH
R^2 = HC(O), MeC(O), PhC(O) etc

Scheme 10

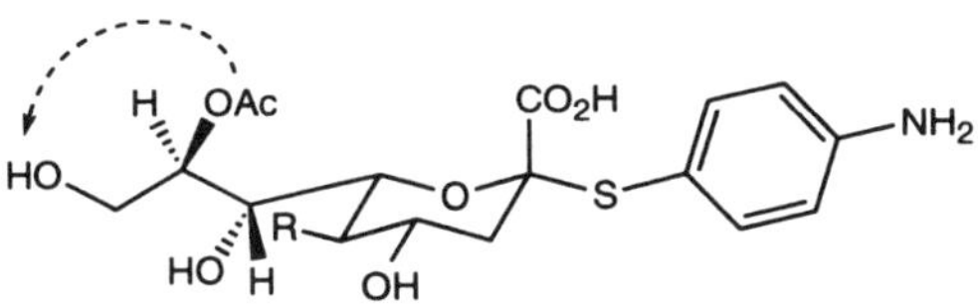

(**58**) R = NHAc

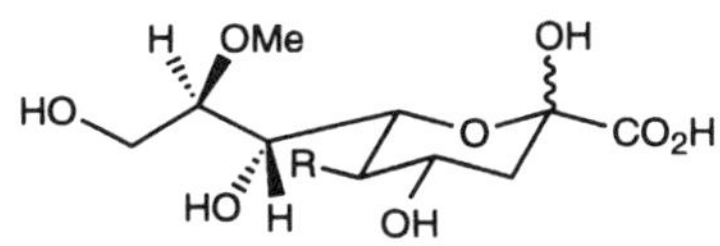

(**59**) R = NHAc

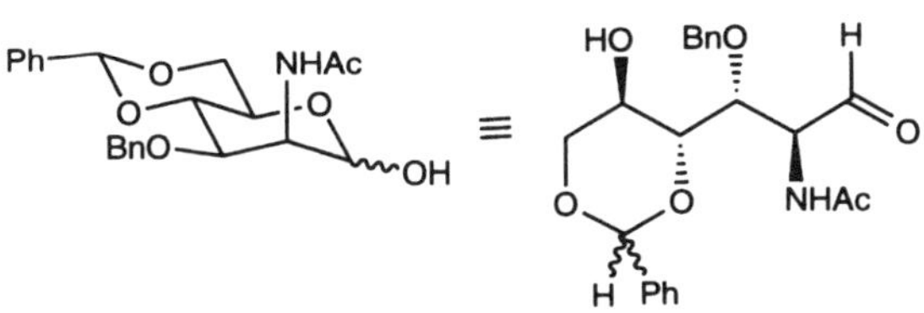

(**60**)

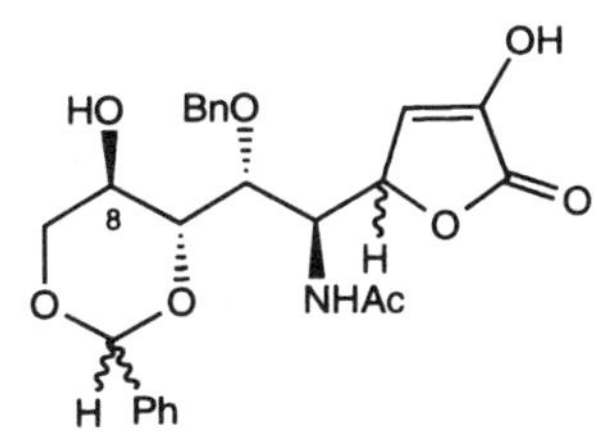

(**61**)

vatives mono-*O*-acetylated at C-4 [116–118], C-7 [117, 119, 120], and C-9 [116–118], as well as various di- and tri-*O*-acetylated compounds [117, 118, 120] have been prepared. The use of ortho esters allows highly regioselective *O*-acytylation at C-9, on the unprotected *N*-acetyl- and *N*-glycolyl-neuraminic acids, to give **57** [116] (Scheme 10). The 8-*O*-acetylated derivative **58** of a Neu5Ac 2-α-thioglycoside was found to be unstable, isomerising to the 9-*O*-acetylated derivative [118]. Naturally occurring 8-methoxy-Neu5Ac (**59**) has been prepared via a chemical synthesis from 3-*O*-benzyl-4,6-*O*-benzylidene-*N*-acetyl-D-mannosamine (**60**) and potassium di-*tert*-butyl oxobutanedioate [74]. The intermediate 6-*O*-benzyl-7,9-*O*-benzylidene-*N*-acetyl-D-neuraminic acid γ-lactone (**61**) was methylated at C-8 and subsequently deprotected to give 8-*O*-methyl-Neu5Ac (**59**).

3.1 Modification of the Glycerol Side-Chain

While many C-9-substituted sialic acids have been synthesised enzymically (see Table 1, Sect. 2.1), a number have also been prepared by manipulation of Neu5Ac itself. Functionalisation of the C-9 hydroxyl group is often possible with only minimal protection of the remainder of the molecule. Regioselective acylation [117, 118, 121], tosylation [54], silylation [112], and tritylation [106] of Neu5Ac glycosides have been reported. The tosylate group of **62** was displaced with azide ion, the product **63** deprotected, and the azide group reduced to give 9-amino-9-deoxy-Neu5Ac (**64**) [54] (Scheme 11). The amine has been converted to a number of *N*-acyl derivatives [122], including both the 9-acetamido **65** [122, 123] and thioacetamido **66** [54] derivatives, which are of interest as substrates for influenza C viral sialidase [124, 125]. Fluorine has been introduced at C-9 by reaction of the partially protected derivative **67** with diethylaminosulfur trifluoride [106]. Oxidation of the C-9 hydroxyl group to give the C-9 carboxy derivative **68** has also been reported (in [101]).

All possible epimeric [126–128] and deoxy [114, 115] side-chain analogues of Neu5Ac have been prepared by Zbiral and coworkers [4]. Using suitable protecting group manipulation, the C-7 and C-8 hydroxyls can each be isolated. They have both been separately oxidised [126, 127, 129, 130] and reduced [126, 127]. The C-7 ketone of **69** was reduced with a 95:5 stereoselectivity for the 7-*epi* compound **70** [126], while reduction of the C-8 ketone of **71** preferentially gave the 8-*epi* derivative **72** (20:1 stereoselectivity) [127]. An interesting 'migration' of the ketone from C-7 to C-8 was observed under basic conditions [130]. The C-7:C-8 epoxy derivatives, **73** and **74**, are known [112, 126, 131], and have been used to prepare the epimeric side-chain compounds (e.g. **75**) [126], as well

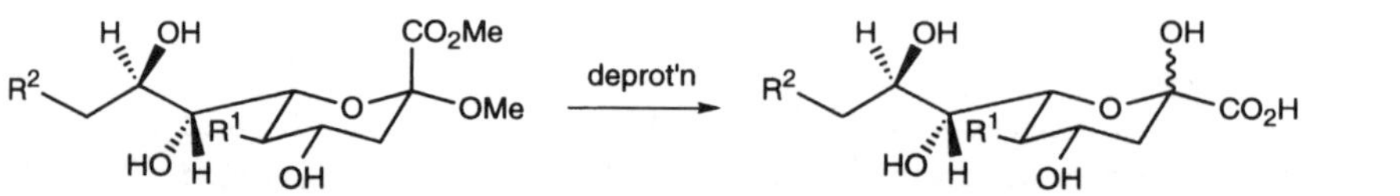

R^1 = NHAc R^1 = NHAc

R^2 = OH → (**62**) OTs → (**63**) N_3 R^2 = N_3 → (**64**) NH_2 → (**65**) NHC(O)CH_3
(**66**) NHC(S)CH_3

Scheme 11

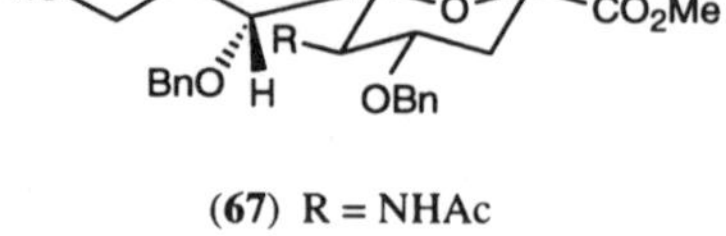

(**67**) R = NHAc

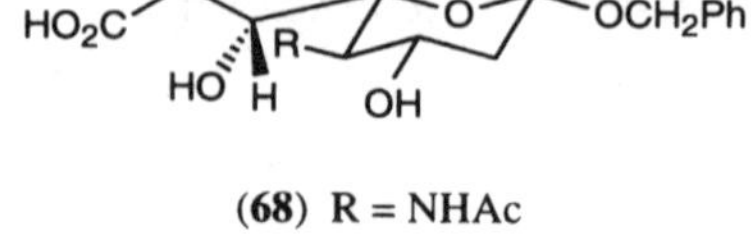

(**68**) R = NHAc

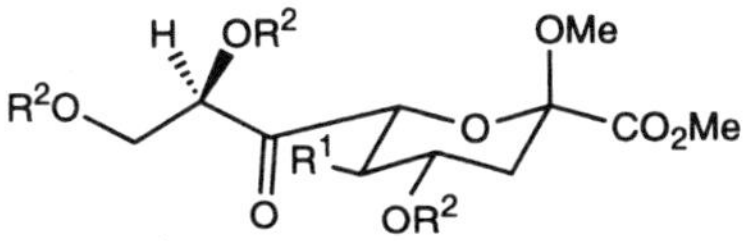

(**69**) R^1 = NHAc, R^2 = Si(*t*-Bu)Me_2

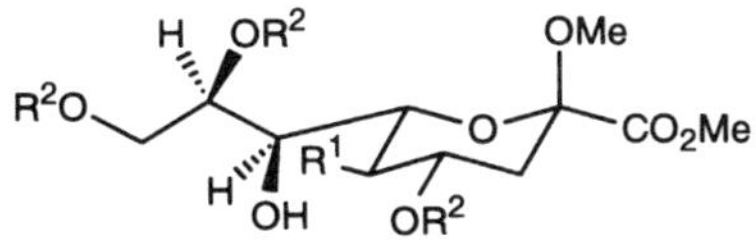

(**70**) R^1 = NHAc, R^2 = Si(*t*-Bu)Me_2

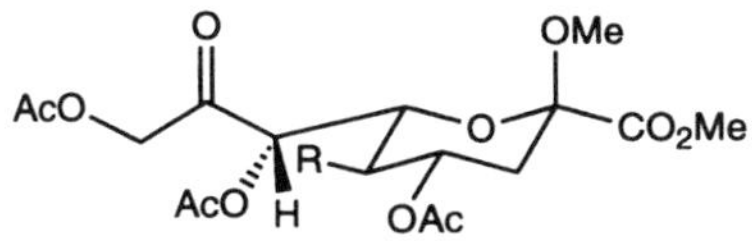

(**71**) R = NHAc

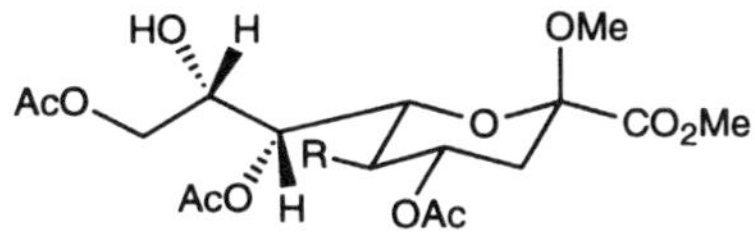

(**72**) R = NHAc

as to introduce halide [131], and azide [112] substituents at C-8 (e.g. **76**) (Scheme 12). Azide has been introduced at C-7 via an enzymic aldol condensation [22, 40]. It is generally accepted that nucleophilic displacements at the C-7 position are difficult as a result of steric constraints. Low reactivity of the C-8 hydroxyl group, in attempted syntheses of the disaccharide Neu5Acα(2$\rightarrow$8)Neu5Ac, has been attributed to interaction with the C-5 acetamido group via the formation of a hydrogen bond when the hexopyranose ring is in the chair-like $^2C_5$ conformation [132]. To avoid this interaction, reactions were carried out using the 1,7-lactone of Neu5Ac **77**, which has the $^5C_2$ conformation [132].

The side-chain of Neu5Ac has been truncated to C-8 (**10**) and to C-7 (**78**) by periodate oxidation followed by borohydride reduction (using 1 and 2 mole equivalents of $NaIO_4$/borohydride respectively) [76]. The 8-carbon analogue **10**

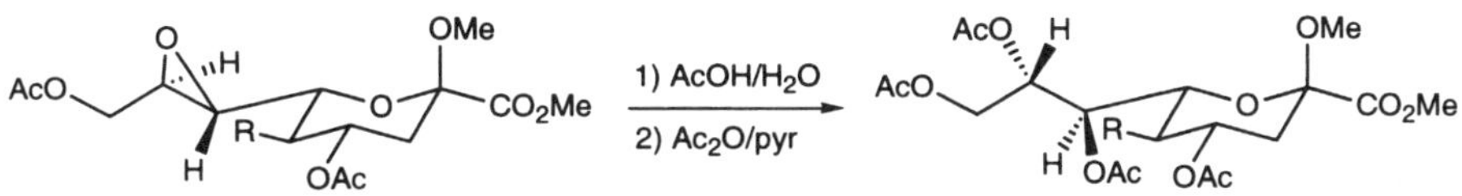

(**73**) R = NHAc (**75**) R = NHAc

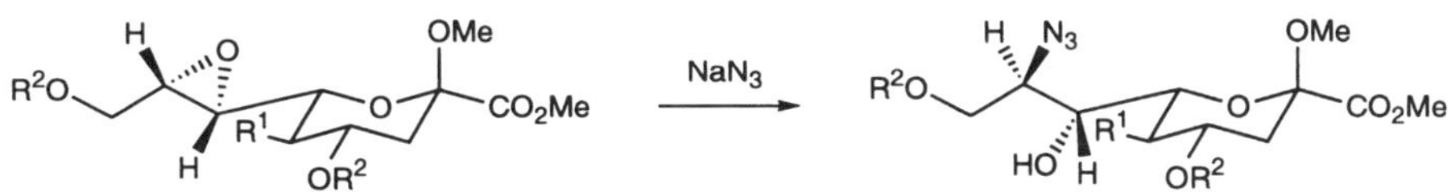

(**74**) R^1 = NHAc, R^2 = Si(*t*-Bu)Me_2 (**76**) R^1 = NHAc, R^2 = Si(*t*-Bu)Me_2

Scheme 12

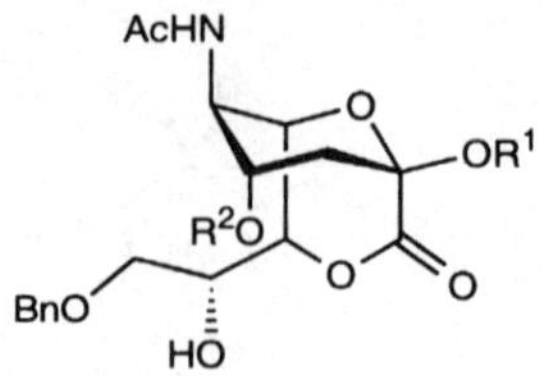

$R^1 = R^2 = Piv$
$R^1 = Me$, $R^2 = Piv$
$R^1 = Me$, $R^2 = Ac$

(**77**) $^5C_2$ conformation

has also been prepared both enzymically [43], and chemically [75, 76] from 2-acetamido-2-deoxy-D-lyxose (**9**). An enzymic aldol condensation with D-lyxose (**79**), led to the formation of the truncated, 8-carbon KDN analogue **80** [15–17]. The 7-carbon KDN analogue **81** has been derived from acetonide-protected L-arabinose (**82**) by reaction with the ylid derived from **24** [82]. A C-7 carboxylic acid derivative of Neu5Ac **83** has also been reported [133].

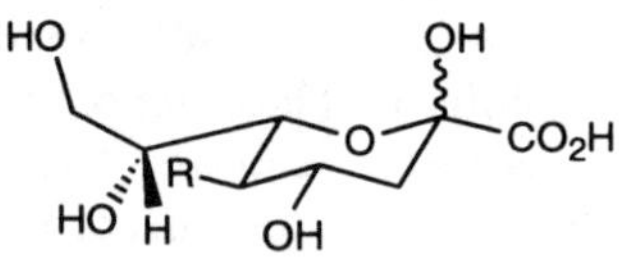

(**10**) R = NHAc
(**80**) R = OH

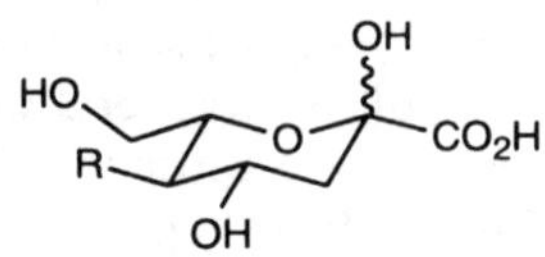

(**78**) R = NHAc
(**81**) R = OH

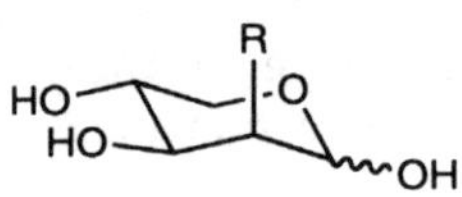

(**9**) R = NHAc
(**79**) R = OH

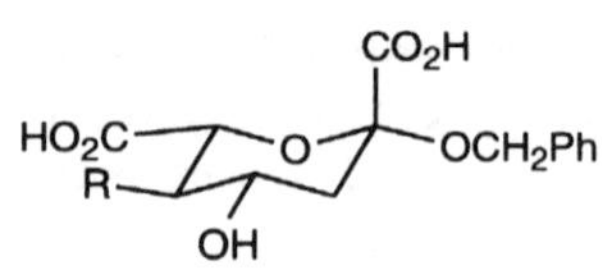

(**83**) R = NHAc

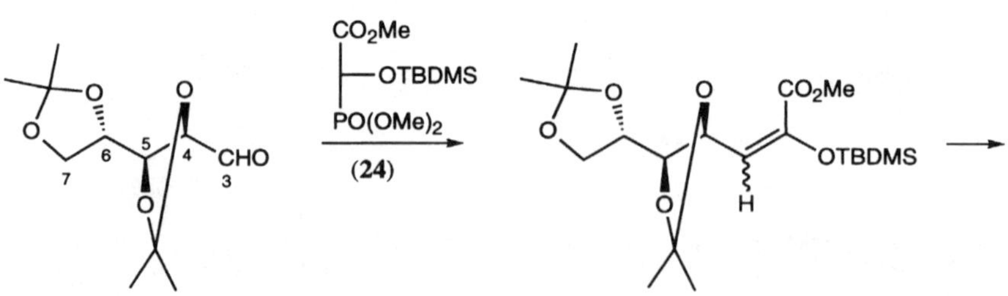

(**81**) R = OH

(**82**) Numbering is indicative of the final derivative

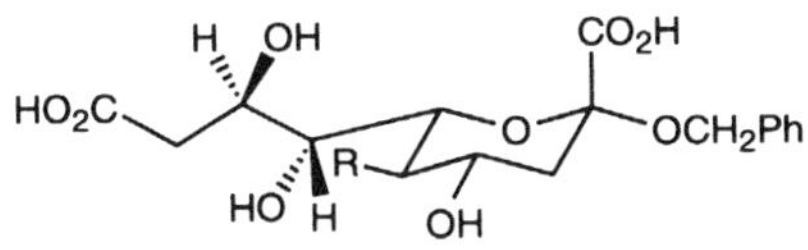

(**84**) R = NHAc

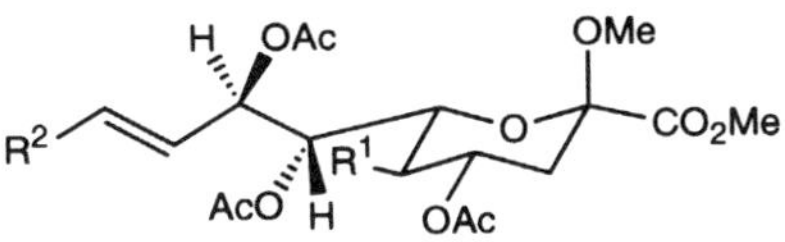

(**85**) R^1 = NHAc, R^2 = CHO
(**86**) R^1 = NHAc, R^2 = CO_2Me

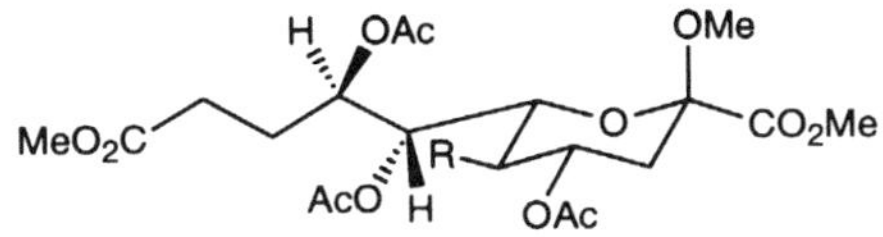

(**87**) R = NHAc

The glycerol side-chain has been extended by one carbon to the C-10 carboxy compound **84** (in [101]). Recently, extension by two carbon atoms to 11- carbon derivatives with aldehyde (**85**) or ester (**86, 87**) functionality at C-11 has been reported [134].

3.2 Modification at C-6

Derivatives at C-6 include both substitutions at the carbon atom and exchange of the oxygen involved in the 2,6-anhydro bond for sulfur, nitrogen, or carbon. These derivatives have all been synthesised from hexose sugars. Chemical syntheses of Neu5Ac analogues containing a methyl **49**, or hydroxymethyl **50** substituent at C-6 from a 1-nitro-D-GlcNAc derivative have been reported by Vasella and coworkers [97] (see Sect. 2.2). 6-Thio-Neu5Ac (**88**) has been prepared by aldol condensation of the 2-acetamido-3-thio-mannofuranose derivative **89** with nickel(II)/oxaloacetic acid [78]. 6-Amino-2,6-dideoxy analogues of Neu5Ac, **90**, **91**, and **92**, were prepared beginning with the condensation of a C-3 azido 1-nitro-D-mannosamine derivative **93** with *tert*-butyl 2-bromomethyl acrylate and subsequent manipulation [90]. Synthesis of the C-7 and C-6 side-chain truncated analogues of 6-amino-2,6-dideoxy-Neu5Ac, **94** and **95**, have also been carried out by Vasella and coworkers, beginning with GlcNAc [135]. While the 6-amino-2,6-dideoxy-Neu5Ac derivatives were found to be inhibitors of both bacterial and viral sialidases, the truncated derivatives were only weak inhibitors, indicating the importance of the glycerol side-chain for recognition by these enzymes [135]. Pyrrolidine analogues of Neu5Ac, **96**, with either a carboxymethyl or a phosphonomethyl group on the ring nitrogen, showed inhibition of a bacterial sialidase [136]. The synthesis of a carbocyclic analogue of Neu5Ac (**97**) using Diels-Alder chemistry has also been reported [99].

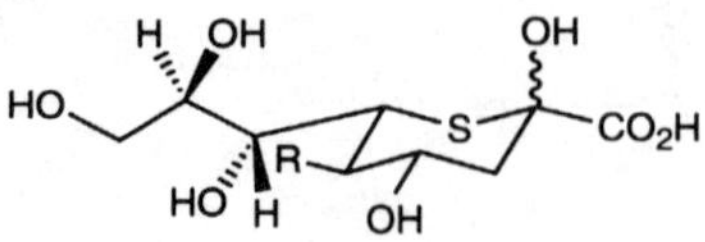

(**88**) R = NHAc

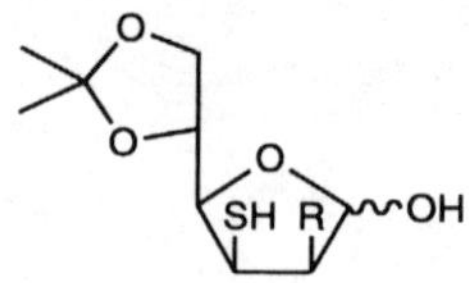

(**89**) R = NHAc

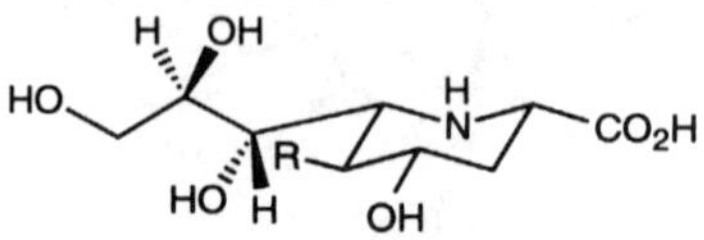

(**90**) R = NHAc

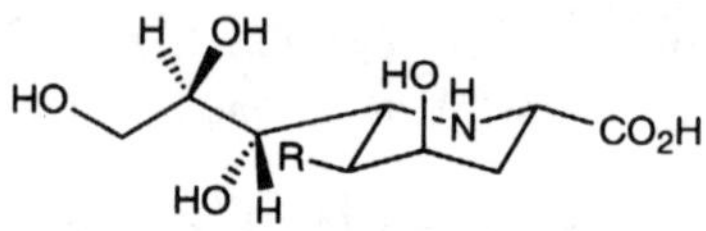

(**91**) R = NHAc

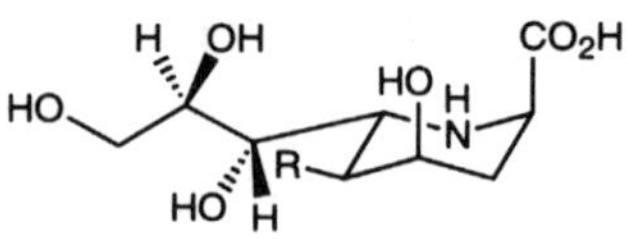

(**92**) R = NHAc

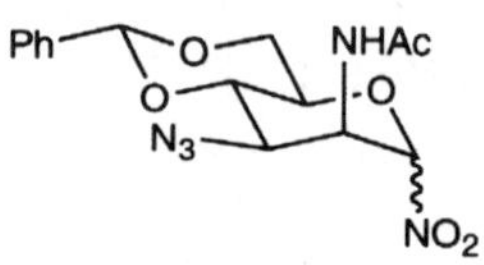

(**93**)

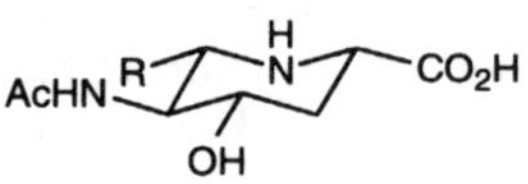

(**94**) R = CH_2OH
R = CH_2F
R = CH_3

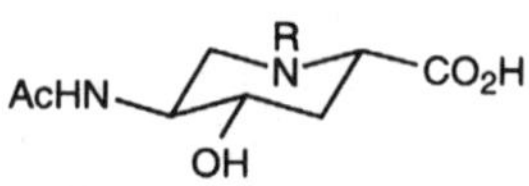

(**95**) R = H
R = CH_2CH_2OH

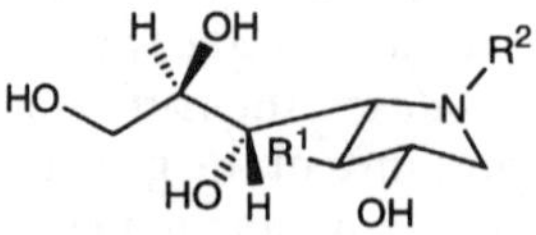

(**96**) R^1 = NHAc
R^2 = CH_2CO_2H, $CH_2PO_3H_2$, $COCO_2H$

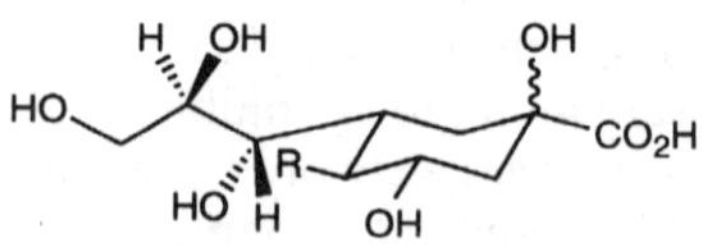

(**97**) R = NHAc

3.3 Modification at C-5

Both Neu5Ac [137] and KDN [138] exist naturally in the 2,6-anhydro hexopyranose form. A furanose analogue of 5-*epi*-KDN has been prepared which contains a 2,5-anhydro linkage. Chemical condensation of D-glucose with oxaloacetic acid followed by treatment with HCl in methanol gave a mixture of 5 isomers, which included two in the furanose form **98**, in a combined yield of 15% (the pyranose form of 5-*epi*-KDN1Me β-methyl glycoside was isolated in only 3% yield) [139]. The furanose products were used to prepare analogues of 5,6-bis-*epi*-KDN containing an acetamido group at C-6. The derivative **99** which contained the C-4 hydroxyl group in the α-configuration, the same stereochemistry as in Neu5Ac and KDN, showed inhibitory activity against influenza virus sialidases [139].

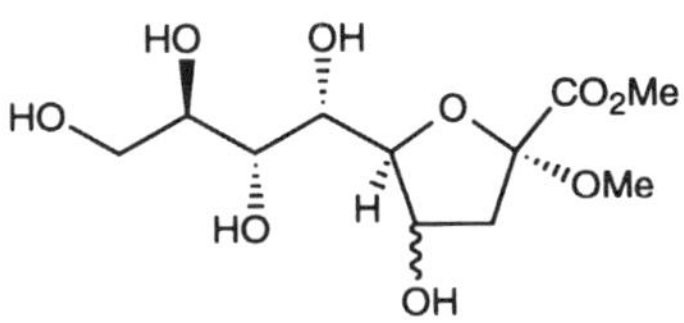

(**98**)

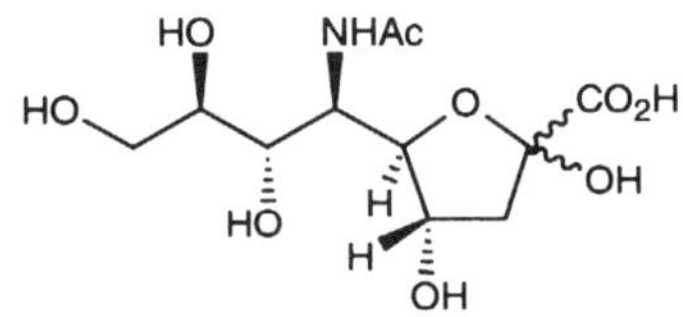

(**99**)

There have been a wide range of substitutions made at C-5 of Neu5Ac. While many derivatives have been prepared from 2-substituted-D-mannose sugars, either enzymically with Neu5Ac aldolase (especially with non-nitrogen substituents, see Table 1 in Sect. 2.1), or chemically [63, 72, 73], they have also been synthesised by acylation of the C-5 amine of neuraminic acid α-glycosides (**100**). Preparation of the C-5 amine has been achieved through basic de-*N*-acetylation of the C-5 acetamido group [7, 54], by base treatment of the imine in **101** [140], or by hydrogenation of a 5-azido (**102**) [55, 56] or 5-benzyloxycarbonylamino (5-NHCBz) (**103**) [56] group in a neuraminic acid analogue. Re-*N*-acylation with various reagents has given a wide range of C-5 acylamido derivatives which have been used to probe interactions with a number of sialic acid-recognising proteins [54, 56].

In a different approach, treatment of peracetylated Neu5Ac1Me (**104**) with nitrosyl acetate gave a mixture of *anti*/*syn* *N*-acetyl-*N*-nitroso derivatives

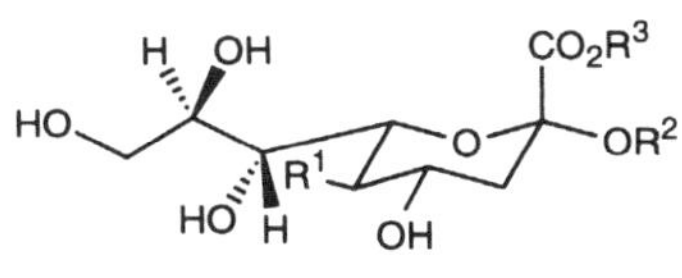

(**100**) $R^1 = NH_2$, R^2 = alkyl or allyl, R^3 = H or Me

(**102**) $R^1 = N_3$, $R^2 = R^3 = Me$

(**103**) R^1 = NHCBz, $R^2 = R^3 = Me$

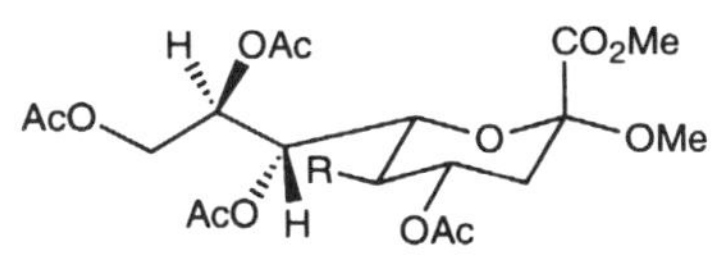

(**101**) R = N=C(OMe)Me

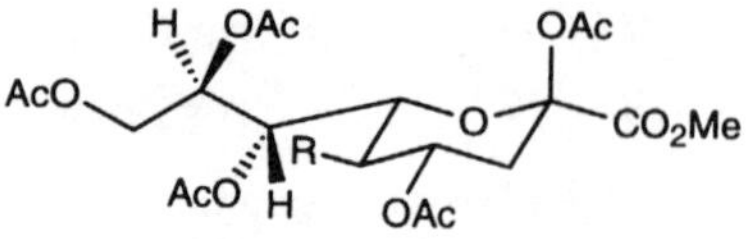

(**104**) R = NHAc
(**105**) R = N(Ac)N=O

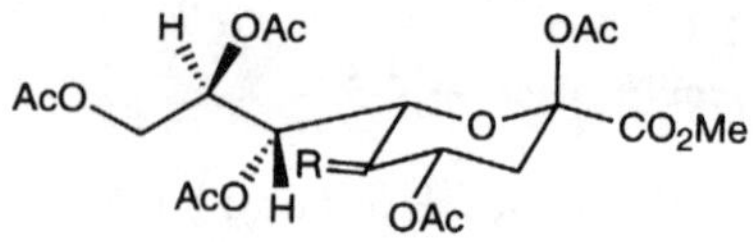

(**106**) R = N_2

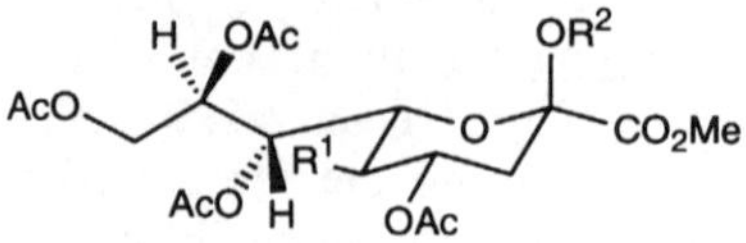

(**107**) $R^1 = N_3$, R^2 = Ac
(**108**) R^1 = OAc, R^2 = Ac
(**109**) R^1 = H, R^2 = Ac
(**110**) R^1 = OAc, R^2 = Me

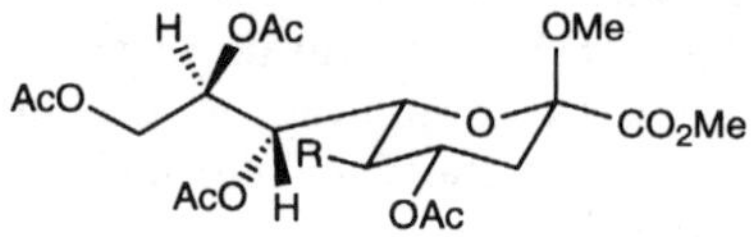

(**111**) R = N(Ac)N=O $\xrightarrow{\Delta}$ (**110**)

105 which were transformed into a C-5 diazo compound **106** [141]. Further reactions of **106** led ultimately to the C-5 azido (**107**), *O*-acetyl (KDN) (**108**), and de-amino (5-deoxy-KDN) (**109**) sialic acids [141]. The acetylated KDN glycoside **110** has been prepared (in somewhat lower yield) by thermal rearrangement of a C-5 *N*-acetyl-*N*-nitroso derivative **111** [142].

3.4
Modification at C-4

Substitution at C-4 of Neu5Ac is of particular interest due to the resistance of the α-glycosides of a number of these analogues to various sialidases ([4, 10, 143] and references therein). Analogues of Neu5Ac with alternative substituents at the C-4 position have not been, and probably cannot be, prepared by enzymic aldol condensation. Chemical synthesis, using various methods, has produced the Neu5Ac derivatives with 4-methoxy (**15**) [50, 84, 85], 4-*epi*-OH (**13**) [88], and 4-deoxy (**12**) [81, 89] substitutions, and the KDN analogue **35** with 4-acetamido [87] substitution. Apart from these examples, derivatisation at the C-4 position has been approached through manipulation of Neu5Ac itself. Protection of the side-chain C-9 and C-8 hydroxyls with an isopropylidene group isolates the C-7 and C-4 hydroxyls. The greater reactivity of the C-4 hydroxyl to oxidation [127, 130, 131, 144], to acylation [114, 116, 117], or to mesylation [47] allows further reactions to be carried out with the C-7 hydroxyl group unprotected. The C-4 position is also open to neighbouring group participation from the C-5 acetamido group. This can either be used to advantage or can cause difficulties in the introduction of substituents at C-4.

Activation of the C-4 hydroxyl of **112** under typical Mitsunobu conditions using triphenylphosphine/diethylazodicarboxylate (TPP/DEAD) [131] resulted in the formation of the 4,5-oxazoline derivative **113** by participation of the C-5 acetamido group. The oxazoline could be opened under acidic conditions to give the epimeric C-4 hydroxyl (**114**) [131]. Participation of the C-5 acetamido group in an attempted iodide displacement of the C-4 mesylate in **115** resulted in the displacement proceeding with overall retention of configuration at C-4 to give the 4-iodo derivative **116** [47]. Hydrogenation of **116** and subsequent deprotection, including hydrolysis of the *α*-methyl glycoside by a viral sialidase, gave 4-deoxy-Neu5Ac (**12**) [47].

Oxidation of the C-4 hydroxyl group of **112** has been carried out with either ruthenium tetroxide [127], pyridinium chlorochromate [130], or pyridinium dichromate/acetic anhydride [144] to give protected 4-oxo-Neu5Ac **117**. Reduction of the ketone with borane-ammonia complex gave rise to a 10:1 mixture of the two epimeric alcohols, the major product having the *epi* configuration at C-4 (**114**) [127]. Reaction of **114** with hydrazoic acid under Mitsunobu conditions led to introduction of azide in the equatorial position at C-4 (**118**) [145]. The azido group was then further manipulated, leading to 4-acetamido-4-deoxy-Neu5Ac (**16**).

The 4-oxo derivative **117** has allowed entry into an interesting range of C-4 mono- and di-substituted derivatives. Reaction with methoxylamine gave the oxime **119** [144]. Reaction of the ketone with methyl zirconium compounds led to epimeric mixtures of the disubstituted C-4 methyl derivatives **120** [146]. Methylenation of the ketone could be achieved by treatment with CH_2I_2-ZnCp_2-$ZrCl_2$ [146]. The methylene compound **121** was readily hydrogenated with palladium on carbon to yield a 3:2 mixture of equatorial and *epi* 4-methyl-4-deoxy derivatives **122** [146]. Epoxidation of the methylene double bond with *m*-chloroperbenzoic acid gave a single isomer of the C-4:C-4′ epoxide **123**, resulting from attack of the oxidant on the '*β*-face' of the double bond [147]. The epoxide has been opened with a number of nucleophiles, with attack occurring at the C-4′ carbon, to give C-4 disubstituted Neu5Ac derivatives **124**–**127** [147].

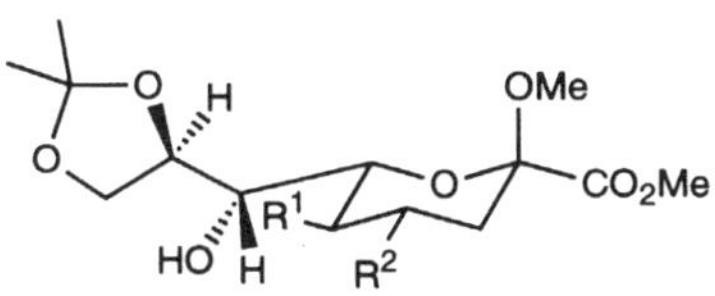

(**112**) R^1 = NHAc, R^2 = OH
(**118**) R^1 = NHAc, R^2 = N_3

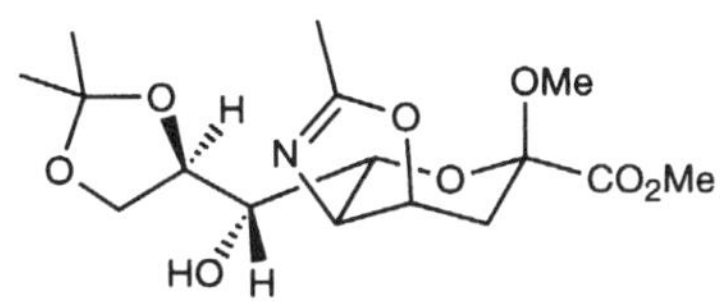

(**113**) R = NHAc

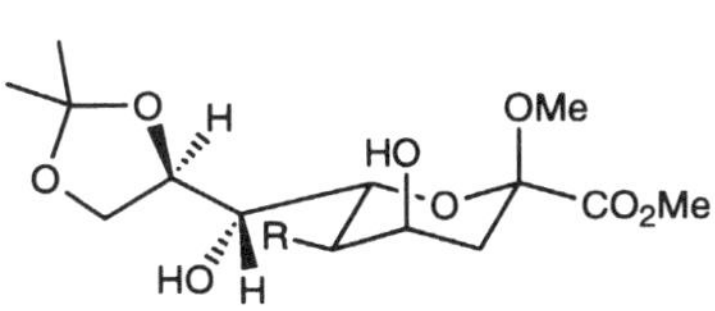

(**114**) R = NHAc

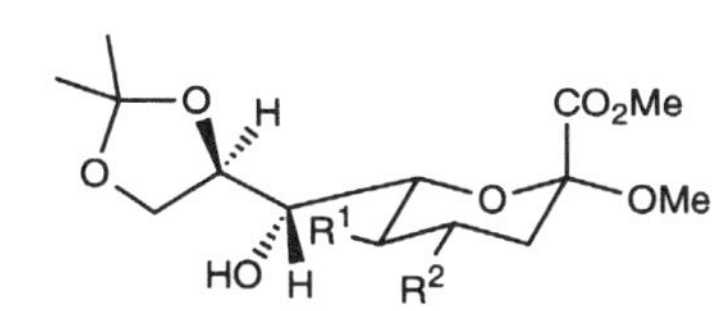

(**115**) R^1 = NHAc, R^2 = OMs
(**116**) R^1 = NHAc, R^2 = I

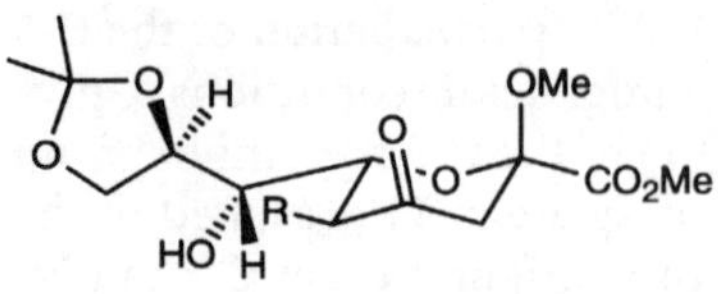

(**117**) R = NHAc

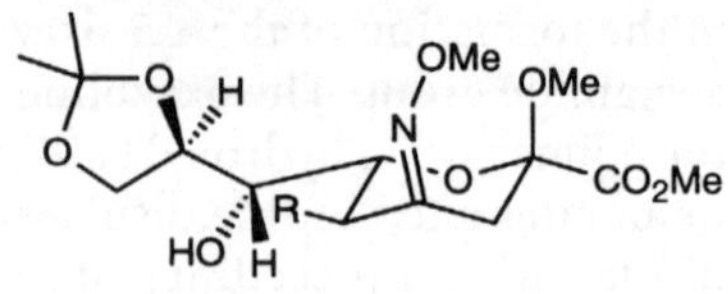

(**119**) R = NHAc

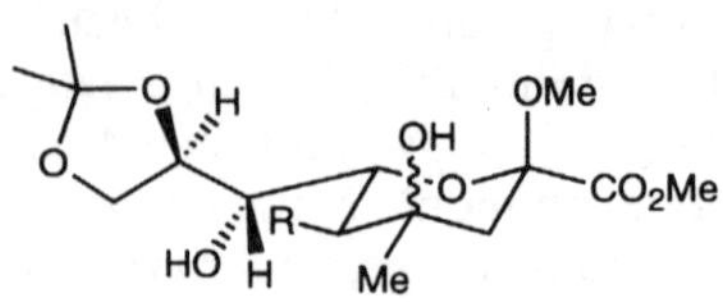

(**120**) R = NHAc

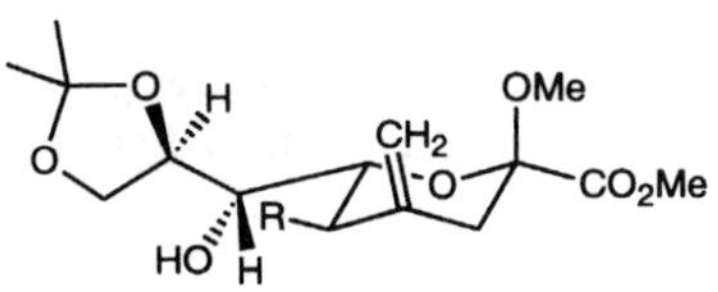

(**121**) R = NHAc

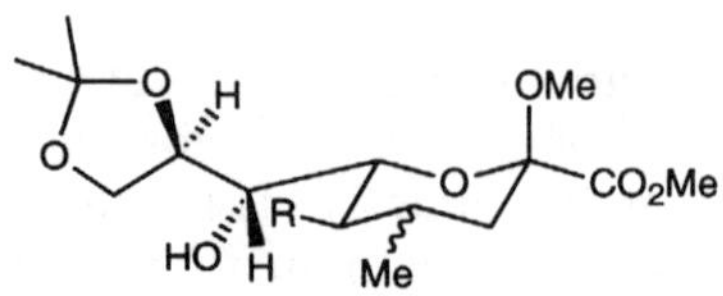

(**122**) R = NHAc

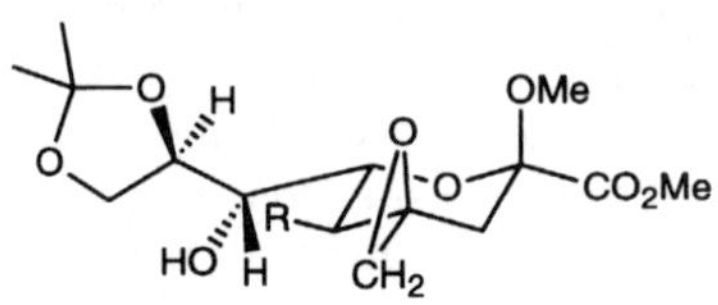

(**123**) R = NHAc

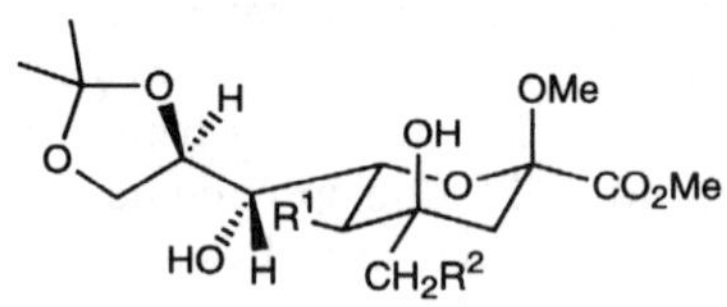

(**124**) R^1 = NHAc, R^2 = CN
(**125**) R^1 = NHAc, R^2 = Cl
(**126**) R^1 = NHAc, R^2 = N_3
(**127**) R^1 = NHAc, R^2 = OMe

3.5 Modification at C-3

The protons at C-3 of Neu5Ac are readily exchanged for deuterium under basic conditions [148, 149]. At a pD of 12.4, both protons were found to exchange (**128**) [148], while under milder conditions (pD 9.0) the axially-oriented proton could be specifically exchanged (**129**) [149]. For obvious reasons, this exchange does not take place in glycosides of Neu5Ac [149]. The α-glycosides prepared from the C-3 deuterated sialic acids have been used in kinetic isotope experiments investigating the mechanism of hydrolysis of the glycosides by sialidase from various sources [150–152].

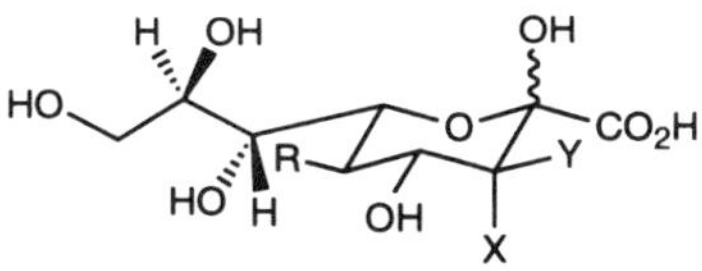

(**128**) R = NHAc, X = Y = D
(**129**) R = NHAc, X = D, Y = H

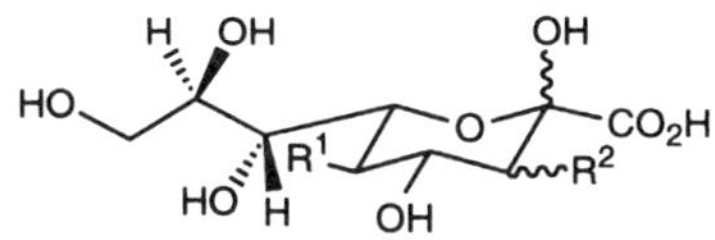

(**130**) R^1 = NHAc, R^2 = OH
(**131**) R^1 = NHAc, R^2 = F

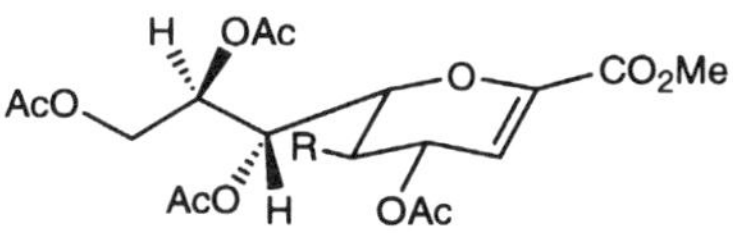

(**46**) R = NHAc

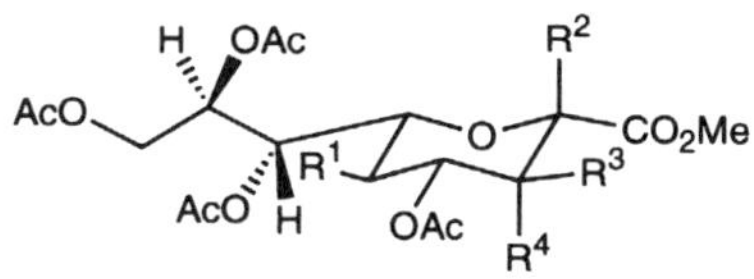

(**132**) R^1 = NHAc, R^2 = OAc, R^3 = F, R^4 = H
(**133**) R^1 = NHAc, R^2 = OAc, R^3 = H, R^4 = F
(**134**) R^1 = NHAc, R^2 = F, R^3 = F, R^4 = H

Neu5Ac derivatives with a hydroxyl (**130**) or fluorine (**131**) substituent at C-3 have been synthesised by an aldol condensation of hydroxypyruvate or bromopyruvate [71] or 3-fluoropyruvate [70] respectively with ManNAc. The addition of halogenating agents across the 2,3-double bond of 2-deoxy-2,3-didehydro-Neu5Ac derivatives (e.g. **46**) produces mixtures of the axially and equatorially substituted C-3 halogen derivatives [153]. The substituent introduced concomitantly at C-2 can be varied by the reagents and reaction conditions chosen [153]. The C-3 fluoro derivatives (**132**–**134**) have been prepared by addition of fluorine to **46** in acetic acid [154]. The derivative **132** with the C-3 fluorine in the equatorial configuration showed activity as an inhibitor of bacterial and viral sialidases [155]. The main use of substituents at C-3 has been as directing auxiliaries for glycosylation at C-2 (Sect. 5.1). In this respect, the substituents used at C-3 (in the equatorial configuration) are the hydroxyl, phenylthio, and phenylselenyl groups [104].

3.6 Modification at C-2 and C-1

Undoubtedly, the most critical area of the sialic acid structure for biological activity is the C-2 centre, which is involved in glycosidic linkages and is substituted with the C-1 carboxylic acid group. 2-Deoxy derivatives are of biological interest [4, 103], in particular the 2-deoxy-2,3-didehydro derivatives, which show significant inhibitory activity against sialidases [4, 156–158].

A range of 2-deoxy-2-H derivatives of Neu5Ac have been synthesised [4, 159, 160]. 2-Deoxy-Neu5Ac itself (**135, 136**) can be obtained through catalytic hydrogenation of peracetylated 2-β-chloro-2-deoxy-Neu5Ac1Me (**55**) [159, 161, 162], by reduction of **55** with tributyltin hydride [161], or by catalytic hydrogenation

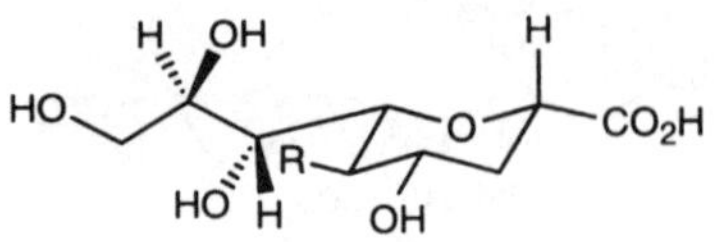

(**135**) R = NHAc

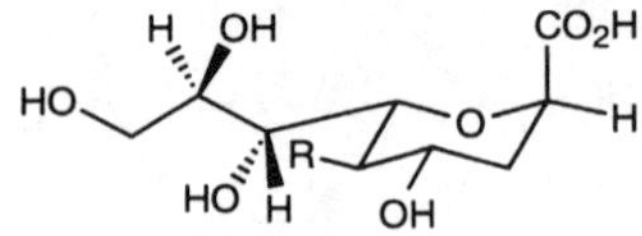

(**136**) R = NHAc

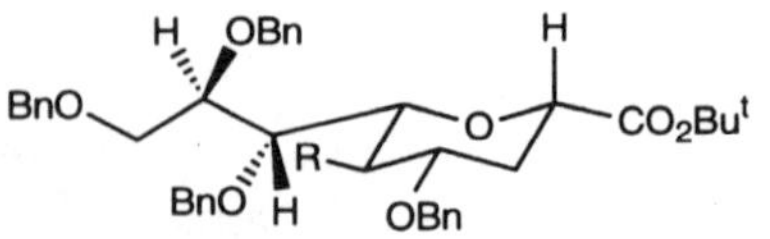

(**137**) R = NHAc

of Neu5Ac2en (protected or as the free acid) [156, 160, 161]. The first procedure is reported to give a mixture of epimers at C-2 (**135** and **136**), while the other methods give only the epimer with an equatorial carboxyl group (**135**) [161]. The C-2 position of a benzyl-protected 2-deoxy-Neu5Ac derivative (**137**) has been epimerised via the enolate anion [163].

2-Deoxy-Neu5Ac analogues with modifications to the glycerol side-chain (epimeric and deoxy) have been prepared either by hydrogenation of the appropriate 2-β-chloro-2-deoxy-Neu5Ac derivatives or, for certain analogues, by manipulation of 2-deoxy-Neu5Ac itself [162]. A series of C-4-substituted analogues were also prepared by manipulation of 2-deoxy-Neu5Ac [159, 164]. These derivatives have been used to investigate structure-activity relationships with influenza virus haemagglutinin [4, 165]. (Other 2-deoxy derivatives are the sulfur, nitrogen, and carbon glycosides, which will be discussed below in Sect. 5.)

A number of modifications have been made to the C-1 position [63, 101]. The carboxylic acid group of both α- and β-glycosides of Neu5Ac has been transformed into a carboxamide (e.g. **138**) [7, 166], or reduced with borohydride to give a hydroxymethyl group (e.g. **139**) [7, 105, 166]. Removal of the carboxylic acid group altogether has been achieved by lead tetraacetate-mediated decarboxylation of the 2-deoxy-Neu5Ac derivative **140**, which produced a mixture of the epimeric acetates **141** [160]. Similar reaction of lead tetraacetate with peracetylated Neu5Ac (**142**) led to a mixture of the epimeric acetates **143** and the α,β-unsaturated lactone **144** [167]. A derivative with homologation at the C-1 position (**145**) has been prepared by chemical condensation of a ManNAc derivative with disodium acetonedicarboxylate [$CO(CH_2COONa)_2$] and subsequent esterification [79]. The free acid was unstable, spontaneously decarboxylating upon de-esterification.

Both the 2-deoxy-2-H- (**146**) [160] and 2-deoxy-2,3-didehydro- (**147**) [168] Neu5Ac analogues with a phosphonic acid group in place of the carboxyl group have been synthesised by Vasella and coworkers. Steps in the synthesis of the 2-deoxy-2,3-didehydro analogue **147** [168] are shown in Scheme 13. The initial step was the bromohydroxylation of protected 2-deoxy-2,3-didehydro Neu5Ac.

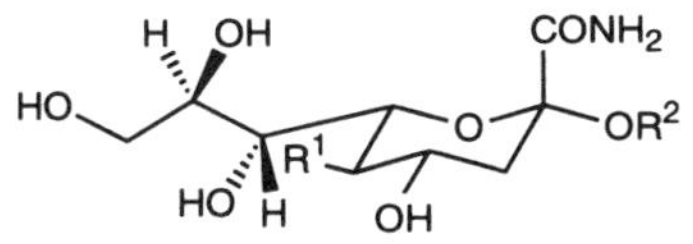

(**138**) R^1 = NHAc, R^2 = CH_2Ph, allyl

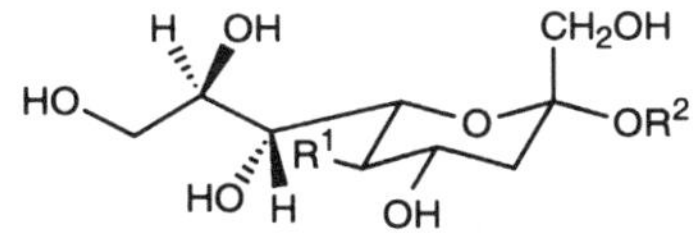

(**139**) R^1 = NHAc, R^2 = CH_2Ph, Me, allyl

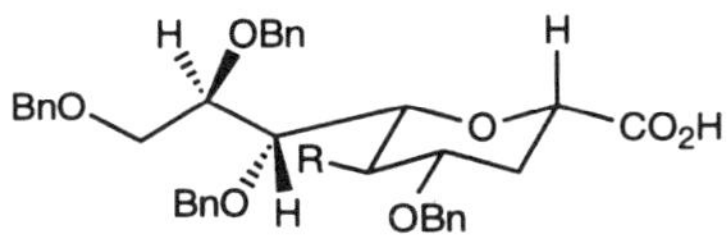

(**140**) R = NHAc

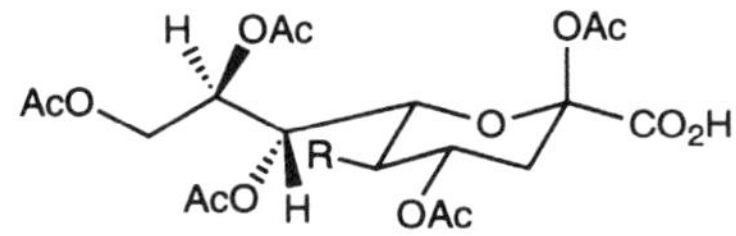

(**142**) R = NHAc

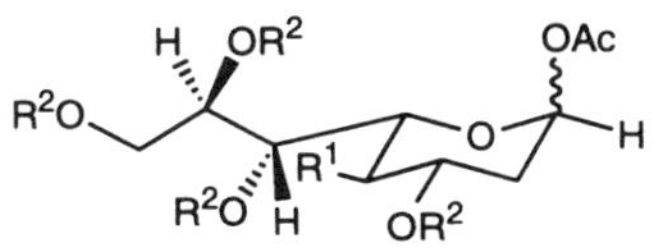

(**141**) R^1 = NHAc, R^2 = Bn
(**143**) R^1 = NHAc, R^2 = Ac

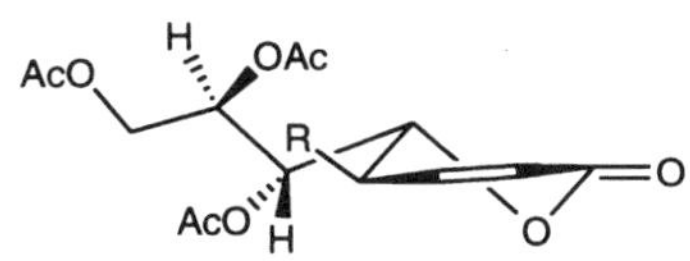

(**144**) R = NHAc

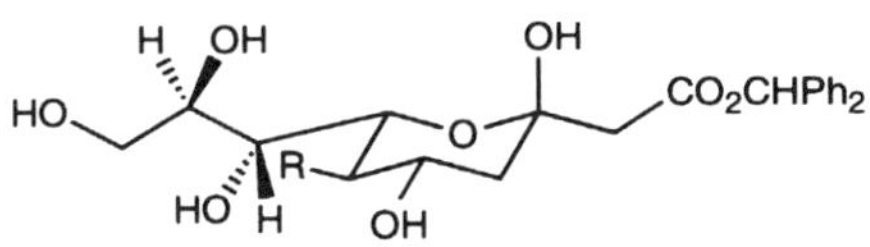

(**145**) R = NHAc

Formation of the 2,3-epoxide, and then epoxide opening in the presence of $TiBr_4$, was followed by oxidative decarboxylation at C-2. A phosphonate was introduced at the anomeric centre by displacement of a trichloroacetimidate. Photobromination at C-1 followed by reductive elimination then gave the 2,3-didehydro derivative **147**.

Modification of sialic acids already incorporated into glycoconjugates has generally been limited to truncation of the side-chain by mild periodate oxidation to the C-8 or C-7 aldehyde and subsequent reduction to the hydroxy function [10, 76, 169]. Lanne et al. [170] have modified the carboxyl group of Neu5Ac on the

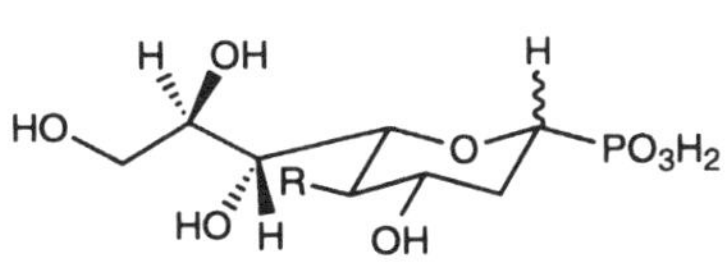

(**146**) R = NHAc

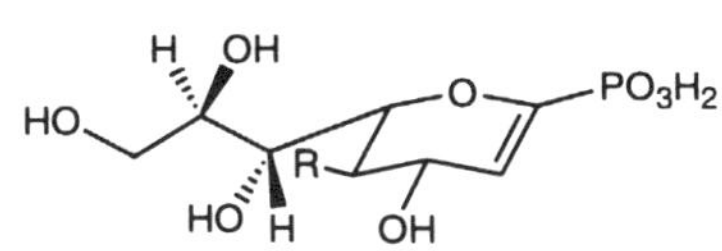

(**147**) R = NHAc

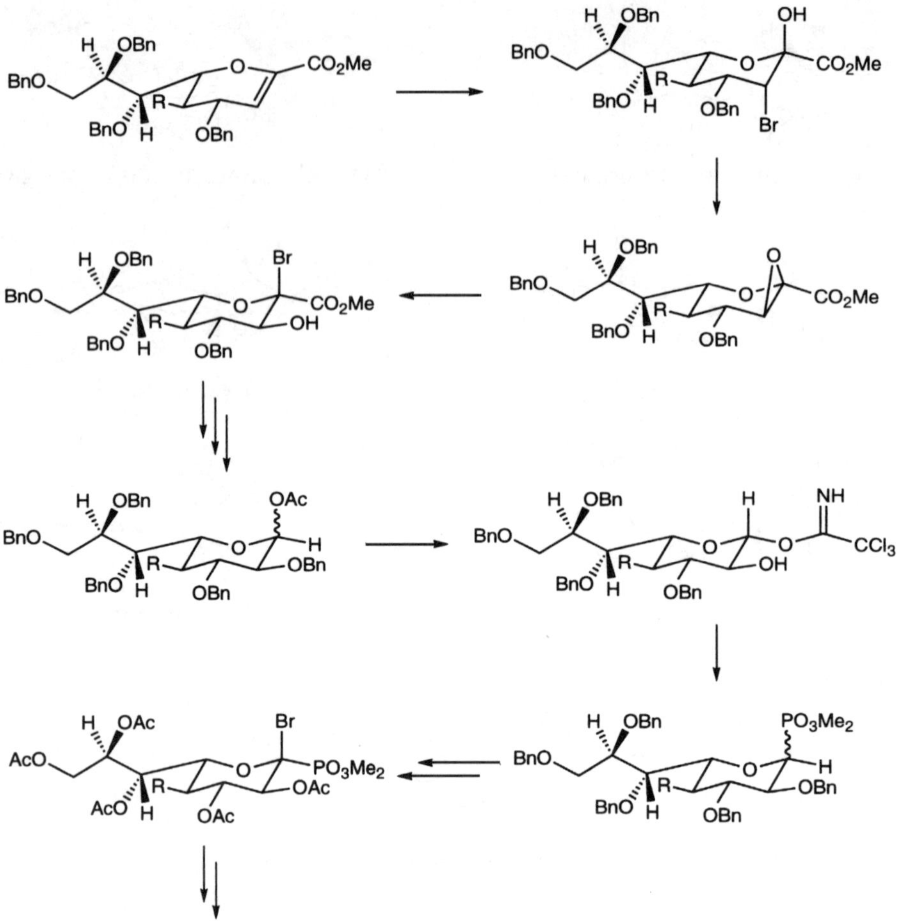

Scheme 13 (**147**) R = NHAc

ganglioside GM_1 by methylation, reduction, and amide formation to investigate the role of the carboxyl group in the binding of *Vibrio cholerae* toxin. In an interesting study, it was shown that Neu5Ac could be modified in the *N*-acyl moiety in vivo by administration of *N*-propanoyl-D-glucosamine or mannosamine [171]. The C-5-modified Neu5Ac derivatives were found to be incorporated into glycoproteins.

4 2-Deoxy-2,3-didehydro Sialic Acids

2-Deoxy-2,3-didehydro-Neu5Ac (Neu5Ac2en, **148**) has long been known to be an inhibitor of sialidases from both bacterial and viral sources [156, 157]. Derivatisation of the parent compound has been pursued with a view to understanding and improving its inhibitory potency.

Neu5Ac2en (**148**) can be prepared in a number of ways [63, 102]. Reaction of peracetylated 2-β-chloro-2-deoxy-Neu5Ac1Me (**55**) with dehydrohalogenating agents such as triethylamine [172], Ag_2CO_3 [172], or diazabicycloundecene (DBU) [153] gives peracetylated Neu5Ac2en1Me (**46**). Reaction of peracetylated Neu5Ac1Me (**104**) with a catalytic amount of trimethylsilyltriflate (TMS-triflate) at room temperature [173], or with 2 equivalents of TMS-triflate at 0 °C [161] is reported to give the unsaturated product (**46**) in greater than 90% yield. The use of TMS-triflate for the synthesis of 2,3-didehydro derivatives has recently been reinvestigated [174]. It was found that the choice of solvent, reaction temperature, and reaction time dramatically influenced the outcome of the reaction. A third method for the preparation of Neu5Ac2en involves the reaction of Neu5Ac1Me (**53**) with sulfuric acid and acetic anhydride ('acetolysis' conditions) [175]. Under these conditions, a mixture of C-4 epimeric acetates **46** and **149** was obtained, postulated to be due to intermediate formation of the oxazoline **150** [175]. Methodology has been developed which readily converts both α- and β-methyl glycosides of Neu5Ac1Me into peracetylated 4-*epi*-Neu5Ac2en1Me (**149**) or the oxazoline **150** under acetolysis-type conditions [176]. The nature of the products formed was dependent on the conditions of workup. The generality and implications of this reaction have also been demonstrated by its application to a range of 4-substituted-Neu5Ac1,2Me_2 derivatives

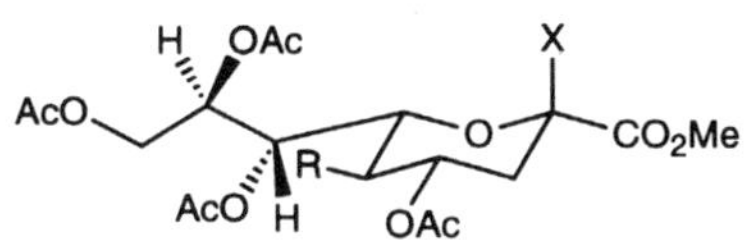

(**55**) R = NHAc, X = Cl
(**104**) R = NHAc, X = OAc

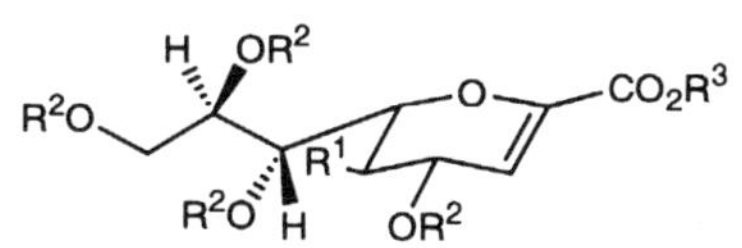

(**46**) R^1 = NHAc, R^2 = Ac, R^3 = Me
(**148**) R^1 = NHAc, R^2 = R^3 = H

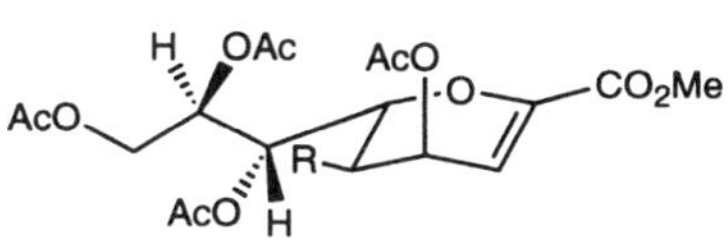

(**149**) R = NHAc

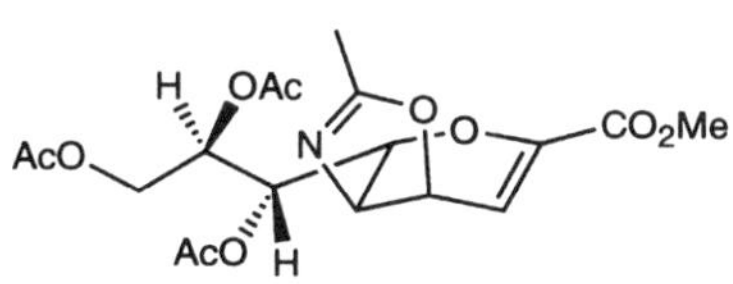

(**150**) R = NHAc

to yield the corresponding 4-substituted-Neu5Ac2en analogues and is discussed in more detail later on in this section [176].

Neu5Ac2en (**148**) has also been prepared from the activated nucleotide sugar CMP-Neu5Ac (**151**) at a pH above 7 [177], although the route for synthesis of Neu5Ac2en in vivo is uncertain [178, 179]. It is conceivable that cleavage of sialyl glycosides by sialidase might give a small amount of Neu5Ac2en [178]. The enzymic hydrolysis is postulated to proceed via an oxocarbocation intermediate [150, 180] which requires only the elimination of a proton from C-3 to form Neu5Ac2en. Indeed, the sialidase from influenza B virus has been shown to convert Neu5Ac to Neu5Ac2en at a low rate [180].

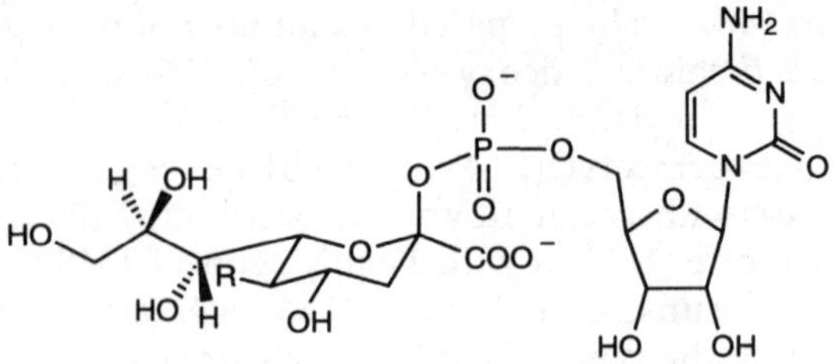

(**151**) R = NHAc CMP-Neu5Ac

In many cases, the methods described above for the synthesis of Neu5Ac2en can be applied to already functionalised Neu5Ac compounds to prepare the corresponding 2,3-didehydro analogues. Derivatives can also be approached by manipulation of Neu5Ac2en itself. C-9 amino and acylamido derivatives (**152**) of *N*-trifluoroacetyl-Neu2en have been reported [60]. Side-chain epimeric [181] and deoxy [182] analogues of Neu5Ac have been converted to the unsaturated derivatives by treatment with TMS-triflate. The glycerol side-chain of Neu5Ac2en has been truncated to the C-7 aldehyde and hydroxy compounds [183]. The side-chain-truncated 7-carbon analogue of KDN2en (**153**) was prepared to determine the effect of both the glycerol side-chain and the C-5 substituent on influenza virus sialidase binding, where it showed significantly weaker inhibition compared to Neu5Ac2en [82].

C-6 methyl- and hydroxymethyl-substituted Neu5Ac2en derivatives **154** [97], and the 6-thio-Neu5Ac2en derivative **155** [78] have been synthesised.

The synthesis of C-5-substituted Neu5Ac2en analogues has been approached both from the saturated analogues [172] and by de-*N*-acetylation of Neu5Ac2en (**148**) with hydrazine [184] or by hydrogenolytic cleavage of Neu5CBz2en (**156**)

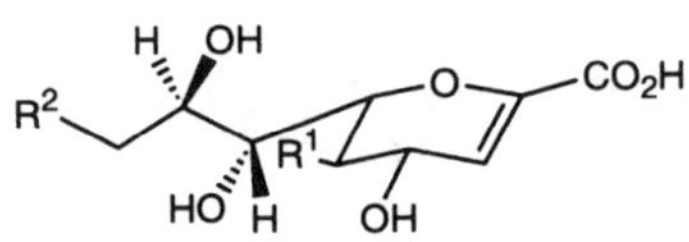

(**152**) $R^1 = NHC(O)CF_3$
$R^2 = NH_2$, $NHC(O)CH_3$, $NHC(O)CF_3$

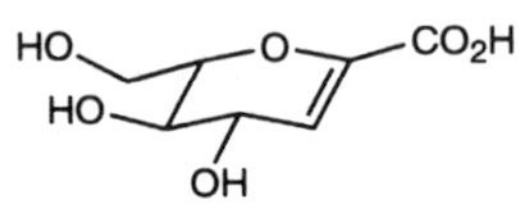

(**153**)

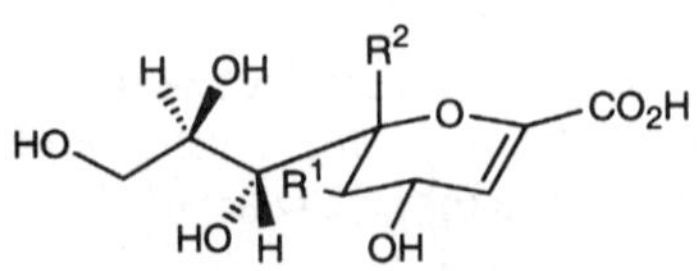

(**154**) $R^1 = NHAc$
$R^2 = CH_3$, CH_2OH

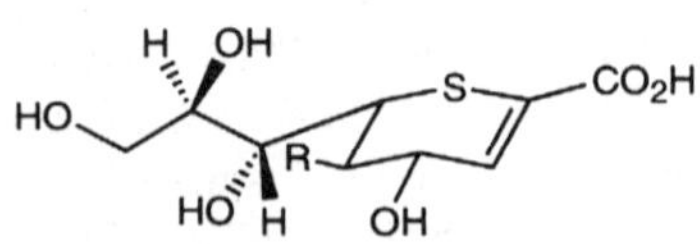

(**155**) R = NHAc

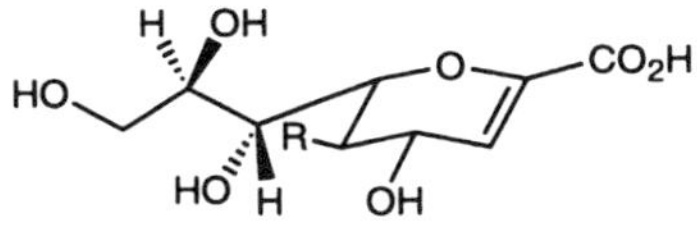

(**156**) R = NHC(O)OCH$_2$Ph

(**157**) R = NH$_2$

(**158**) R = NHC(O)CH$_2$OH

(**159**) R = OH

(**160**) R = N$_3$

[185] to give Neu2en (**157**), and subsequent re-*N*-acylation. Using the latter procedure, Meindl and Tuppy prepared a wide range of *N*-acylated- Neu2en derivatives [185], which were used to examine the effect of the C-5 substituent on inhibition of sialidases [157]. Naturally occurring *N*-glycolyl-Neu2en (**158**) has been prepared by a number of methods [172, 179]. KDN2en (**159**) and 5-azido-5-deoxy-KDN2en (**160**) were prepared by reaction of the saturated derivatives with TMS-triflate [184].

4-Deoxy-Neu5Ac2en (**161**) has been prepared from the saturated analogue by reaction with TMS-triflate [82], by hydrogenolytic cleavage of the 4,5-oxazoline of **150** [186], and as a by-product from a glycosylation using the β-glycosyl chloride of 4-deoxy-Neu5Ac [187]. 4-Oxo-Neu5Ac2en (**162**) was obtained by oxidation of the C-4 hydroxyl of a Neu5Ac2en derivative with manganese dioxide [188]. As previously mentioned, an acetolysis-type reaction has been used to convert the methyl glycosides of protected 4-oxo- (**117**), and 4-methyl-4-deoxy-Neu5Ac1Me (**122**) to the corresponding peracetylated Neu5Ac2en analogues [176]. This method offers a convenient entry into functionalised 2,3-didehydro derivatives directly from the methyl glycosides normally used in the original modification of Neu5Ac.

A number of C-4-substituted 2,3-didehydro derivatives show significant inhibition of influenza A and B virus sialidase. The C-4 amino derivative (**163**) was originally targeted as a potential inhibitor of influenza virus sialidase based on molecular modelling studies of the X-ray crystal structure of influenza A virus sialidase [189–191].

The 4,5-oxazoline derivative of Neu5Ac2en **150**, produced by reaction of peracetylated Neu5Ac2en1Me (**46**) with BF$_3\cdot$etherate [192] or of peracetylated Neu5Ac1Me (**104**) with 2 equivalents of TMS-triflate at 50 °C [186, 193], has been used as an intermediate for the introduction of substituents at C-4. Nucleophilic

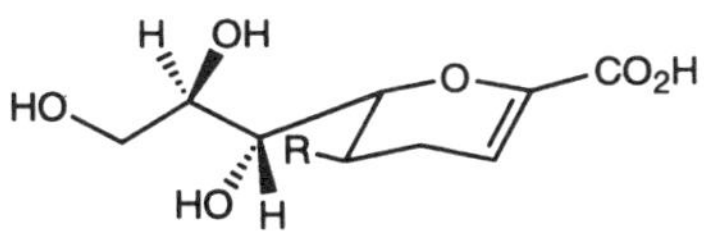

(**161**) R = NHAc

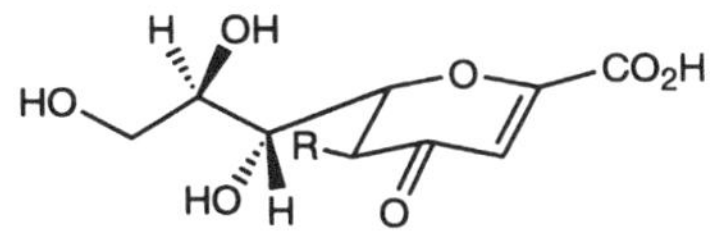

(**162**) R = NHAc

attack by azide or thioacetate under acidic conditions gave the C-4-substituted compounds **164** and **165**, respectively, in good yield [192]. Only small amounts (<5%) of the 4-*epi* compounds were observed. The azide **164** has also been prepared by reaction of the 4-*epi*-Neu5Ac2en derivative **166** with hydrazoic acid under Mitsunobu conditions [186]. However, this reaction resulted in formation of both epimers at C-4, the ratio being dependent on the solvent chosen. It was postulated that the 4-*epi*-azido derivative (**167**) was formed as a result of a competing attack of azide ion at the C-2 position and subsequent rearrangement. The C-4 azide in **164** can be reduced to an amino substituent in the presence of the double bond by using triphenylphosphine [186] or by hydrogenation in the presence of palladium on carbon in toluene, methanol and glacial acetic acid [194, 195]. The 4-amino substituent has been acylated [186, 195], alkylated [195] (see ref. [158]), and converted to the more basic 4-guanidino derivative **168** [194, 195]. Both the 4-amino-4-deoxy- and 4-deoxy-4-guanidino-Neu5Ac2en derivatives showed significant increase (up to 100,000 fold) in inhibition of influenza A and B virus sialidase compared to Neu5Ac2en itself [158, 189, 196]. The synthesis of the 4-guanidino compound **168** has since been reported by other groups [193, 197].

The potent and selective inhibition of influenza virus sialidase by 4-amino-4-deoxy- (**163**) and 4-deoxy-4-guanidino-Neu5Ac2en (**168**) [158] has prompted the synthesis of a range of structural variants of these compounds. The sulfur isosteres **169** have been synthesised from 6-thio-Neu5Ac and show comparable activity to their oxygen-containing counterparts [198]. The side-chain has been truncated to the C-8 (**170**) and C-7 (**171**) hydroxy derivatives [98]. Derivatives with no glycerol side-chain at all (**172**) were based on a template built up from GlcNAc [98]. Different acyl groups have been introduced at C-5 [199], and the 5-desacetamido derivatives **173** have been prepared beginning with an aldolase-catalysed condensation of 2-deoxy-D-glucose with sodium pyruvate [200]. The guanidine substituent at C-4 has been modified by introduction of substituents

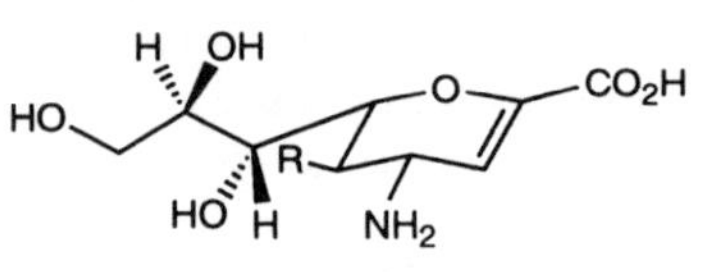

(**163**) R = NHAc

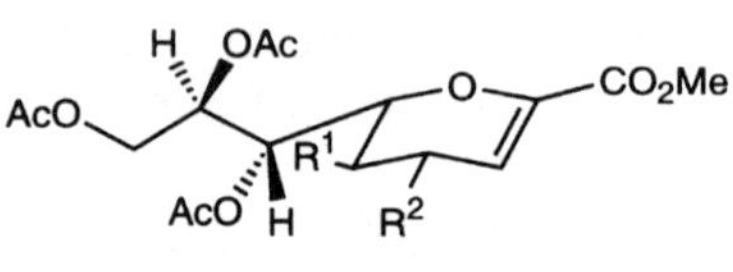

(**164**) R^1 = NHAc, R^2 = N_3
(**165**) R^1 = NHAc, R^2 = SAc

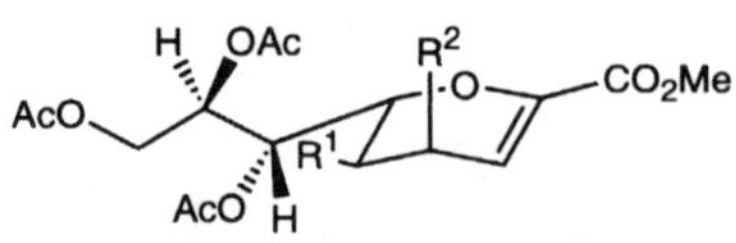

(**166**) R^1 = NHAc, R^2 = OH
(**167**) R^1 = NHAc, R^2 = N_3

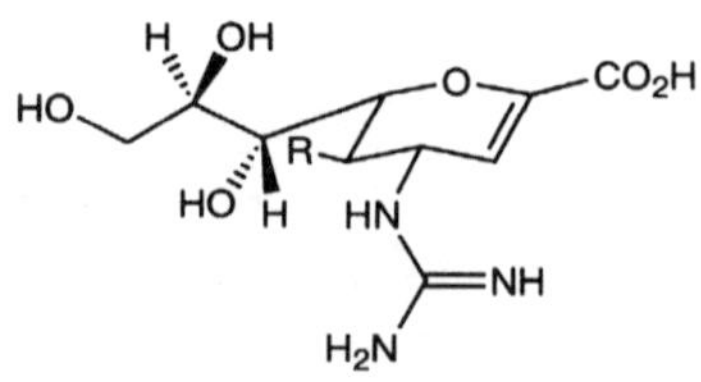

(**168**) R = NHAc

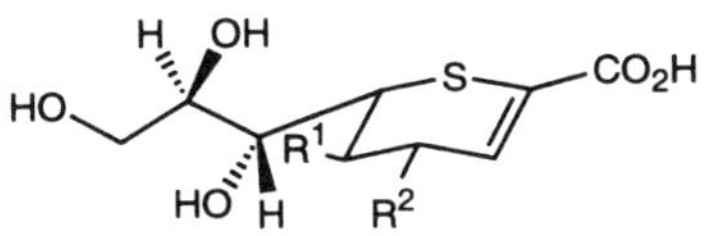

(**169**) R^1 = NHAc
R^2 = NH_2, NHC(NH)NH_2

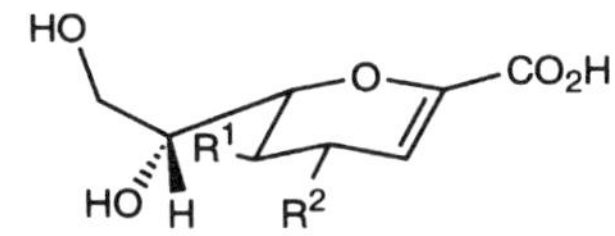

(**170**) R^1 = NHAc
R^2 = NH_2, NHC(NH)NH_2

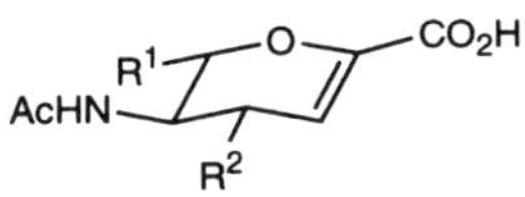

(**171**) R^1 = CH_2OH
(**172**) R^1 = H
R^2 = NH_2, NHC(NH)NH_2

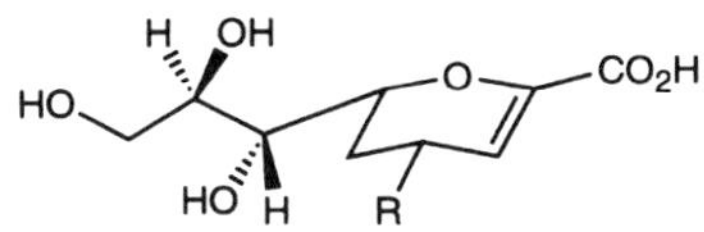

(**173**) R = NH_2, NHC(NH)NH_2

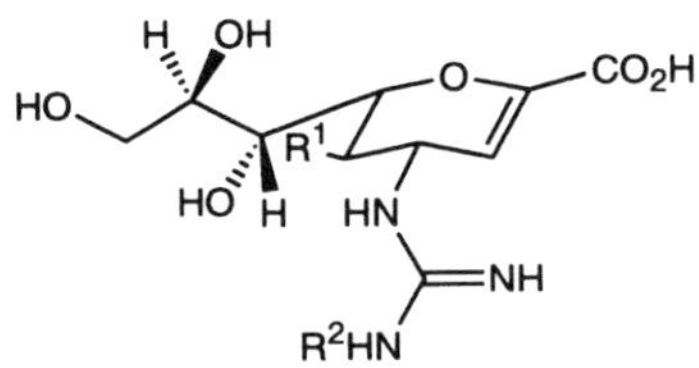

(**174**) R^1 = NHAc
R^2 = Me, OH, NH_2, NO_2, CO_2Et

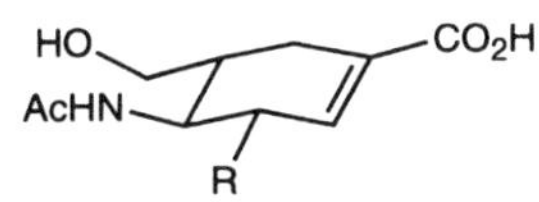

(**175**) R = NH_2, NHC(NH)NH_2

on the 'terminal' amine (**174**) [193]. Carbocyclic analogues **175** of the truncated, 7-carbon compounds have also been prepared [100]. While none of the derivatives showed better inhibitory activity against influenza A virus sialidase than the parent compounds **163** and **168**, a few instances of improvement in plaque reduction assays (indicating inhibition of virus replication) were reported [100, 199]. The varied activities of the derivatives provide insight into the binding requirements of the sialidase enzyme.

The addition of halogenating agents across the double bond of the 4-azido-4-deoxy-Neu5Ac2en compound **164** has been used for the convenient preparation of the corresponding 'saturated' derivatives. The addition of *N*-bromosuccinimide in methanol gave a mixture of the methyl glycosides **176** and **177** of the C-4-substituted 3-bromo-3-dehydro derivative [201]. Reduction of the azide in **176**, with concomitant removal of the C-3 bromine, with tri-*n*-butyltin hydride was followed by guanidination. The 2-α-methyl glycoside of 4-guanidino-4-deoxy-Neu5Ac (**178**) so prepared is currently under investigation as a probe for influenza virus haemagglutinin [201, 202]. Sabesan has reported the addition of HCl across the double bond of compound **164** to give the β-glycoxyl chloride **179** [203]. This was then converted into a series of *O*- and *S*-linked $\alpha(2\rightarrow6)$Gal

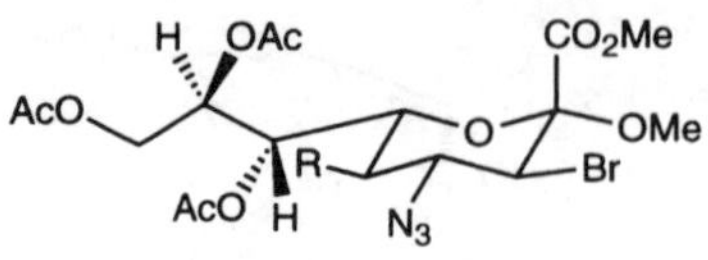

(176) R = NHAc

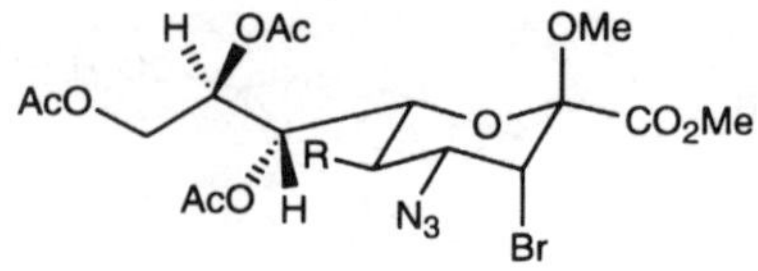

(177) R = NHAc

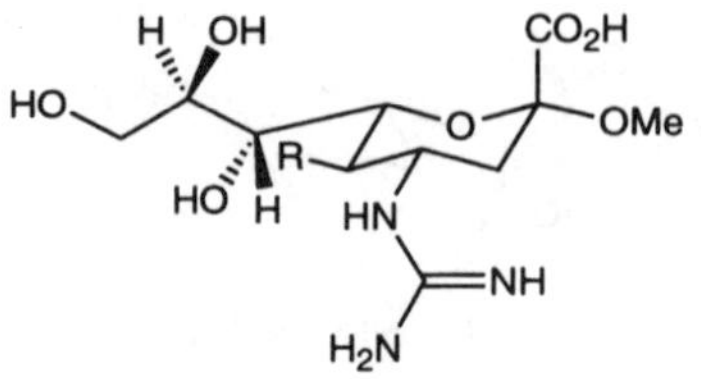

(178) R = NHAc

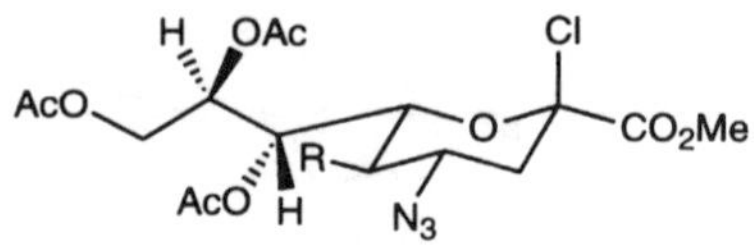

(179) R = NHAc

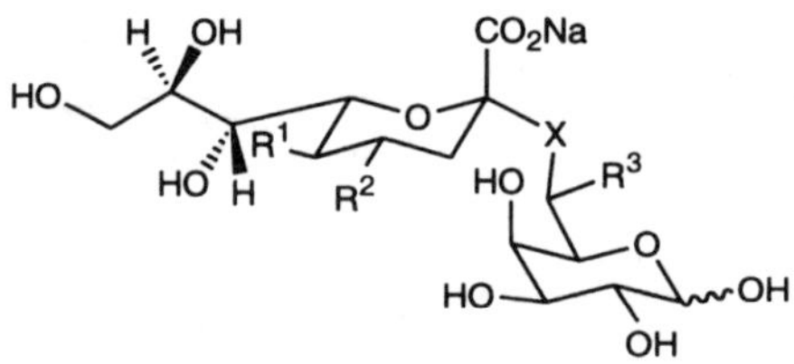

(180) R^1 = NHAc
R^2 = N_3, NH_2, NHAc
R^3 = H, CH_3
X = O, S

derivatives (**180**). Interactions of these disaccharides with influenza virus sialidase were then studied and compared to the 2,3-didehydro derivatives [203]. The addition of HCl across the double bond was limited to the 4-azido-4-deoxy-Neu5Ac2en derivative, reactions with the 4-amino and 4-thioacetyl derivatives being unsuccessful.

The versatility of the addition of halogenating agents to the double bond of 2,3-didehydro analogues depends on the stability of the remaining substituents to the reaction conditions. However, the examples described here show promise of a means of converting potential inhibitors into their substrate analogues. In conjunction with the 'acetolysis' reaction which converts methyl glycosides of the saturated species to their 2,3-didehydro analogues [176], this conversion will be useful for the elucidation of structure-activity relationships. This could lead to a better understanding of the effects of the same substitution pattern on the substrate or inhibitor specificity of an enzyme.

5
Synthesis of Sialyl Glycosides as Biological Probes

Free sialic acids are used by the enzymes of sialic acid metabolism, CMP-sialate synthase and Neu5Ac aldolase. However the majority of sialic acid-metabolising and -recognising proteins interact with glycosidically bound sialic acid. Therefore, sialyl glycosides and sialyloligosaccharides are often used to probe structure-function relationships of sialic acid-recognising proteins. The nature of the glycosidic linkage and/or the structure and functionality of the sialic acid residue can be altered to produce variation in the biological properties of the glycosides.

5.1 Synthesis of Sialyl Glycosides by Chemical Methods

The synthesis of sialic acid glycosides has been reviewed [64, 102, 104, 204] and the synthesis of biologically active sialyl glycosides and sialoglycoconjugates discussed (in refs. [205] and [206]) by several authors. The formation of glycosides of Neu5Ac is generally more difficult than for typical hexopyranose sugars due to steric and electronic factors. The presence of the carboxyl group at the C-2 carbon both sterically restricts glycoside formation and reduces the reactivity of the anomeric centre [64, 104, 204]. As the adjacent carbon C-3 is unsubstituted, there is no possibility of neighbouring group participation to assist in glycoside formation as often seen in the glycosylation of typical hexopyranose sugars [207]. Additionally, when glycosylation is slow or difficult, an intramolecular elimination to give the 2,3-didehydro sugar can and quite often does occur, much to the annoyance of the experimentalist. Although natural sialic acid glycosides have an α-configuration, the anomeric effect favours the formation of the β-anomer [204, 208].

Reaction of the activated glycosyl donor, peracetylated 2-β-chloro-2-deoxy-Neu5Ac1Me (**55**) with sodium alcoholates (in analogy to a Williamson synthesis) [107, 209] or with alcohols using phase-transfer catalysis [210, 211] has been used to produce a range of alkyl and aryl α-glycosides. To react with the less nucleophilic and often sterically hindered sugar alcohols, the system requires further activation such as the addition of a heavy metal promoter (Koenigs-Knorr conditions) [104, 209]. Alternative activating groups at C-2 include a 2-β-acetoxy or phenylcarbonyloxy group (**181**) (promoted by a silver catalyst) [212], 2-β-phosphites (e.g. **182**) [132, 213–215], simple alkyl α-thioglycosides (e.g. **183**) [216–221], and 2-α-xanthates **184** [218, 219, 222, 223]. Activation of the 2-thio substituents with "thiophilic reagents" produces predominantly α-glycosides, with the ratio of α- and β-anomers dependent on the reaction conditions chosen [216, 223, 224]. Reaction of activated glycosyl donors in acetonitrile has been postulated to proceed through intermediate formation of nitrilium cations (discussed in references [214] and [223]).

In attempts to reduce the competing elimination reaction and to introduce some means of stereocontrol in the glycosylation, substituents such as hydroxyl, phenylthio, and phenylselenyl have been introduced at C-3 in the equatorial configuration (**185**) [104, 225]. These substituents can be reductively removed after the glycosylation reaction. The addition of halogenating agents across the 2,3-double bond of protected Neu5Ac2en [153] and subsequent manipulation of the substituents at C-2 and C-3 has provided a range of glycosyl donors. The diaxially substituted dibromide **186** acted as a glycosyl donor to produce only β-glycosides (**187**) because of steric hindrance of the α-face by the C-3 bromine

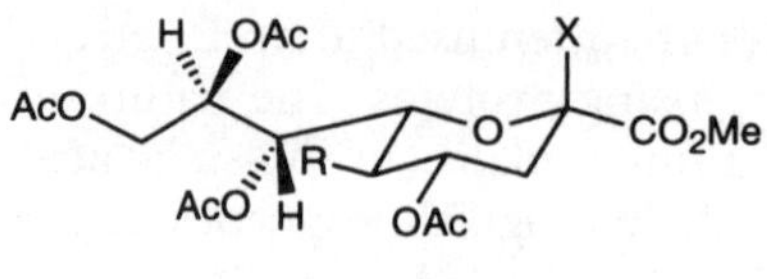

(**55**) R = NHAc, X = Cl
(**181**) R = NHAc, X = OAc, OC(O)OPh
(**182**) R = NHAc, X = $OP(OEt)_2$

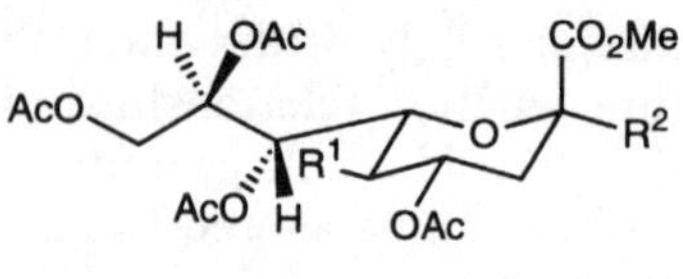

(**183**) R^1 = NHAc, R^2 = SMe
(**184**) R^1 = NHAc, R^2 = SC(S)OEt

[226, 227]. Glycosylation of the 3-α-bromo-Neu5Ac 2-β-phosphite **188** similarly gave, stereoselectively, a β-linked sialyl glycoside [228]. The 2,3-epoxide **189** [229], the 2-β-halo-3-β-hydroxy compounds **190** [230], and 3-β-phenylselenyl [231] and 3-β-phenylthio-substituted compounds [232] have all been used as glycosyl donors. It has been suggested that the C-3 equatorially substituted phenylselenyl and phenylthio groups may assist in directing glycosylation to give the α-glycosides through the intermediacy of an 'episelenonium', or 'episulfonium' ion **191** [225, 231, 233].

Using the glycosylation methods described above, a wide range of sialyl glycosides have been prepared (Fig. 2). Certain aromatic α-glycosides, for example 4-methylumbelliferyl (**192**) and *p*-nitrophenyl (**193**) glycosides, are used as substrates in sialidase assays, where cleavage of the glycosidic linkage results in the

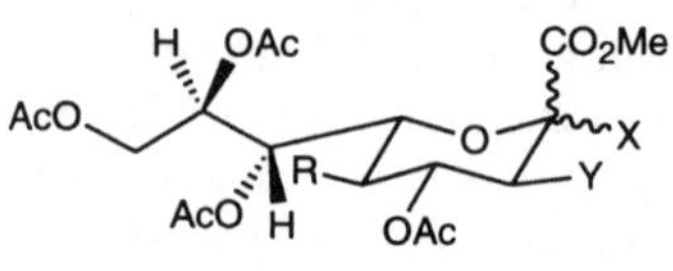

(**185**) R = NHAc
Y = OH, SPh, SePh
X = halide, SMe, SC(S)OEt

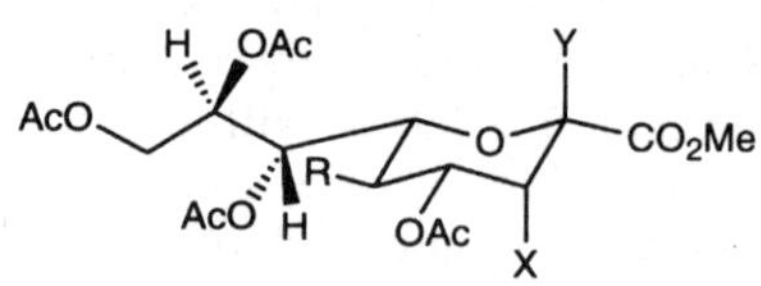

(**186**) R = NHAc, X = Br, Y = Br
(**187**) R = NHAc, X = Br, Y = OR'
(**188**) R = NHAc, X = Br, Y = $OP(OBn)_2$

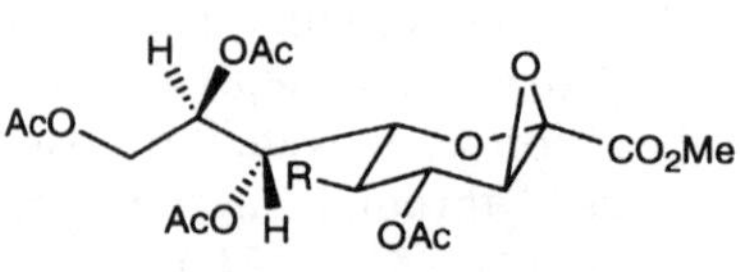

(**189**) R = NHAc

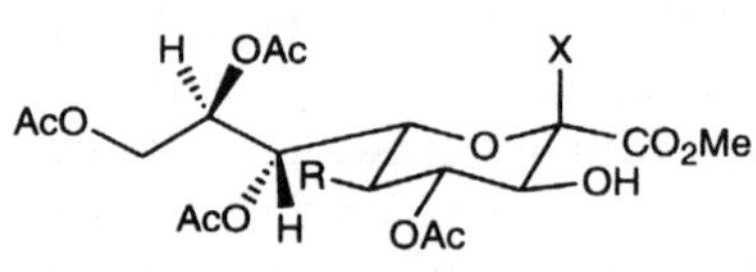

(**190**) R = NHAc, X = Cl, Br, F

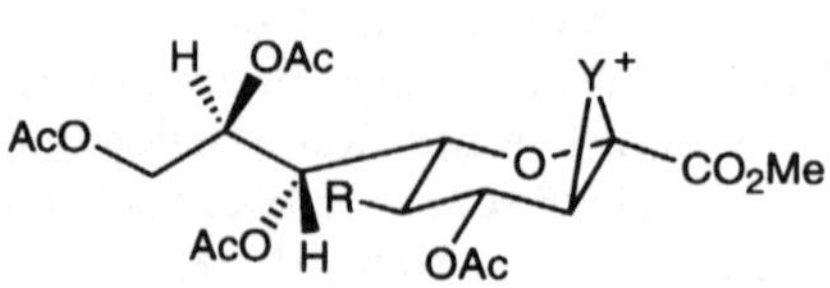

(**191**) R = NHAc, Y = SPh, SePh

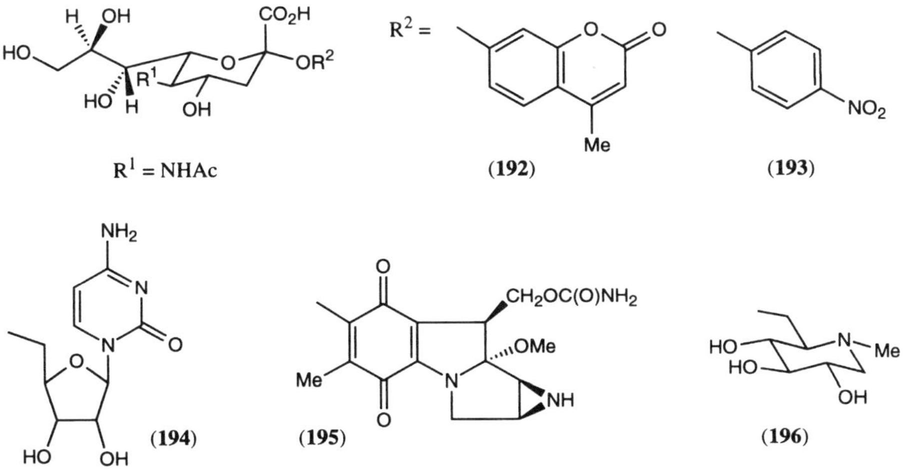

Fig. 2

Neu5Acα(2→3)Galβ(1→4)Glcβ1(→)1'Ceramide Ganglioside GM_3 (**197**)

Neu5Acα(2→3)Galβ(1→4)GlcNAcβ
3
↑
Fucα1

Sialy-Lewis^x tetrasaccharide (**198**)

Gal: D-galactose
Glc: D-glucose
Fuc: L-fucose

release of a detectable product [141, 187, 234–238]. There are reports of the glycosylation of sialic acids with cholesterol [239], uridine [105], inosine and cytidine (**194**) [240], mitomycins (**195**) [209], and 1-deoxynojirimycin derivatives (**196**) [241, 242] to produce potentially biologically active sialyl glycosides. The syntheses of many and varied sialyl disaccharides and oligosaccharides ([204] and references therein), gangliosides (**197**) [224, 243], and structures containing the sialyl-Lewis^x tetrasaccharide unit **198** [243–245] have been accomplished. The majority of reported sialyl glycosylations have been carried out with Neu5Ac itself, and with KDN [209]. Glycosylation of modified sialic acids is usually limited to the formation of aromatic glycosides used for assay purposes. Hasegawa, however, has used a range of modified sialic acids (e.g. side-chain epimeric, deoxy, truncated, etc.) for the synthesis of modified GM_3 gangliosides [224] and modified sialyl-Lewis^x analogues [243].

A novel approach to sialyl disaccharide synthesis is outlined in Scheme 14. The 7-carbon aldehyde **199**, the precursor to the Neu5Ac residue, was built up from ManNAc. Reaction of the aldehyde with phosphonate **200**, which contains the protected aglycon carbohydrate, methyl 2,3,4-tri-*O*-benzyl-*α*-D-glucopyranoside, gave a mixture of the enol ethers **201**, which could be separated. Hydrolysis of the acetonide protecting groups, followed by stereospecific Hg(II) induced cyclisation and then demercuration, gave the *α*- and *β*-glycosides of the

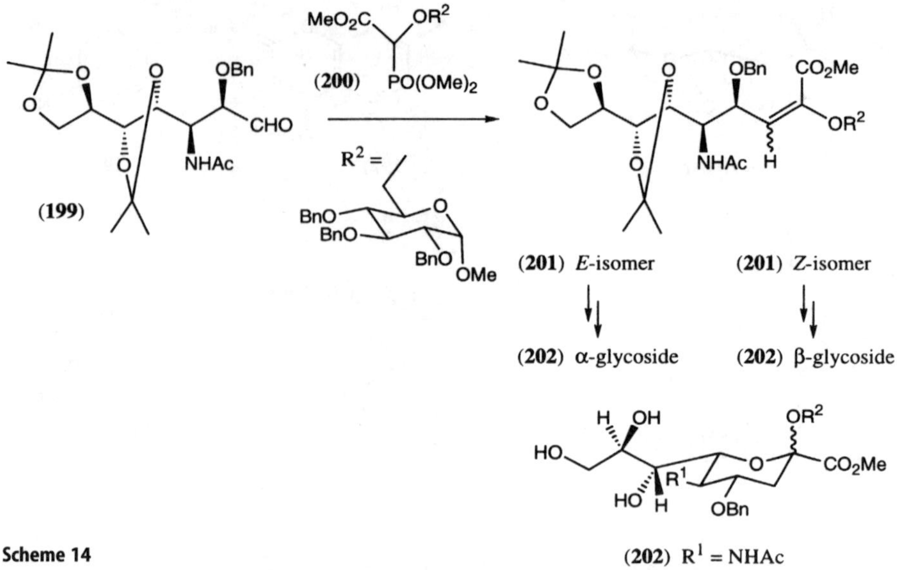

Scheme 14

(2 → 6)-linked sialyl disaccharide **202** from the *E* and *Z* isomers of **201**, respectively [246].

Compounds which contain hydrolytically stable glycosidic linkages are of value in the investigation of sialidase enzymes or systems which contain such enzymes. One example is in the design of sialic acid-based inhibitors of influenza virus haemagglutinin [247–251], where both the haemagglutinin and the sialidase glycoproteins are located on the viral cell coat [252]. Sialic acid glycosides containing sulfur [253–255], nitrogen [253], and carbon [249] glycosidic linkages have all been shown to have resistance to sialidase action. Once again, the majority of these glycosylations have been performed with Neu5Ac itself, with the variation in the glycosidic linkage offering the potential for alteration of biological activity.

A number of routes into the *α*-thioglycosides of Neu5Ac have been reported (reviewed in ref. [103]), including the reaction of peracetylated 2-*β*-chloro-2-deoxy-Neu5Ac1Me (**55**) with sodium thiolates [107] or with thiols under phase-transfer catalysis [211, 256]. An alternative approach is to use a thioglycoside of Neu5Ac as the glycosyl acceptor. A convenient starting material for these syntheses is peracetylated 2-*α*-thioacetyl-Neu5Ac1Me (**203**), prepared by reaction of **55** with potassium thioacetate under 'normal' [257, 258] or phase-transfer [247, 248] conditions. Selective de-*S*-acetylation of **203** with diethylamine [259], hydrazinium acetate [260], or a molar equivalent of NaOMe [257, 258, 261], followed by reaction with halides or sulfonates, has been used to prepare a range of *S*-linked glycosides of Neu5Ac. Using these procedures, alkyl [107, 256, 258, 262], aryl [211, 256, 257], cytidine [263, 264], and sugar thioglycosides [224, 255, 259, 265], including gangliosides [224], have been prepared. Dendritic *α*-thioglycosides of Neu5Ac have been prepared as inhibitors of influenza A virus haemagglutinin [247, 248]. A number of the sulfur-linked ganglioside analogues were reported to show inhibition of influenza virus sialidase [254, 261]. Recently a totally thio-linked analogue of sialyl-Lewis^x has been reported [266].

There are fewer reports of *N*-glycosides of sialic acids [63, 104]. The *p*-nitrophenyl aminoglycoside **204**, prepared by reaction of peracetylated 2-β-chloro-2-deoxy-Neu5Ac1Me (**55**) with *p*-nitroaniline, was not cleaved by either bacterial or viral sialidases [253]. Reaction of peracetylated Neu5Ac1Me (**104**) with trimethylsilyluracil (**205**, R=H), in the presence of $SnCl_4$, produced a mixture of the α- and β-aminogylcosides of uracil (**206**) [267]. When **55** was reacted with trimethylsilyluracil under Koenigs-Knorr conditions, only the β-aminogylcoside was isolated. The 2-α-azido-2-deoxy-Neu5Ac derivative **207** has been prepared by reaction of **55** with sodium azide under 'normal' [268] or phase-transfer [269] conditions. The β-anomer has been prepared by reaction of peracetylated Neu5Ac1Me (**104**) with trimethylsilylazide [268]. While **207** was found to be a good substrate for sialidases [268], the corresponding 2-α-azido glycoside of 6-thio-Neu5Ac (**208**) was hardly cleaved by sialidase and was also a sialidase inhibitor [78]. Both the α- and β-anomers of 2-azido-2-deoxy-Neu5Ac have been converted to a mixture of anomers of 2-amino-2-deoxy-Neu5Ac [268].

Sialyl glycosides with a carbon linkage have also been prepared. Reaction of peracetylated 2-β-chloro-2-deoxy-Neu5Ac1Me (**55**) with allyl(tri-*n*-butyl) stannane, utilising radical coupling conditions, gave an epimeric mixture of the

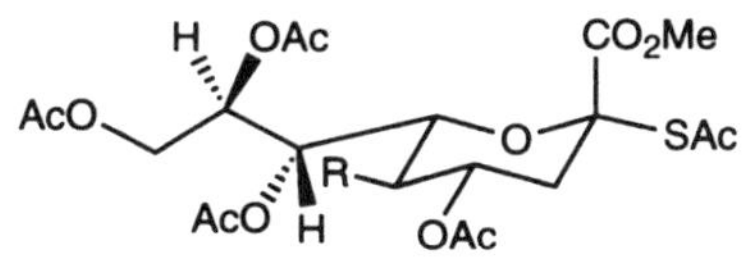

(**203**) R = NHAc

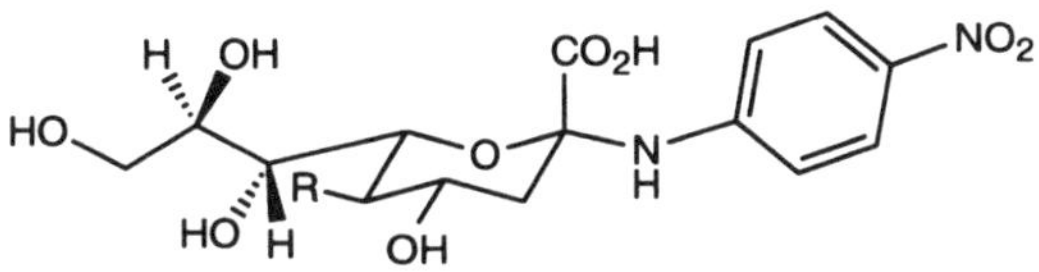

(**204**) R = NHAc

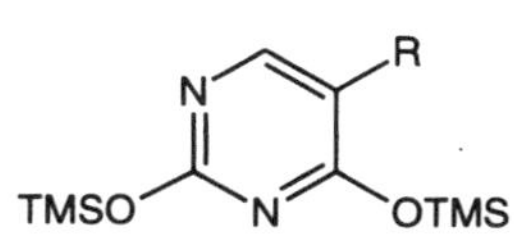

(**205**) R = H, F

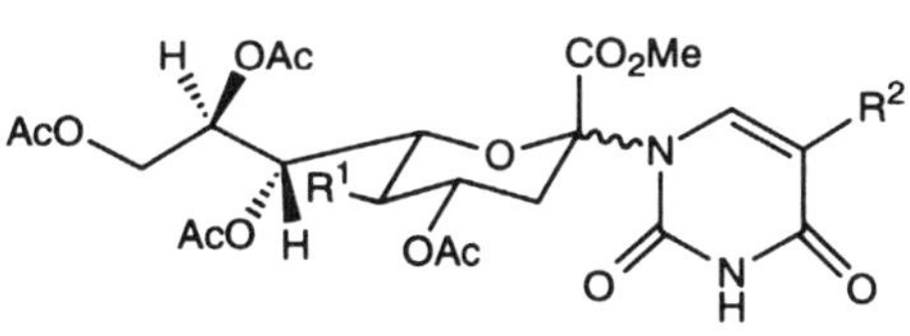

(**206**) R^1 = NHAc, R^2 = H, F

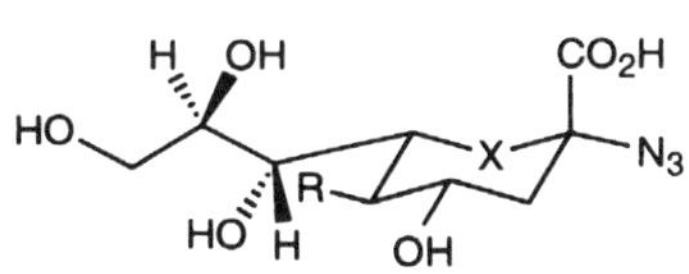

(**207**) R = NHAc, X = O

(**208**) R = NHAc, X = S

2-allyl glycosides (**209**) [270, 271]. Reaction at the allyl double bond has been used to attach the sialyl glycoside to polymeric supports to produce multivalent sialic acid species [249, 250]. Potent inhibitors of influenza virus haemagglutinin have been developed using multivalent *C*-linked sialosides [251]. *C*-Glycosides have also been prepared by hydroxymethylation of a lithium ester enolate derived from 2-deoxy-Neu5Ac derivative **137** [163]. This gave a 3:1 mixture of the α- and β-anomers of the 2-hydroxymethyl derivative **210.** The 2-α-aminomethyl derivative **211** was prepared by mesylation of the C-1′ hydroxyl group of the α-anomer of **210**, displacement with azide, and subsequent reduction to the amine.

5.2 Synthesis of Sialyl Glycosides by Enzymic Methods

The fact that the sialic acid residue is predominantly in a terminal position on natural oligosaccharide or glycoconjugate structures (except in polysialyl oligosaccharides) allows the use of a chemo-enzymic approach to the synthesis of these compounds. The asialo-oligosaccharide is built up using chemical or enzymic methodology, and the sialic acid transfer is carried out using a sialyltransferase (reviewed in ref. [204]). The enzymic transfer of the sialic acid from its activated form, CMP-Neu5Ac (**151**), proceeds with strict regio- and stereospecificity based on the sialyltransferase chosen.

Difficulties in the enzymic methodology are the need for CMP-activated sialic acids, a need to recycle the nucleotide, and the restricted substrate specificity of the sialyltransferase. The majority of work in sialyl oligosaccharide synthesis to date has involved the synthesis of Neu5Ac-containing oligosaccharides, using CMP-Neu5Ac which can be prepared from Neu5Ac either enzymically using CMP-sialate synthase [30, 42, 272] or chemically [215, 273]. The problem of regeneration of the nucleotide sugar has been overcome by Wong and coworkers, who have developed a system for enzymic recycling [274, 275].

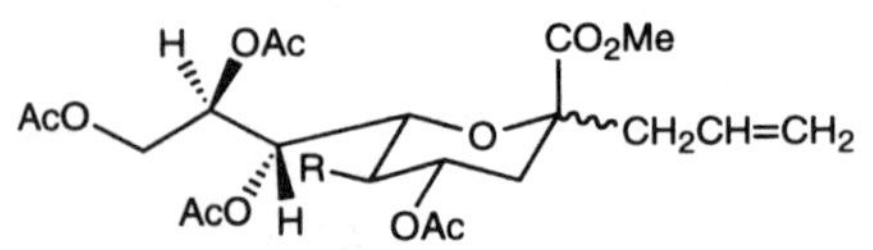

(**209**) R = NHAc

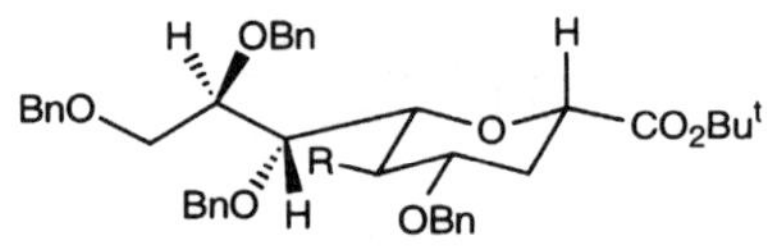

(**137**) R = NHAc

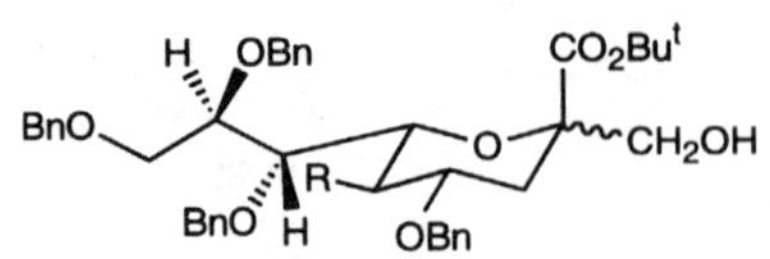

(**210**) R = NHAc

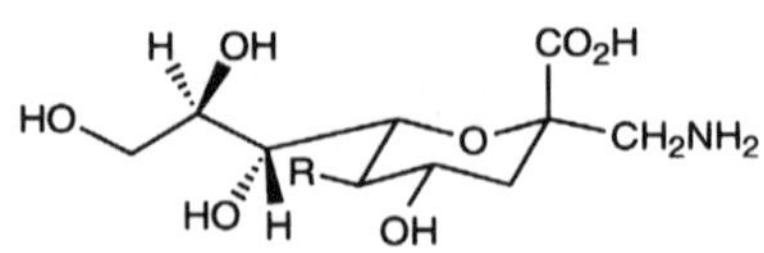

(**211**) R = NHAc

The extension of this methodology towards the totally enzymic synthesis of sialyl oligosaccharides has been growing with the isolation and purification of glycosyltransferases, their increased production using recombinant technology [276, 277], and the determination of their substrate specificities [278]. The use of multi-enzyme systems, in combination with cofactor (CTP, ATP etc.) regeneration, for the synthesis of sialyl oligosaccharides is now possible [279, 280], and has been applied to the synthesis of a number of sialyl oligosaccharides, including sialyl-Lewis^x and its analogues [281, 282].

The transfer of modified sialic acids by enzymic methods is dependent upon the substrate specificity of the CMP-sialate synthase and sialyltransferase enzymes chosen. The substrate specificity of CMP-sialate synthase has been examined ([4, 10, 26, 42] and references therein), and has been found to show some differences depending on the source of the enzyme [42]. The enzyme from rat liver [4] was found to activate most side-chain deoxy and epimeric Neu5Ac derivatives with varying efficiency. However, there was a sensitivity to alteration at the C-8 position. For the enzymes cloned from the *E. coli* system, a number of C-5 analogues of Neu5Ac, including KDN, were not substrates [42]. Several C-9 (**212**) [122, 123] and C-5 (**213**) [122, 283] substituted Neu5Ac derivatives have been activated. The enzyme is also tolerant of some modifications at the C-4 position, activating 4-acetamido-4-deoxy- (**16**) [145] and 4-deoxy-Neu5Ac (**12**) [284], although 4-*epi*-Neu5Ac (**13**) showed no activation [48].

The transfer of the modified sialic acids from their activated form to saccharides using sialyltransferases has been reported for 4-deoxy-Neu5Ac (**12**) [284] and for a number of C-9 [122, 123, 169] and C-5 [122, 169, 283] analogues. Cell surface glycoconjugates sialylated with the C-9 acetyl, acetamido, and thioacetamido Neu5Ac analogues have been investigated as receptors for influenza C virus [124, 125]. CMP-Neu5Ac derivatives substituted at the C-9 (Fig. 3) or C-5 position with large fluorescent dyes, for example CMP-9-fluoresceinyl-Neu5Ac (**214**), or photoactivatable groups, e.g. **215**, have been prepared from the appropriate CMP-Neu derivative (e.g. **216**, R^2=H) [169]. The sialic acid derivatives could be transferred onto asialo-glycoconjugates using specific sialyltransferases, enabling evaluation of the expression of certain cell surface glycoconjugates.

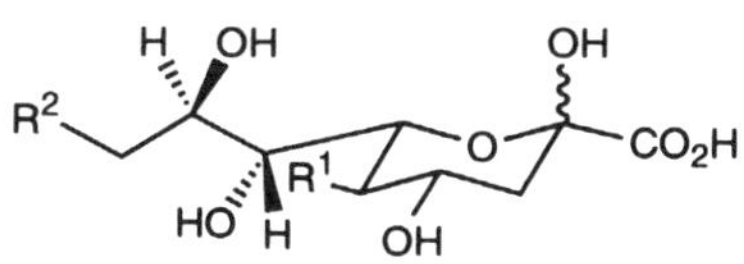

(**212**) R^1 = NHAc
R^2 = N_3, NH_2, NHAc, NHC(S)Me, NHC(O)$(CH_2)_4CH_3$, NHC(O)Ph

(**213**) R^1 = NHC(O)H, NHC(O)CH_2NH_2, NHC(S)Me
R^2 = OH

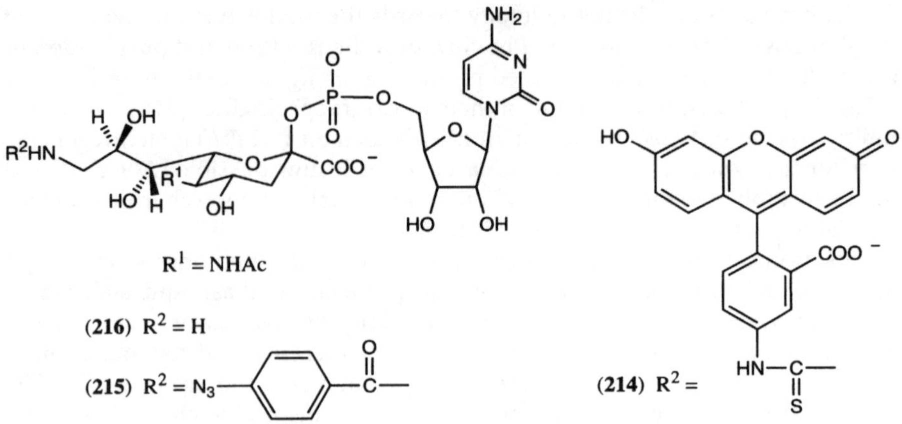

Fig. 3

6 Conclusion and Future Prospects

Since the first preparation of Neu5Ac in 1957, the synthesis, manipulation and glycosylation of sialic acids has been intensely studied. Today, there are procedures available for the modification of each position of the neuraminic acid skeleton. The growing awareness that sialic acids play significant roles in a number of biological processes has led to investigation of the proteins which recognise and/or metabolise them. The advent of structure determination by both X-ray crystallographic and NMR spectroscopic methods of some of these proteins permits detailed examination at an atomic level of the binding site as well as ligand-protein interactions. The information collected from these studies identifies specific key residues within the protein to be targeted by appropriately functionalised sialic acid-containing templates for improvement in binding affinity, and hence possible intervention of normal protein function. The development of a Neu5Ac2en derivative as a potential anti-influenza drug [189] is an example of the structure-based design of biological probes that is possible when structural information is available. However, in the absence of structural information, the use of modified sialic acids as biological probes is the only means of elucidating the key requirements for binding (and, in the case of enzymes, catalysis) of the proteins. With the knowledge of functional group manipulation that is now available, proteins can be probed for a wide range of structure-activity relationships.

A number of carbohydrates have therapeutic applications [285, 286], and there are now reports of several potential therapeutic applications for sialic acid-containing compounds [189, 287–289]. The field of sialic acid chemistry is ever-expanding, and most of those working in this challenging area of chemistry are now closely allied with glycobiologists. The future for this field is bright, and much is yet to be discovered.

Acknowledgements. The authors would like to thank Barry Matthews, Milton Kiefel, and Jeff Dyason for helpful discussions during the preparation of this review.

References

1. Schauer R, Kelm S, Reuter G, Roggentin P, Shaw L (1995) Biochemistry and role of sialic acids. In: Rosenberg A (ed) Biology of the sialic acids. Plenum, New York, p 7
2. Schauer R (1982) Adv Carbohydr Chem Biochem 40:131
3. Reutter W, Köttgen E, Bauer C, Gerok W (1982) Biological significance of sialic acids. In: Schauer R (ed) Sialic acids - Chemistry, metabolism and function. Springer, Berlin Heidelberg New York, p 263
4. Zbiral E (1992) Synthesis of sialic acid analogs and their behaviour towards the enzymes of sialic acid metabolism and hemagglutinin X-31 of influenza A-virus. In: Ogura H, Hasegawa A, Suami T (eds) Carbohydrates. Synthetic methods and applications in medicinal chemistry. VCH, Weinheim, p 304
5. Schauer R, Buscher H-P (1974) Biochim Biophys Acta 338:369
6. Czarniecki MF, Thornton ER (1977) J Am Chem Soc 99:8273
7. Roy R, Laferrière CA (1990) Can J Chem 68:2045
8. Blix G (1962) Meth Carbohydr Chem 1:246
9. Schauer R, Corfield AP (1982) Isolation and purification of sialic acids. In: Schauer R (ed) Sialic acids - Chemistry, metabolism and function. Springer-Verlag, Wien, p 51
10. Corfield AP, Schauer R (1982) Metabolism of sialic acids. In: Schauer R (ed) Sialic acids - Chemistry, metabolism and function. Springer, Berlin Heidelberg New York, p 195
11. Comb DG, Roseman S (1960) J Biol Chem 235:2529
12. Brunetti P, Jourdian GW, Roseman S (1962) J Biol Chem 237:2447
13. Augé C, David S, Gautheron C, Malleron A, Cavayé B (1988) New J Chem 12:733
14. Augé C, Gautheron C (1987) J Chem Soc, Chem Commun:859
15. Kim M-J, Hennen WJ, Sweers HM, Wong C-H (1988) J Am Chem Soc 110:6481
16. Augé C, Bouxom B, Cavayé B, Gautheron C (1989) Tetrahedron Lett 30:2217
17. Augé C, Gautheron C, David S, Malleron A, Cavayé B, Bouxom B (1990) Tetrahedron 46:201
18. David S, Malleron A, Cavayé B (1992) New J Chem 16:751
19. Lin C-H, Sugai T, Halcomb RL, Ichikawa Y, Wong C-H (1992) J Am Chem Soc 114:10138
20. Kragl U, Gödde A, Wandrey C, Lubin N, Augé C (1994) J Chem Soc, Perkin Trans 1:119
21. Halcomb RL, Fitz W, Wong C-H (1994) Tetrahedron: Asymmetry 5:2437
22. Kong DCM, von Itzstein M (1995) Tetrahedron Lett 36:957
23. Fitz W, Schwark J-R, Wong C-H (1995) J Org Chem 60:3663
24. David S, Augé C, Gautheron C (1991) Adv Carbohydr Chem Biochem 49:175
25. Drueckhammer DG, Hennen WJ, Pederson RL, Barbas IIIrd CF, Gautheron CM, Krach T, Wong C-H (1991) Synthesis:499
26. Wong C-H, Whitesides GM (1994) Enzymes in organic synthesis. Elsevier, Oxford
27. Kragl U, Gygax D, Ghisalba O, Wandrey C (1991) Angew Chem Int Ed Engl 30:827
28. Bednarski MD, Chenault HK, Simon ES, Whitesides GM (1987) J Am Chem Soc 109:1283
29. Spivak CT, Roseman S (1959) J Am Chem Soc 81:2403
30. Simon ES, Bednarski MD, Whitesides GM (1988) J Am Chem Soc 110:7159
31. Nees S, Schauer R, Mayer F, Ehrlich K (1976) Hoppe-Seyler's Z Physiol Chem 357:839
32. Uchida Y, Tsukada Y, Sugimori T (1984) J Biochem 96:507
33. Schauer R, Stoll S, Zbiral E, Schreiner E, Brandstetter HH, Vasella A, Baumberger F (1987) Glycoconjugate J 4:361
34. Zbiral E, Kleineidam RG, Schreiner E, Hartmann M, Christian R, Schauer R (1992) Biochem J 282:511
35. Baumann W, Freidenreich J, Weisshaar G, Brossmer R, Friebolin H (1989) Biol Chem Hoppe-Seyler 370:141
36. Schauer R, Wember M (1971) Hoppe-Seyler's Z Physiol Chem 352:1517
37. Deijl CM, Vliegenthart JFG (1983) Biochem Biophys Res Commun 111:668
38. Izard T, Lawrence MC, Malby RL, Lilley GG, Colman PM (1994) Structure 2:361
39. Kong DCM, Dyason JC, von Itzstein M (unpublished work)

40. Kong DCM, von Itzstein M (1994) Chemoenzymatic synthesis of a novel sialic acid analogue using membrane-enclosed enzyme catalysis. XVIIth International Carbohydrate Symposium (Ottawa) p 246
41. Fitz W, Wong C-H (1994) J Org Chem 59:8279
42. Liu JL-C, Shen G-J, Ichikawa Y, Rutan JF, Zapata G, Vann WF, Wong C-H (1992) J Am Chem Soc 114:3901
43. Suttajit M, Urban C, McLean RL (1971) J Biol Chem 246:810
44. Shukla AK, Schauer R (1986) Anal Biochem 158:158
45. Kiefel MJ, Bennett S, von Itzstein M (unpublished work)
46. Warren L (1964) Biochim Biophys Acta 83:129
47. Hagedorn H-W, Brossmer R (1986) Helv Chim Acta 69:2127
48. Gross HJ, Kovac A, Rose U, Watzlawick H, Brossmer R (1988) Biochemistry 27:4279
49. Gross HJ, Brossmer R (1988) FEBS Lett 232:145
50. Beau J-M, Schauer R (1980) Eur J Biochem 106:531
51. Sugai T, Shen G-J, Ichikawa Y, Wong C-H (1993) J Am Chem Soc 115:413
52. Brossmer R, Rose U, Kasper D, Smith TL, Grasmuk H, Unger FM (1980) Biochem Biophys Res Commun 96:1282
53. Kong DCM, von Itzstein M (unpublished work)
54. Isecke R, Brossmer R (1994) Tetrahedron 50:7445
55. Schrell A, Whitesides GM (1990) Liebigs Ann Chem:1111
56. Sparks MA, Williams KW, Lukacs C, Schrell A, Priebe G, Spaltenstein A, Whitesides GM (1993) Tetrahedron 49:1
57. Koppert K, Brossmer R (1992) Tetrahedron Lett 33:8031
58. Koppert K, Brossmer R (1992) 5-C-Acetamidomethyl-5-deoxy-KDN, a branched homologue of *N*-acetylneuraminic acid, and additional KDN analogues modified at C-5. XVIth International Carbohydrate Symposium (Paris) p 296
59. Wu W-Y, Kok GB, Jin B, von Itzstein M (unpublished work)
60. Murakami M, Ikeda K, Achiwa K (1996) Carbohydr Res 280:101
61. Danishefsky SJ, DeNinno MP (1986) J Org Chem 51:2615
62. Danishefsky SJ, DeNinno MP, Chen S-H (1988) J Am Chem Soc 110:3929
63. Tuppy H, Gottschalk A (1972) The structure of sialic acids and their quantitation. In: Gottschalk A (ed) Glycoproteins. Their composition, structure and function. Elsevier, Amsterdam, p 403
64. DeNinno MP (1991) Synthesis:583
65. Cornforth JW, Daines ME, Gottschalk A (1957) Proc Chem Soc Lond:25
66. Cornforth JW, Firth ME, Gottschalk A (1958) Biochem J 68:57
67. Carroll PM, Cornforth JW (1960) Biochim Biophys Acta 39:161
68. How MJ, Halford MDA, Stacey M, Vickers E (1969) Carbohydr Res 11:313
69. Kuhn R, Baschang G (1962) Liebigs Ann Chem 659:156
70. Gantt R, Millner S, Binkley SB (1964) Biochem 3:1952
71. DeVries GH, Binkley SB (1972) Arch Biochem Biophys 151:243
72. Meindl P, Tuppy H (1966) Monatsh Chem 97:1628
73. Brossmer R, Nebelin E (1969) FEBS Lett 4:335
74. Khorlin AY, Privalova IM (1970) Carbohydr Res 13:373
75. McLean R, Beidler J (1969) J Am Chem Soc 91:5388
76. McLean RL, Suttajit M, Beidler J, Winzler RJ (1971) J Biol Chem 246:803
77. Shirai R, Ogura H (1989) Tetrahedron Lett 30:2263
78. Mack H, Brossmer R (1987) Tetrahedron Lett 28:191
79. Hagedorn HW, Merten H, Brossmer R (1992) Carbohydr Res 236:89
80. Benzing-Nguyen L, Perry MB (1978) J Org Chem 43:551
81. Estenne G, Saroli A, Doutheau A (1991) J Carbohydr Chem 10:181
82. Driguez P-A, Barrere B, Quash G, Doutheau A (1994) Carbohydr Res 262:297
83. Mirzayanova MN, Davydova LP, Samokhvalov GI (1970) J Gen Chem USSR 40:663
84. Beau J-M, Sinaÿ P (1978) Carbohydr Res 65:1
85. Beau J-M, Sinaÿ P, Kamerling JP, Vliegenthart JFG (1978) Carbohydr Res 67:65

86. Dondoni A, Marra A, Merino P (1994) J Am Chem Soc 116:3324
87. Dondoni A, Boscarato A, Marra A (1994) Tetrahedron: Asymmetry 5:2209
88. Baumberger F, Vasella A (1986) Helv Chim Acta 69:1205
89. Baumberger F, Vasella A (1986) Helv Chim Acta 69:1535
90. Baumberger F, Vasella A, Schauer R (1988) Helv Chim Acta 71:429
91. Gordon DM, Whitesides GM (1993) J Org Chem 58:7937
92. Chan T-H, Lee M-C (1995) J Org Chem 60:4228
93. Chan T-H, Li C-J (1992) J Chem Soc, Chem Commun:747
94. Kim E, Gordon DM, Schmid W, Whitesides GM (1993) J Org Chem 58:5500
95. Gao J, Härter R, Gordon DM, Whitesides GM (1994) J Org Chem 59:3714
96. Julina R, Müller I, Vasella A, Wyler R (1987) Carbohydr Res 164:415
97. Vasella A, Wyler R (1990) Helv Chim Acta 73:1742
98. Bamford MJ, Pichel JC, Husman W, Patel B, Storer R, Weir NG (1995) J Chem Soc, Perkin Trans 1:1181
99. Ogawa S, Yoshikawa M, Taki T, Yokoi S, Chida N (1995) Carbohydr Res 269:53
100. Chandler M, Conroy R, Cooper AWJ, Lamont RB, Scicinski JJ, Smart JE, Storer R, Weir NG, Wilson RD, Wyatt PG (1995) J Chem Soc, Perkin Trans 1:1189
101. Holmquist L (1975) FOA Reports 9:1
102. Vliegenthart JFG, Kamerling JP (1982) Synthesis of sialic acids and sialic acid derivatives. In: Schauer R (ed) Sialic acids - Chemistry, metabolism and function. Springer, Berlin Heidelberg New York, p 59
103. von Itzstein M, Kiefel MJ (1996) Sialic acid analogues as potential anti-microbial agents. In: Witczak ZJ Nieforth KA (eds) Carbohydrates: Targets for rational drug design. MDI, New York (in press)
104. Okamoto K, Goto T (1990) Tetrahedron 46:5835
105. Ogura H, Furuhata K, Itoh M, Shitori Y (1986) Carbohydr Res 158:37
106. Sharma M, Petrie IIIrd CR, Korytnyk W (1988) Carbohydr Res 175:25
107. Lubineau A, Le Gallic J (1991) J Carbohydr Chem 10:263
108. Meindl P, Tuppy H (1965) Monatsh Chem 96:802
109. Kononov LO, Magnusson G (1994) Easy sialylation without promoter. Rapid and highly stereoselective reaction between acetochloroneuraminic acid and alcohols. International Conference on the Biology and Chemistry of Sialic Acids (Moscow) p 71
110. Meindl P, Tuppy H (1965) Monatsh Chem 96:816
111. Meindl P, Tuppy H (1966) Monatsh Chem 97:990
112. Brandstetter HH, Zbiral E, Schulz G (1982) Liebigs Ann Chem:1
113. Marra A, Sinaÿ P (1989) Carbohydr Res 190:317
114. Schreiner E, Christian R, Zbiral E (1990) Liebigs Ann Chem:93
115. Zbiral E, Brandstetter HH, Schreiner EP (1988) Monatsh Chem 119:127
116. Ogura H, Furuhata K, Sato S, Anazawa K, Itoh M, Shitori Y (1987) Carbohydr Res 167:77
117. Furuhata K, Ogura H (1989) Chem Pharm Bull 37:2037
118. Reinhard B, Faillard H (1994) Liebigs Ann Chem:193
119. Anazawa K, Furuhata K, Ogura H (1988) Chem Pharm Bull 36:4976
120. Hasegawa A, Murase T, Ogawa M, Ishida H, Kiso M (1990) J Carbohydr Chem 9:429
121. Sepulveda-Boza S, Stather U (1994) Bol Soc Chil Quim 39:299
122. Brossmer R, Gross HJ (1994) Methods Enzymol 247:153
123. Gross HJ, Bünsch A, Paulson JC, Brossmer R (1987) Eur J Biochem 168:595
124. Herrler G, Gross H-J, Imhof A, Brossmer R, Milks G, Paulson JC (1992) J Biol Chem 267:12501
125. Brossmer R, Isecke R, Herrler G (1993) FEBS Lett 323:96
126. Salunkhe M, Hartmann M, Schmid W, Zbiral E (1988) Liebigs Ann Chem:187
127. Hartmann M, Zbiral E (1989) Monatsh Chem 120:899
128. Zbiral E, Brandstetter HH (1985) Monatsh Chem 116:87
129. Zbiral E, Phadtare S, Schmid W (1987) Liebigs Ann Chem:39
130. Hartmann M, Christian R, Zbiral E (1990) Liebigs Ann Chem:83
131. Brandstetter HH, Zbiral E (1983) Liebigs Ann Chem:2055

132. Tsvetkov YE, Schmidt RR (1994) Tetrahedron Lett 35:8583
133. Brossmer R, Holmquist L (1974) FEBS Lett 40:250
134. Kiefel MJ, Bennett S, von Itzstein M (1996) J Chem Soc, Perkin Trans 1:439
135. Glänzer BI, Györgydeák Z, Bernet B, Vasella A (1991) Helv Chim Acta 74:343
136. Czollner L, Kuszmann J, Vasella A (1990) Helv Chim Acta 73:1338
137. Friebolin H, Kunzelmann P, Supp M, Brossmer R, Keilich G, Ziegler D (1981) Tetrahedron Lett 22:1383
138. Nakamura M, Takayanagi H, Furuhata K, Ogura H (1992) Chem Pharm Bull 40:879
139. Yamamoto T, Kumazawa H, Inami K, Teshima T, Shiba T (1992) Tetrahedron Lett 33: 5791
140. Ogura H, Furuhata K (1981) Tetrahedron Lett 22:4265
141. Schreiner E, Zbiral E (1990) Liebigs Ann Chem:581
142. Shirai R, Nakamura M, Hara S, Takayanagi H, Ogura H (1988) Tetrahedron Lett 29: 4449
143. Saito M, Yu RK (1995) Biochemistry and function of sialidases. In: Rosenberg A (ed) Biology of the sialic acids. Plenum Press, New York, p 261
144. Groves DR, Wilson JC, von Itzstein M (1995) Aust J Chem 48:1217
145. Zbiral E, Schreiner E, Christian R (1989) Carbohydr Res 194:C15
146. Hartmann M, Christian R, Zbiral E (1991) Monatsh Chem 122:111
147. Groves DR, von Itzstein M (1996) J Chem Soc, Perkin Trans 1: (accepted for publication)
148. Friebolin H, Schmidt H, Supp M (1981) Tetrahedron Lett 22:5171
149. Dorland L, Haverkamp J, Schauer R, Veldink GA, Vliegenthart JFG (1982) Biochem Biophys Res Commun 104:1114
150. Chong AKJ, Pegg MS, Taylor NR, von Itzstein M (1992) Eur J Biochem 207:335
151. Sinnott ML, Guo X, Li S-C, Li Y-T (1993) J Am Chem Soc 115:3334
152. Tiralongo J, Pegg MS, von Itzstein M (1995) FEBS Lett 372:148
153. Okamoto K, Kondo T, Goto T (1987) Bull Chem Soc Japan 60:631
154. Nakajima T, Hori H, Ohrui H, Meguro H, Ido T (1988) Agric Biol Chem 52:1209
155. Hagiwara T, Kijima-Suda I, Ido T, Ohrui H, Tomita K (1994) Carbohydr Res 263: 167
156. Meindl P, Tuppy H (1969) Hoppe-Seyler's Z Physiol Chem 350:1088
157. Meindl P, Bodo G, Palese P, Schulman J, Tuppy H (1974) Virology 58:457
158. Holzer CT, von Itzstein M, Jin B, Pegg MS, Stewart WP, Wu W-Y (1993) Glycoconjugate J 10:40
159. Bandgar BP, Zbiral E (1995) Carbohydr Res 270:201
160. Wallimann K, Vasella A (1990) Helv Chim Acta 73:1359
161. Schmid W, Christian R, Zbiral E (1988) Tetrahedron Lett 29:3643
162. Bandgar BP, Hartmann M, Schmid W, Zbiral E (1990) Liebigs Ann Chem 1185
163. Wallimann K, Vasella A (1991) Helv Chim Acta 74:1520
164. Hartmann M, Zbiral E (1991) Liebigs Ann Chem:795
165. Kelm S, Paulson JC, Rose U, Brossmer R, Schmid W, Bandgar BP, Schreiner E, Hartmann M, Zbiral E (1992) Eur J Biochem 205:147
166. Brossmer R, Holmquist L (1971) Hoppe-Seyler's Z Physiol Chem 352:1715
167. Potter JJ, von Itzstein M (1996) Carbohydr Res 282:181
168. Vasella A, Wyler R (1991) Helv Chim Acta 74:451
169. Brossmer R, Gross HJ (1994) Methods Enzymol 247:177
170. Lanne B, Schierbeck B, Karlsson K-A (1994) J Biochem 116:1269
171. Kayser H, Zeitler R, Kannicht C, Grunow D, Nuck R, Reutter W (1992) J Biol Chem 267:16934
172. Meindl P, Tuppy H (1969) Monatsh Chem 100:1295
173. Claesson A, Luthman K (1982) Acta Chem Scand B36:719
174. Kok GB, Mackey BL, von Itzstein M (1996) Carbohydr Res 289:67
175. Kumar V, Kessler J, Scott ME, Patwardhan BH, Tanenbaum SW, Flashner M (1981) Carbohydr Res 94:123
176. Kok GB, Groves DR, von Itzstein M (1996) J Chem Soc, Chem Commun:

177. Beau J-M, Schauer R, Haverkamp J, Kamerling JP, Dorland L, Vliegenthart JFG (1984) Eur J Biochem 140:203
178. Saito M, Rosenberg A (1984) Biochemistry 23:3784
179. Nöhle U, Shukla AK, Schröder C, Reuter G, Schauer R, Kamerling JP, Vliegenthart JFG (1985) Eur J Biochem 152:459
180. Burmeister WP, Henrissat B, Bosso C, Cusack S, Ruigrok RWH (1993) Structure 1:19
181. Zbiral E, Brandstetter HH, Christian R, Schauer R (1987) Liebigs Ann Chem:781
182. Zbiral E, Schreiner E, Christian R, Kleineidam RG, Schauer R (1989) Liebigs Ann Chem: 159
183. Meindl P, Tuppy H (1970) Monatsh Chem 101:639
184. Schreiner E, Zbiral E, Kleineidam RG, Schauer R (1991) Carbohydr Res 216:61
185. Meindl P, Tuppy H (1973) Monatsh Chem 104:402
186. Schreiner E, Zbiral E, Kleineidam RG, Schauer R (1991) Liebigs Ann Chem:129
187. Baumberger F, Vasella A, Schauer R (1986) Helv Chim Acta 69:1927
188. Kumar V, Tanenbaum SW, Flashner M (1982) Carbohydr Res 103:281
189. von Itzstein M, Wu W-Y, Kok GB, Pegg MS, Dyason JC, Jin B, Phan TV, Smythe ML, White HF, Oliver SW, Colman PM, Varghese JN, Ryan DM, Woods JM, Bethell RC, Hotham VJ, Cameron JM, Penn CR (1993) Nature 363:418
190. von Itzstein M, Dyason JC, Oliver SW, White HF, Wu W-Y, Kok GB, Pegg MS (1996) J Med Chem 39:388
191. Taylor NR, von Itzstein M (1994) J Med Chem 37:616
192. von Itzstein M, Jin B, Wu W-Y, Chandler M (1993) Carbohydr Res 244:181
193. Chandler M, Bamford MJ, Conroy R, Lamont B, Patel B, Patel VK, Steeples IP, Storer R, Weir NG, Wright M, Williamson C (1995) J Chem Soc, Perkin Trans 1:1173
194. von Itzstein M, Wu W-Y, Jin B (1994) Carbohydr Res 259:301
195. von Itzstein M, Wu W-Y, Phan TV, Danylec B, Jin B (1991) Patent App (Biota Holdings Pty Ltd) WO 91 16,320: Chem Abstr 117:49151y
196. Pegg MS, von Itzstein M (1994) Biochem Mol Biol Int 32:851
197. Scheigetz J, Zamboni R, Bernstein MA, Roy B (1995) Org Prep Proc Int 27:637
198. Kok GB, Campbell M, Mackey B, von Itzstein M (1996) J Chem Soc, Perkin Trans 1 (accepted for publication)
199. Smith PW, Starkey ID, Howes PD, Sollis SL, Keeling SP, Cherry PC, von Itzstein M, Wu W-Y, Jin B (1996) Eur J Med Chem 31:143
200. Starkey ID, Mahmoudian M, Noble D, Smith PW, Cherry PC, Howes PD, Sollis SL (1995) Tetrahedron Lett 36:299
201. Ciccotosto S, von Itzstein M (1995) Tetrahedron Lett 36:5405
202. Watowich SJ, Skehel JJ, Wiley DC (1994) Structure 2:719
203. Sabesan S (1994) Bioorg Med Chem Lett 4:2457
204. Ito Y, Gaudino JJ, Paulson JC (1993) Pure Appl Chem 65:753
205. Ogura H, Hasegawa A, Suami T (eds) (1992) Carbohydrates. Synthetic methods and applications in medicinal chemistry. VCH and Kodansha, Weinheim and Tokyo
206. Kováč P (ed) (1994) Synthetic oligosaccharides. Indispensable probes for the life sciences. ACS Symposium Series 560. American Chemical Society, Washington
207. Whitfield DM, Douglas SP (1996) Glycoconjugate J 13:5
208. Garegg PJ, Lindberg AA (1988) The synthesis of oligosaccharides for biological and medical purposes. In: Kennedy JF (ed) Carbohydrate chemistry. Clarendon Press, Oxford, p 500
209. Ogura H (1992) Sialic acid derivatives as glycolipoids. In: Ogura H, Hasegawa A, Suami T (eds) Carbohydrates. Synthetic methods and applications in medicinal chemistry. VCH, Weinheim, p 282
210. Rothermel J, Faillard H (1990) Carbohydr Res 196 : 29
211. Roy R, Andersson FO, Harms G, Kelm S, Schauer R (1992) Angew Chem Int Ed Engl 31:1478
212. Mukaiyama T, Sasaki T, Iwashita E, Matsubara K (1995) Chem Lett:455

213. Martin TJ, Schmidt RR (1992) Tetrahedron Lett 33:6123
214. Martin TJ, Brescello R, Toepfer A, Schmidt RR (1993) Glycoconjugate J 10:16
215. Kondo H, Ichikawa Y, Wong C-H (1992) J Am Chem Soc 114:8748
216. Hasegawa A, Ogawa M, Ishida H, Kiso M (1990) J Carbohydr Chem 9:393
217. Hasegawa A, Ohki H, Nagahama T, Ishida H, Kiso M (1991) Carbohydr Res 212:277
218. Sabesan S, Neira S, Davidson F, Duus JØ, Bock K (1994) J Am Chem Soc 116:1616
219. Ercégovic T, Magnusson G (1995) J Org Chem 60:3378
220. Ercégovic T, Magnusson G (1996) J Org Chem 61:179
221. Sugata T, Higuchi R (1996) Tetrahedron Lett 37:2613
222. Marra A, Sinaÿ P (1990) Carbohydr Res 195:303
223. Martichonok V, Whitesides GM (1996) J Org Chem 61:1702
224. Hasegawa A, Kiso M (1992) Systematic synthesis of gangliosides toward the elucidation and biomedical application of their biological functions. In: Ogura H, Hasegawa A, Suami T (eds) Carbohydrates. Synthetic methods and applications in medicinal chemistry. VCH, Weinheim, p 243
225. Ito Y, Ogawa T (1990) Tetrahedron 46:89
226. Okamoto K, Kondo T, Goto T (1986) Chem Lett:1449
227. Okamoto K, Kondo T, Goto T (1987) Tetrahedron 43:5909
228. Kondo H, Aoki S, Ichikawa Y, Halcomb RL, Ritzen H, Wong C-H (1994) J Org Chem 59:864
229. Okamoto K, Kondo T, Goto T (1987) Bull Chem Soc Japan 60:637
230. Okamoto K, Kondo T, Goto T (1988) Tetrahedron 44:1291
231. Ito Y, Ogawa T (1987) Tetrahedron Lett 28:6221
232. Ito Y, Ogawa T (1988) Tetrahedron Lett 29:3987
233. Nakahara Y, Iijima H, Ogawa T (1994) Stereocontrolled approaches to *O*-glycopeptide synthesis. In: Kováč P (ed) Synthetic oligosaccharides. Indispensable probes for the life sciences. ACS Symposium Series 560. American Chemical Society, Washington, p 249
234. Myers RW, Lee RT, Lee YC, Thomas GH, Reynolds LW, Uchida Y (1980) Anal Biochem 101:166
235. Baggett N, Marsden BJ (1982) Carbohydr Res 110:11
236. Eschenfelder V, Brossmer R (1987) Carbohydr Res 162:294
237. Nakamura M, Furuhata K, Ogura H (1989) Chem Pharm Bull 37:821
238. Zbiral E, Schreiner E, Salunkhe MM, Schulz G, Kleineidam RG, Schauer R (1989) Liebigs Ann Chem:519
239. Sato S, Fujita S, Furuhata K, Ogura H, Yoshimura S, Itoh M, Shitori Y (1987) Chem Pharm Bull 35:4043
240. Sato S, Furuhata K, Itoh M, Shitori Y, Ogura H (1988) Chem Pharm Bull 36:914
241. Kiso M, Ando K, Furui H, Ishida H, Hasegawa A (1993) J Carbohydr Chem 12:985
242. Kiso M, Ando K, Inagaki H, Ishida H, Hasegawa A (1995) Carbohydr Res 272:159
243. Hasegawa A (1994) Synthesis of sialo-oligosaccharides and their ceramide derivatives as tools for elucidation of biologic functions of gangliosides. In: Kováč P (ed) Synthetic oligosaccharides. Indispensable probes for the life sciences. ACS Symposium Series 560. American Chemical Society, Washington, p 184
244. Nicolaou KC, Hummel CW, Bockovich NJ, Wong C-H (1991) J Chem Soc, Chem Commun: 870
245. Nicolaou KC, Hummel CW, Iwabuchi Y (1992) J Am Chem Soc 114:3126
246. Paquet F, Sinaÿ P (1984) Tetrahedron Lett 25:3071
247. Roy R, Zanini D, Meunier J, Romanowski A (1993) J Chem Soc, Chem Commun: 1869
248. Roy R, Zanini D, Meunier SJ, Romanowska A (1994) Synthesis and antigenic properties of sialic acid based dendrimers. In: Kováč P (ed) Synthetic oligosaccharides. Indispensable probes for the life sciences. ACS Symposium Series 560. American Chemical Society, Washington, p 104

249. Nagy JO, Wang P, Gilbert JH, Schaefer ME, Hill TG, Callstrom MR, Bednarski MD (1992) J Med Chem 35:4501
250. Spevak W, Nagy JO, Charych DH, Schaefer ME, Gilbert JH, Bednarski MD (1993) J Am Chem Soc 115:1146
251. Mammen M, Dahmann G, Whitesides GM (1995) J Med Chem 38:4179
252. Herrler G, Hausmann J, Klenk H-D (1995) Sialic acid as receptor determinant of ortho- and paramyxoviruses. In: Rosenberg A (ed) Biology of the sialic acids. Plenum Press, New York, p 315
253. Khorlin AY, Privalova IM, Zakstelskaya LY, Molibog EV, Evstigneeva NA (1970) FEBS Lett 8:17
254. Suzuki Y, Sato K, Kiso M, Hasegawa A (1990) Glycoconjugate J 7:349
255. Kiefel MJ, Beisner B, Bennett S, Holmes ID, von Itzstein M (1996) J Med Chem 39: 1314
256. Cao S, Meunier SJ, Andersson FO, Letellier M, Roy R (1994) Tetrahedron: Asymmetry 5:2303
257. Warner TG, Lee LA (1988) Carbohydr Res 176:211
258. Hasegawa A, Nakamura J, Kiso M (1986) J Carbohydr Chem 5:11
259. Bennett S, von Itzstein M, Kiefel MJ (1994) Carbohydr Res 259:293
260. Park WKC, Meunier SJ, Zanini D, Roy R (1994) Selective deprotection of thiolacetates with hydrazinium acetate. XVIIth International Carbohydrate Symposium (Ottawa) p 306; (1995) Carbohydr Lett 1:179
261. Hasegawa A, Morita M, Ito Y, Ishida H, Kiso M (1990) J Carbohydr Chem 9:369
262. Angus DI, von Itzstein M (1995) Carbohydr Res 274:279
263. Smalec B, von Itzstein M (1995) Carbohydr Res 266:269
264. Kanie O, Nakamura J, Kiso M, Hasegawa A (1987) J Carbohydr Chem 6:105
265. Kanie O, Nakamura J, Itoh Y, Kiso M, Hasegawa A (1987) J Carbohydr Chem 6:117
266. Eisele T, Toepfer A, Kretzschmar G, Schmidt RR (1996) Tetrahedron Lett 37:1389
267. Ogura H, Fujita H, Furuhata K, Itoh M, Shitori Y (1986) Chem Pharm Bull 34:1479
268. Supp M, Rose U, Brossmer R (1980) Hoppe-Seyler's Z Physiol Chem 361:338
269. Tropper FD, Andersson FO, Braun S, Roy R (1992) Synthesis: 618
270. Paulsen H, Matschulat P (1991) Liebigs Ann Chem: 487
271. Nagy JO, Bednarski MD (1991) Tetrahedron Lett 32:3953
272. Corfield AP, Schauer R, Wember M (1979) Biochem J 177:1
273. Kajihara Y, Ebata T, Koseki K, Kodama H, Matsushita H, Hashimoto H (1995) J Org Chem 60:5732
274. Ichikawa Y, Shen G-J, Wong C-H (1991) J Am Chem Soc 113:4698
275. Ichikawa Y, Liu JL-C, Shen G-J, Wong C-H (1991) J Am Chem Soc 113:6300
276. Schachter H (1994) Molecular cloning of glycosyltransferase genes. In: Fukuda M, Hindsgaul O (eds) Molecular glycobiology. Oxford University Press, Oxford, p 88
277. Williams MA, Kitagawa H, Datta AK, Paulson JC, Jamieson JC (1995) Glycoconjugate J 12:755
278. Khan SH, Hindsgaul O (1994) Chemical synthesis of oligosaccharides. In: Fukuda M, Hindsgaul O (eds) Molecular glycobiology. Oxford University Press, Oxford, p 206
279. Ichikawa Y, Wong C-H (1994) Topochemistry and inhibition of selectin-mediated cell adhesion: Chemical-enzymatic synthesis of inhibitors related to E-selectin recognition. In: Bock K, Clausen H (eds) Complex carbohydrates in drug research. Structural and functional aspects. Munksgaard, Copenhagen, p 118
280. Wong C-H (1995) Topochemistry and inhibition of carbohydrate-mediated cell adhesion. In: Jolles G, Dean PM (eds) New perspectives in drug design. Academic Press, London, p 35
281. Ichikawa Y, Lin Y-C, Dumas DP, Shen G-J, Garcia-Junceda E, Williams MA, Bayer R, Ketcham C, Walker LE, Paulson JC, Wong C-H (1992) J Am Chem Soc 114:9283
282. DeFrees SA, Kosch W, Way W, Paulson JC, Sabesan S, Halcomb RL, Huang D-H, Ichikawa Y, Wong C-H (1995) J Am Chem Soc 117:66
283. Gross HJ, Brossmer R (1995) Glycoconjugate J 12:739
284. Gross HJ, Brossmer R (1987) Glycoconjugate J 4:145

285. Musser JH (1992) Ann Rep Med Chem 27:301
286. Petitou M (1994) Drugs based on carbohydrates. Past and future. In: Kováč P (ed) Synthetic oligosaccharides. Indispensable probes for the life sciences. ACS Symposium Series 560. American Chemical Society, Washington, p 19
287. von Itzstein M, Smalec B (1994) Today's Life Science 6:22
288. Mulligan MS, Paulson JC, DeFrees S, Zheng Z-L, Lowe JB, Ward PA (1993) Nature 364:149
289. Borman S (1993) Chem. Eng. News June 28:27

Synthesis of Oligosaccharides of Bacterial Origin Containing Heptoses, Uronic Acids and Fructofuranoses as Synthetic Challenges

Stefan Oscarson

Department of Organic Chemistry, Arrhenius Laboratory, Stockholm University, S-106 91 Stockholm, Sweden

In this article, syntheses of bacterial oligosaccharides containing additional synthetic challenges are presented. In the first part, syntheses of L-*glycero*-D-*manno*-heptopyranosyl-containing oligosaccharides are reported. Synthesis of the heptose trisaccharide structures from the core region of lipopolysaccharides from *Salmonella* and *Haemophilus* bacteria are described together with larger fragments containing hexoses as well. In the second part, development of reactive β-selective glucuronic acid thioglycoside donors is presented. These donors, promoted by DMTST, are used to prepare disaccharide structures corresponding to the repeating unit of the capsular polysaccharide from *Streptococcus pneumoniae* type 3 and to parts of the capsular polysaccharide of *Cryptococcus neoformans*. In the third and last part, stereoselective synthesis of α- and β-D-fructofuranosides using thioglycoside donors are discussed. With participating benzoyl groups and DMTST as promoter, excellent yields of α-linked fructofuranosyl disaccharides are obtained. Application of the internal aglycon delivery approach, with the aglycon tethered to the β-face of the fructofuranosyl thioglycoside donor as part of a 3-*O*-*p*-methoxybenzylidene acetal, produced stereospecifically high yields of β-linked fructofuranosyl disaccharides, inter alia, structures from the *Haemophilus influenzae* type e capsular polysaccharide, after activation of the tethered intermediates with DMTST.

Table of Contents

Topics in Current Chemistry, Vol. 186

1 General Introduction

Because of the polyhydroxy nature of saccharides, together with the many stereogenic centres present, inter alia, the one in the anomeric centre which give rise to α- and β-glycosidic bonds, carbohydrate structures are very complex. In man, mainly eight monosaccharide building blocks, galactose, glucose, fucose, mannose, 2-amino-2-deoxy-glucose, 2-amino-2-deoxy-galactose, glucuronic acid and neuraminic acid, are used to form all carbohydrate structures, but a vast number of permutations and thereby biological information are still possible even in short sequences compared to proteins and nucleic acids. From a synthetic point of view, the limited size of oligosaccharides necessary for biological activity is an advantage, but the inherent complexity which governs these limited sizes makes the synthesis of oligosaccharides a real challenge, for which a common automated approach using solid phase synthesis is still not available. In carbohydrates of microbial origin, a further complexity is introduced, namely the presence of an almost endless variety of monosaccharide moieties which are not commercially available, and, in a synthesis, these have to be constructed and their protecting group and glycosylation properties investigated prior to the assembly of the oligosaccharide. Various substituents such as phosphates, sulfates, acyl and acetal groups add further complexity to carbohydrates from all sources. In this article, the syntheses of oligosaccharides containing various types of additional challenges to the already difficult task of synthesising oligosaccharides are presented. In the first part, synthesis of L-*glycero*-D-*manno*-heptopyranose-containing oligosaccharides is reported, in the second part synthesis of glucuronic acid-containing oligosaccharides is described, and, in the last part, synthesis with fructofuranosidic donors is discussed. The target oligosaccharides are used in biological experiments to investigate their interaction with various proteins, e.g. receptors, lectins and antibodies. The results are used in the characterisation of bacteria, in the making of vaccines, and to explore other ways to avoid bacterial diseases.

2 Synthesis of L-*glycero*-D-*manno*-Heptopyranose-Containing Oligosaccharides

2.1 Introduction

Glycero-D-*manno*-heptose is a monosaccharide residue found in bacterial polysaccharides, predominantly in the inner core region of lipopolysaccharides of Gram-negative bacteria. The L-*glycero* form is the most abundant, but the D-*glycero* isomer has also been found [1, 2].

The number of syntheses of heptose-containing oligosaccharides described in the literature is rather limited and comes almost exclusively from four different groups, Paulsen et al., Zamojski et al., van Boom et al. and ours, and have been discussed in a number of dissertations from these groups [3–9]. In Table 1, a summary of synthesised oligosaccharide structures containing two or more L-*glycero*-D-*manno*-heptopyranose residues is given.

The syntheses are performed in work mainly connected with the bacteria *Salmonella*, *Neisseria* and *Haemophilus*, and the synthesised structures correspond to lipopolysaccharide core structures of these bacteria (Figs. 1–3).

Table 1. Synthesised structures containing two or more L-*glycero*-D-*manno*-heptopyranosyl residues

Structure	Ref.
L-α-D-Hep*p*-(1 → 3)-L-α-D-Hep*p*	[10, 11]
L-α-D-Hep*p*-(1 → 6)-L-α-D-Hep*p*	[12]
L-α-D-Hep*p*-(1 → 7)-L-α-D-Hep*p*	[13, 14]
L-α-D-Hep*p*-(1 → 3)-L-α-D-Hep*p*-(1 → 5)-α-Kdo*p*	[7, 15]
L-α-D-Hep*p*-(1 → 3)-L-α-D-Hep*p*-(1 → 5)-β-Kdo*p*	[7]
L-α-D-Hep*p*-(1 → 3)-L-α-D-Hep*p*-(1 → 5)-α-Kdo*p*-(2 → 6)-β-D-Glc*p*Nhm-(1 → 6))-D-Glc*p*Nhm	[7]
α-D-Glc*p*NAc*p*-(1 → 2)-L-α-D-Hep*p*-(1 → 3)-L-α-D-Hep*p*	[16]
α-D-Glc*p*NAc*p*-(1 → 2)-[α-D-Glc*p*-(1 → 3)]-L-α-D-Hep*p*-(1 → 3)-L-α-D-Hep*p*	[5]
α-D-Glc*p*-(1 → 3)-[L-α-D-Hep*p*-(1 → 7)]-L-α-D-Hep*p*	[13, 18]
L-α-D-Hep*p*-(1 → 7)-L-α-D-Hep*p*-(1 → 3)-L-α-D-Hep*p*	[7, 19, 20]
L-α-D-Hep*p*-(1 → 2)-L-α-D-Hep*p*-(1 → 3)-L-α-D-Hep*p*	[9]
L-α-D-Hep*p*-(1 → 3)-[L-α-D-Hep*p*-(1 → 7)]-L-α-D-Hep*p*	[18]
β-D-Gal*p*-(1 → 2)-L-α-D-Hep*p*-(1 → 2)-L-α-D-Hep*p*-(1 → 3)-L-α-D-Hep*p*	[9]
α-D-Gal*p*-(1 → 2)-L-α-D-Hep*p*-(1 → 2)-L-α-D-Hep*p*-(1 → 3)-L-α-D-Hep*p*	[9]
α-D-Glc*p*-(1 → 3)-[L-α-D-Hep*p*-(1 → 7)]-L-α-D-Hep*p*-(1 → 3)-L-α-D-Hep*p*	[21]
α-D-Gal*p*-(1 → 3)-α-D-Glc*p*-(1 → 3)-[L-α-D-Hep*p*-(1 → 7)]-L-α-D-Hep*p*-(1 → 3)-L-α-D-Hep*p*	[21]
α-D-Gal*p*-(1 → 3)-[α-D-Gal*p*-(1 → 6)]-α-D-Glc*p*-(1 → 3)-[L-α-D-Hep*p*-(1 → 7)]-L-α-D-Hep*p*-(1 → 3)-L-α-D-Hep*p*	[21]

```
                                                            P              PPOCH2CH2NH2
                                                            |              |
                                                            4              4
α-D-Glcp-(1→2)-α-D-Galp-(1→3)-α-D-Glcp-(1→3)-α-Hepp-(1→3)-α-Hepp-(1→5)-Kdo
   2                              6                      7                        4
   ↑                              ↑                      ↑                        ↑
   1                              1                      1                        2
α-D-GlcpNAc                   α-D-Galp                α-Hepp          α-Kdo-(2→4)-α-Kdo
```

Fig. 1. Structure of the *Salmonella* Ra core

```
     R1             R2        β-D-Glcp-(4←1)-R3
     1              1               1
     ↓              ↓               ↓                 R1, R2, R3 = short
                                                      oligosaccharides
     2              3               4
α-Hepp-(1→2)-α-Hepp-(1→3)-α-Hepp-(1→5)-Kdo
```

Fig. 2. Generalised structure of the dephosphorylated *Haemophilus influenzae* and *ducreyi* lipooligosaccharide without the lipid A part

```
                       R1       β-D-Glcp-(4←1)-R2
                       1               1
                       ↓               ↓               R1, R2 = short
                                                       oligosaccharides
                       3               4
α-D-GlcpNAc-(1→2)-α-Hepp-(1→3)-α-Hepp-(1→5)-Kdo
```

Fig. 3. Generalised structure of the dephosphorylated *Neisseria meningitidis* oligosaccharide part of the LPS

The oligosaccharides are generally synthesised as glycosides at the reducing end, sometimes as methyl glycosides but most often as spacer glycosides, to be able to form neo-glycoconjugates of the oligosaccharides. This allows their further use in a variety of biological experiments apart from inhibition experiments, e.g. in affinity chromatography and immunisation. Paulsen et al. use the allyl glycoside as a linking arm. The allyl group disqualifies benzyls as protecting groups, but is smoothly converted by addition, to the double bond, of aminomercaptoethanol into an amino-derivative [22], which then can be further coupled to different carriers. We and van Boom et al. use spacers already from the beginning containing an amino group in the structure, a 2-(4-aminophenyl)ethanol or a 3-aminopropanol, respectively.

2.2 Synthesis of L-*glycero*-D-*manno*-Heptopyranose Monosaccharide Derivatives

Through the development of effective methods for the stereoselective synthesis of L-*glycero*-D-*manno*-heptopyranose, the synthesis of oligosaccharide structures, including this residue, became possible. Several good methods appeared at about the same time.

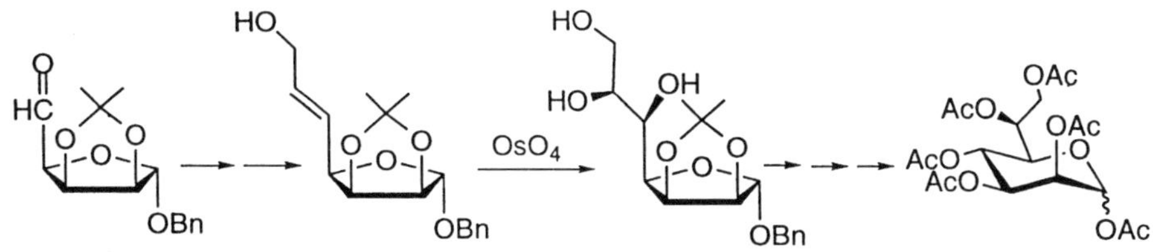

Scheme 1

Brimacombe and Kabir [23] used a two-carbon Wittig-type elongation of a lyxose derivative, followed by a *cis*-hydroxylation with osmium tetroxide according to Kishi's empirical rule [24] (or a Sharpless epoxidation followed by epoxide-opening with hydroxide [25]) to give predominantly the L-*glycero*-D-*manno* stereoisomer (Scheme 1). The stereoselectivity was high (7:1) and the synthesis easily performed on a large scale. Since a two-carbon elongation is made, the synthesis has to start from a furanose system, and therefore a transformation in a number of steps into a pyranosidic form is necessary before use in the oligosaccharide synthesis.

A more direct approach to the heptopyranose is the one-carbon elongation, using various Grignard reagents, at C-6 of a mannopyranose derivative (Scheme 2). Complexation of the reagent with the ring oxygen ensures a high stereoselectivity in the reaction, to give predominantly the L-*glycero* configuration in the product according to an anti-Cram addition [26].

The use of allyl or benzyl chloromethyl ether as the precursor gives directly the 7-*O*-allyl or benzyl-protected derivative [20, 27, 28]. These alkoxy methyl Grignard reagents are, however, very labile, due to the possibility of internal α-elimination [29], and the reproducibility of the reaction can be a problem. A two-step procedure, using (isopropoxy- or phenyldimethylsilyl)methylmagnesium chloride as Grignard reagents followed by an oxidative cleavage of the carbon-silicon bond with peroxides or peracids to give the hydroxyl compound, has been developed by Tamao et al. [30] and Fleming et al. [31], and was

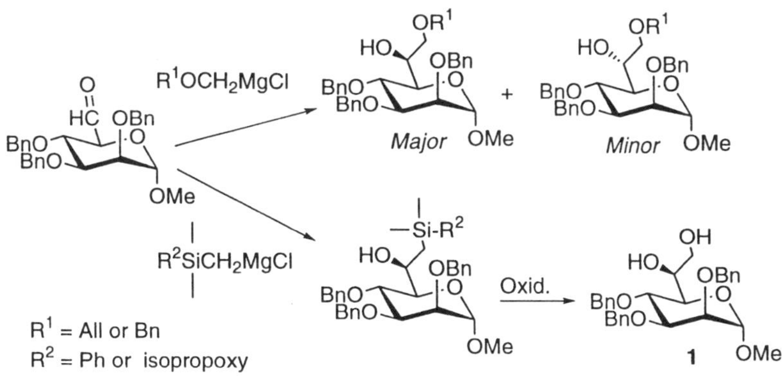

Scheme 2

introduced into the synthesis of heptoses by van Boom et al. [12, 32]. These reagents, when applied on the mannose precursor methyl-2,3,4-tri-*O*-benzyl-α-D-manno-hexodialdo-1,5-pyranoside, give high yields of almost exclusively the L-*glycero*-α-D-*manno*-heptopyranoside stereoisomer (Scheme 2). These reagents have also been further developed by van Boom into protecting groups in carbohydrates [33] and into a new type of reagent which allows an alternative oxidative introduction of the hydroxyl group [34].

Protecting group manipulations, to allow the formation of heptopyranosyl donors and acceptors, can be performed either before the carbon-elongation on mannose derivatives, as long as protecting groups compatible with the Grignard and the oxidation reaction conditions are used, or after the elongation, on heptopyranose derivatives. Both approaches have been used, the former having the advantage of fewer manipulations at the heptose level, the latter the advantage of the need of stereoselective formation and determination of configuration of only one heptose derivative, since the heptopyranoside derivative **1**, easily obtained in multigram-scale from the two-step Grignard reaction, can conveniently be transformed into almost any acceptor or donor necessary (see below).

2.3 Synthesis of *Salmonella* Ra Core Structures

Many syntheses in Table 1 are of structures from the core of the lipopolysaccharide from *Salmonella*, which contains a linear triheptoside structure L-α-D-Hep*p*-(1$\rightarrow$7)-L-α-D-Hep*p*-(1$\rightarrow$3)-L-α-D-Hep*p* linked to a Kdo residue at the reducing end and branched with a glucose moiety in the 3-position of the middle heptose residue (Fig. 1, p. 174) [35]. Similar inner core structures with (1$\rightarrow$3)- and (1$\rightarrow$7)-linked heptopyranosyl residues have also been proposed for *Citrobacter* [36].

In our syntheses of *Salmonella* structures [10,13,19,21], the heptose derivative **1** was used as a precursor for all of the heptose residues found in the target molecules. An acceptor precursor for later formation of the (1$\rightarrow$7)-linkage was obtained by regioselective silylation of the primary hydroxyl group followed by benzoylation of OH-6 (Scheme 3). Van Boom et al. postpone the oxidation step

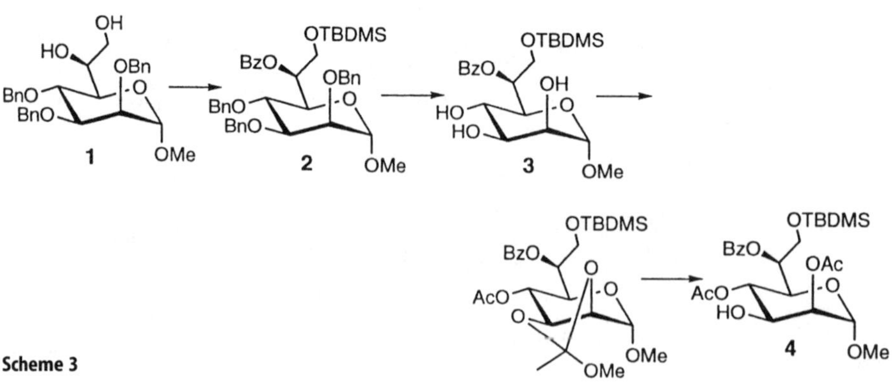

Scheme 3

and use the carbon-silica group as a temporary 7-*O*-protecting group [18]. This is possible with the phenyl derivative but not with the isopropoxy derivative, which is too labile and not compatible with glycosylation conditions.

To obtain acceptors with a free 3-OH, an approach exploiting the regioselective opening of cyclic orthoesters was used. Debenzylation of **2** afforded the 2,3,4-triol **3**, from which the 2,3-cyclic orthoester was formed. This was followed by acetylation of the 4-hydroxyl group to give a fully protected compound. Acidic opening of the orthoester then gave exclusively the 3-OH acceptor **4** (Scheme 3).

The regioselective opening of five-membered ring orthoesters formed from *cis-vic*-diols to give the axial ester-equatorial hydroxyl products was first studied by Lemieux and Drigeuz [37]. With the use of chlorinated orthoesters [38] to give the axial chloroacetate removable in the presence of other acyl groups, a very flexible protecting group scheme, inter alia for the formation of branched oligosaccharides, can be constructed. With six-membered ring (4, 6-) orthoesters, most of the regioselectivity is lost [39], but the easy procedure and the formation of only monoesters still makes it an alternative as a regioselective esterification method of 4,6-diols [40, 41].

Other ways to obtain 3-OH acceptors are the use of mannose precursors with a temporary 3-*O*-allyl protecting group in the Grignard reaction [16, 18, 20], or phase transfer benzylation [42] of a heptopyranosyl 2,3-diol, or benzylation of a tin-activated [43] heptofuranosyl 2,3-diol, both last methods giving preferentially the 2-*O*-benzyl derivative [27, 15]. However, if the tin-assisted alkylation is performed on a heptopyranosyl 2,3-diol, the 3-*O*-alkyl derivative is the main product [5, 28].

As heptosyl donor, thioglycoside derivatives were preferentially chosen. In part because of our familiarity with this type of donors, but mainly because of the flexibility of thioglycosides as donors or as donor precursors [44]. A large number of promoters are known to activate thioglycosides, and, if all these

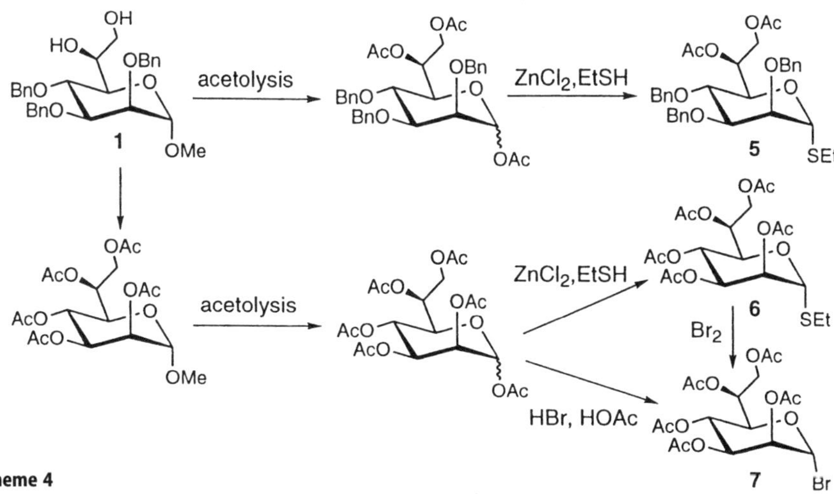

Scheme 4

fail, the thioglycosides are conveniently (using mild procedures) transformed into other type of donors, e.g. halides, sulfoxides and trichloroacetimidates for further testings. Bromide, chloride and trichloroacetimidate heptosyl donors have also been used. Acetolysis of methyl glycoside derivatives, e.g. **1**, followed by treatment with a Lewis acid and mercaptoethanol, smoothly gave ethyl thioglycosides [19] (Scheme 4). Thioglycoside derivatives cannot be used as the precursor in the formation of heptoses using the silyl Grignard reagents because of the oxidative conditions necessary to cleave the silicon-carbon bond. However, ethyl thiomannopyranosides have been used as precursors in reactions with alkoxymethyllithium reagents to give thioheptopyranosides, but then most of the stereoselectivity in the carbon elongation reaction was lost, and about equal amounts of the D-and L-*glycero*-configuration were obtained [5].

With the supply of both heptose acceptors and donors, the assembly of oligosaccharide structures could be accomplished. Glucosylation of acceptor **4** with methyl 2,3,4,6-tetra-*O*-benzyl-1-thio-β-D-glucopyranoside as donor and methyl triflate as promoter, using diethyl ether as solvent to improve α-selectivity, gave the α-(1$\rightarrow$3)-linked disaccharide (Scheme 5). Removal of the silyl group and a silver triflate-promoted glycosylation of the resulting 7-OH with acetobromoheptopyranose (**7**) gave the branched trisaccharide, which was deprotected to give the target trisaccharide **8** [13]. The fully acetylated heptosyl donors are prone to form the anomeric orthoester, and special care to avoid this by-product must be taken, e.g., in the coupling above the right amount of collidine was essential for the outcome of the reaction. They are also unreactive, e.g., if the thioglycoside **6** was used as donor with methyl triflate

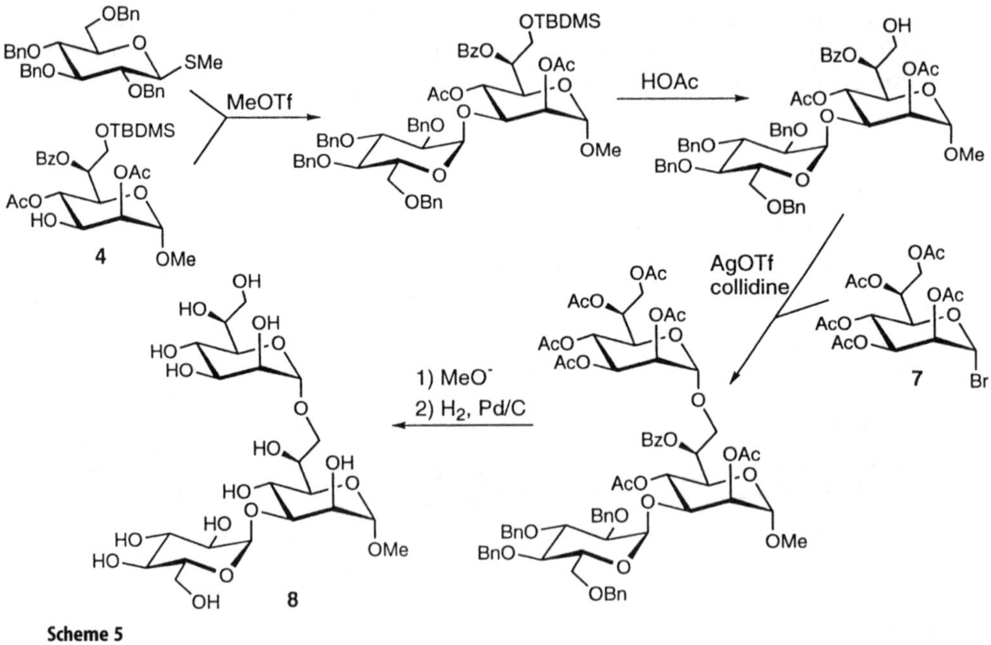

Scheme 5

as promoter the main product was the 7-*O*-methylated disaccharide. To improve the reactivity and lessen the tendency for orthoester formation, Paulsen and Heitmann introduced the use of 2-*O*-acetyl-3,4,6,7-tetra-*O*-benzyl-L-*glycero*-α-D-*manno*-heptopyranosyl chloride as a more effective α-selective donor [14]. This donor was also used by van Boom et al. in their assembly of the same trisaccharide **8** [18].

Using the same methodology, syntheses of larger fragments of the core structure were then performed, now as spacer derivatives [19, 21]. The spacer arm was introduced by an NIS-promoted coupling between the thioglycoside **5** and 2-(*p*-trifluoroacetamidophenyl)ethanol. Although the donor **5** lacks a 2-*O*-participating group, the *manno* configuration ensures the preferential formation of α-linked products. With hindered aglycons, generally only the α-product is formed, but with primary alcohols the β-product can also be found. With the spacer alcohol, a substantial amount of the β-linked derivative was formed, wherefore the α/β-mixture obtained was treated with BF_3-etherate in acetonitrile to allow the anomerisation of the β-form into the more stable α-form [45] to give an equilibrium with almost exclusively α-linked product **9**. Debenzylation and the orthoester procedure then gave the 3-OH acceptor **10**, after which coupling with thioethyl donor **11**, with a temporary 7-*O*-silyl protecting group, using DMTST as promoter in diethyl ether, gave exclusively the α-linked disaccharide **12** (Scheme 6). Removal of the silyl ether followed by silver triflate-promoted coupling with benzobromoheptopyranose then gave the α-(1 → 7)-linked trisaccharide **13**. This time the problem with orthoester forma-

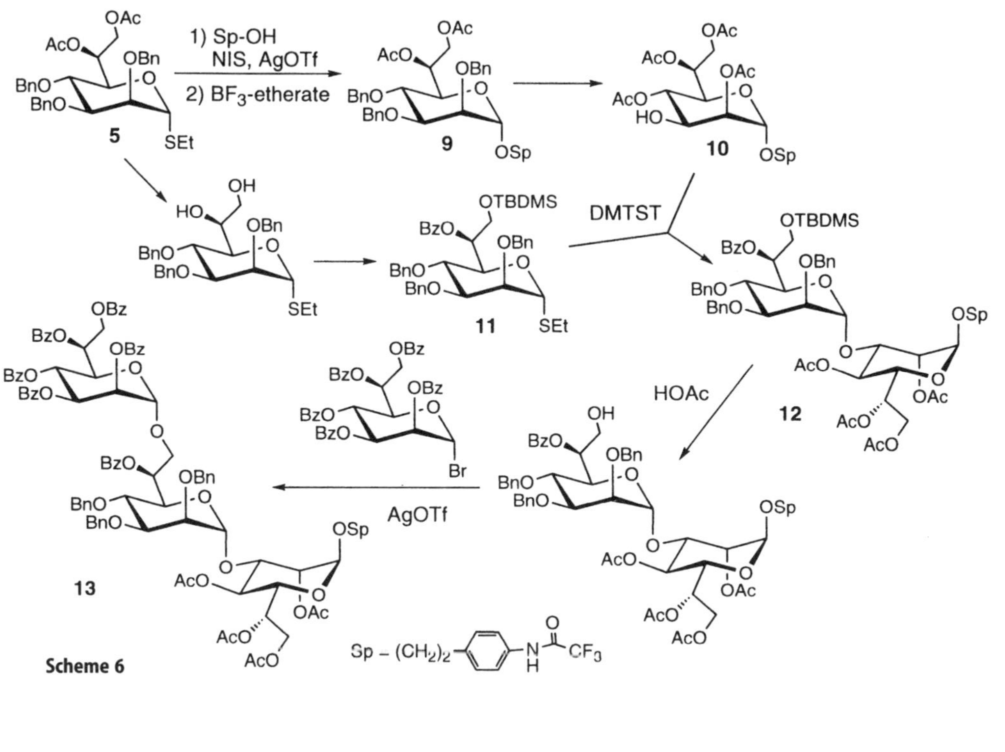

Scheme 6

tion during glycosylation found for the acetylated donors was avoided by the use of a benzoylated donor, as found earlier [46].

Derivative **13** was deprotected to give the triheptoside backbone of the *Salmonella* core [19], but the protecting pattern in **13** was also chosen to allow further synthesis of core structures. Thus, debenzylation of **13** to give the 2′,3′,4′-triol, followed by the orthoester protocol, once more gave a 3′-OH acceptor (**14**), which was reacted with methyl 2,3,4,6-tetra-*O*-benzyl-1-thio-*β*-D-glucopyranoside and DMTST to give the tetrasaccharide **15** (Scheme 7).

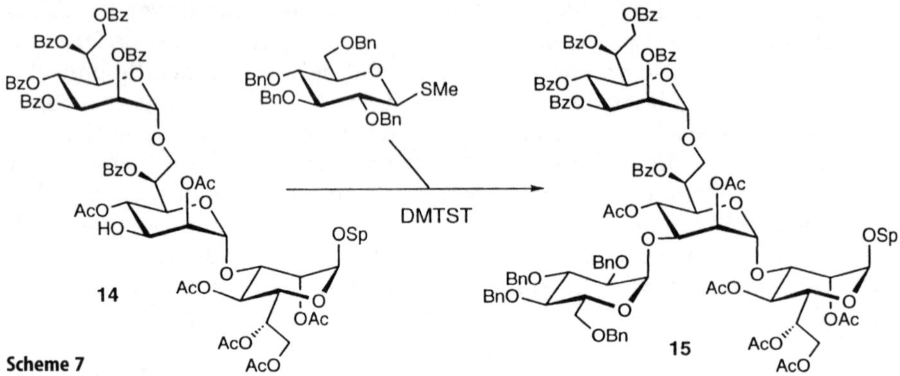

Scheme 7

A trisaccharide, corresponding to the hexose part linked to the heptose part in the Ra core, had earlier been synthesised as its methyl thioglycoside in our laboratory [47]. This trisaccharide was now tried as donor in a DMTST-promoted coupling to **14** (Scheme 8). A major product was isolated but was found not to be a hexasaccharide but a tetrasaccharide. The same tetrasaccharide was formed if

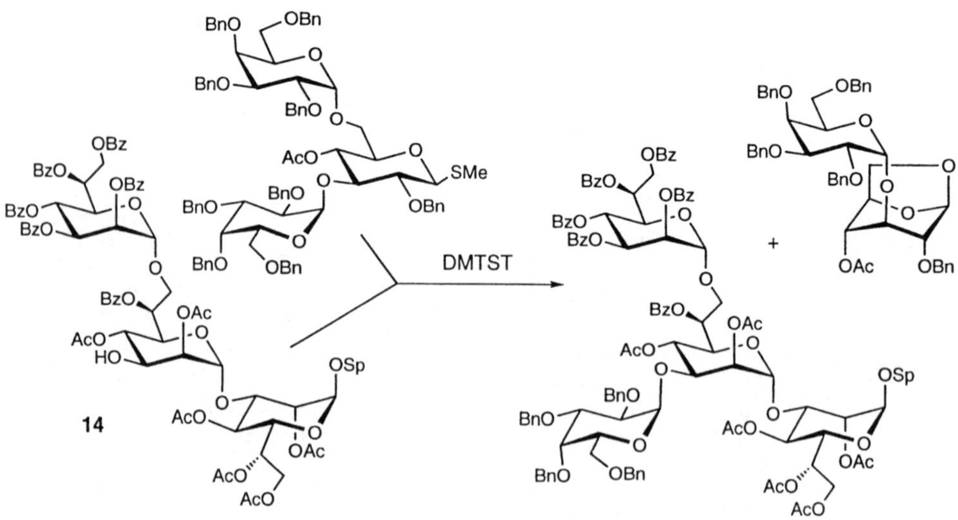

Scheme 8

the disaccharide methyl 6-(2,3,4,6-tetra-*O*-benzyl-α-D-galactopyranosyl)-4-*O*-acetyl-2,3-di-*O*-benzyl-1-thio-α-D-glucopyranoside was used as donor. A by-product from these reactions was isolated, analysed, and found to be the 1,6-anhydro derivative of the glucopyranosyl residue. This is a rather common by-product from donors with a conceivable leaving group at O-6, in this reaction the galactopyranosyl unit, which thus became an activated galactosyl donor and coupled to **14**.

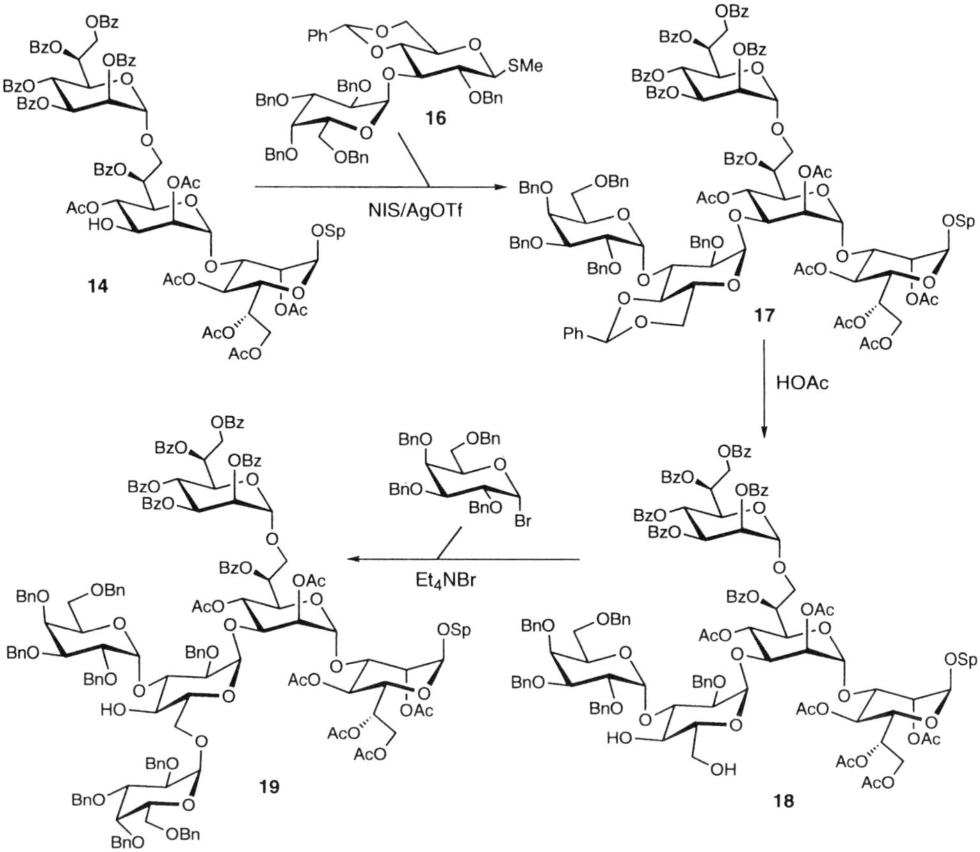

Scheme 9

To prevent this decomposition of the donor derivative, compound **16**, with a 4,6-*O*-benzylidene acetal that restrains the 6-oxygen from interaction with the anomeric centre but allows the later introduction of the 6-*O*-linked galactosyl moiety, was chosen as donor. Now, coupling with acceptor **14** gave the desired pentasaccharide **17**, this time using NIS/silver triflate as promoter. Debenzylidenation followed by a halide-assisted coupling between the resulting 4,6-diol **18** and 2,3,4,6-tetra-*O*-benzyl-α-D-galactopyranosyl bromide gave, in a stereo- and regiospecific reaction, the α-(1 $\rightarrow$ 6)-linked hexasaccharide **19** (Scheme 9). Deprotection of compounds **15, 17** and **19** then gave spacer oligosaccharides ready for biological experiments [21].

2.4 Synthesis of *Haemophilus influenzae* and *ducreyi* Lipopolysaccharide Structures

The other synthesised oligosaccharides shown in Table 1 are structures from the lipopolysaccharide core of *Haemophilus influenzae* and *ducreyi* and *Nesseria menigitidis* (Figs. 2 and 3, p. 174). The lipopolysaccharides from these bacteria show a number of similarities. They both lack the polysaccharidic *O*-antigen chain with its repeating units. They also show a lot of microheterogeneity, both in the carbohydrate structure and in the phosphorylation pattern. This phase variation, together with the presence of structures common in human tissues, is an efficient way for the bacteria to elude the immune system during infection. The heterogeneity makes structural elucidation of these structures very complicated, but, with the access to new techniques, both separational and analytical, proposed structures have been reported (Figs. 2 and 3) [48–50]. As can be seen from the figures there are also structural similarities. The tetrasaccharide motif, β-D-Glc*p*-(1→4)-[L-α-D-Hep*p*-(1→3)]-L-α-D-Hep*p*-(1→5)-Kdo, is common to both core structures.

Compared to the *Salmonella* core, a totally different heptose part is found in the *Haemophilus* core, and new heptose acceptors allowing coupling in the 2- and the 4-position, and preferably also branched 2,3- and 3,4-structures, had to be constructed. Debenzylation of spacer derivative **9** (Scheme 6, p. 179) and isopropylidenation of the resulting 2,3,4-triol gave a 4-OH acceptor (**20**) with the capacity to yield branched structures through later selective removal of the acetal. The same approach to 4-*O*-linked heptose derivatives has been used by van Boom et al. [51]. The isopropylidene group can also be introduced already at the hexose level before the carbon elongation reaction [5, 28].

Silver triflate-promoted coupling of acceptor **20** with benzobromolactose or benzobromocellobiose smoothly gave the two trisaccharides **21** and **22** respectively (Scheme 10), which were deprotected to give core structure oligosaccharides [9, 52].

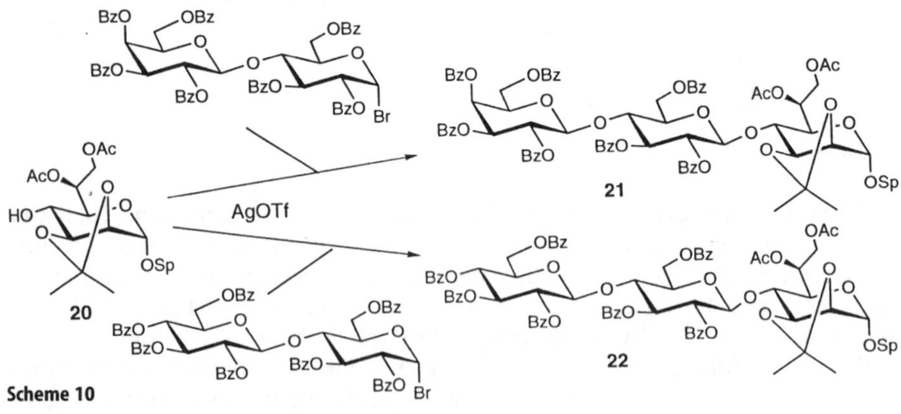

Scheme 10

A latent 2-OH acceptor is the earlier mentioned 2-*O*-acetyl-3,4,6,7-tetra-*O*-benzyl-L-*glycero*-α-D-*manno*-heptopyranosyl chloride obtained from the

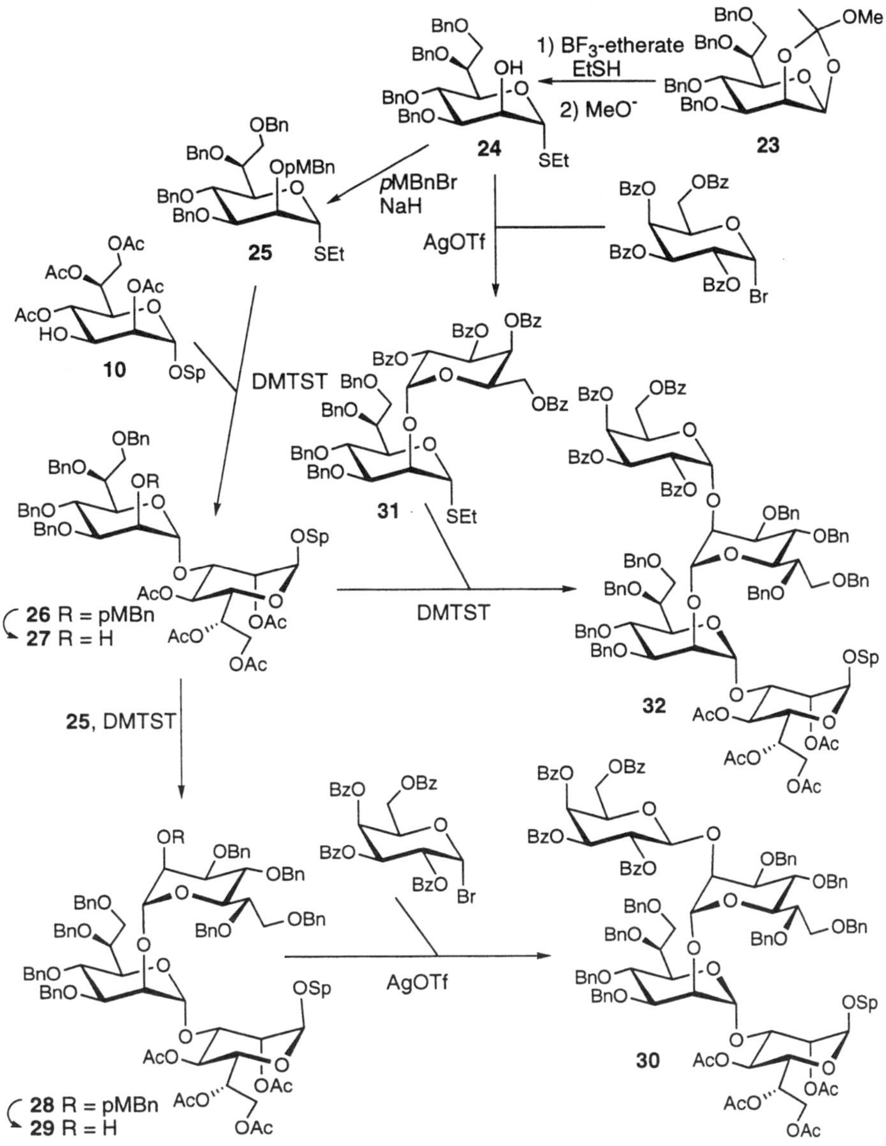

Scheme 11

orthoester **23** [14]. Since permanent acyl protecting groups were already present, another temporary 2-*O*-protecting group had to be introduced. Rearrangement of the orthoester **23** in the presence of excess ethylmercaptan gave the 2-*O*-acetyl ethyl thioglycoside, which was deacetylated and *p*-methoxybenzylated to give **25** (Scheme 11).

DMTST-promoted coupling of donor **25** with the earlier used 3-OH acceptor **10** gave the heptoside disaccharide **26**, from which the *p*-methoxybenzyl group was removed by DDQ-treatment to give the 2-OH acceptor **27** (Scheme 11). Another coupling with the same donor **25** then gave derivative **28**, corresponding to the triheptoside backbone of the *Haemophilus* core. Now, using the same protocol, a further 2-*O*-elongation could be accomplished. DDQ-treatment of compound **28** gave acceptor **29**, which was glycosylated with benzobromogalactose to give the β-(1 $\rightarrow$ 2)-linked tetrasaccharide **30**.

Initially, this tetrasaccharide was designed to be synthesised through a convergent strategy employing acceptor **27** and a disaccharide donor obtained from the coupling of the thioglycoside **24** and benzobromogalactose. However, when this latter glycosylation was attempted, the only product obtained from the rather sluggish reaction was, in spite of the participating benzoyl group in the donor, the α-(1 $\rightarrow$ 2)-linked disaccharide **31** (Scheme 11). Various donors and promoters were tried, but no β-linked product was isolated [9]. It is occasionally found that the 1,2-*cis*-product is formed in preference to the *trans*-form although a 2-*O*-participating group is used in the coupling, especially with galactosyl donors preferring the α-configuration (see ref. [53] for examples). Van Boeckel et al. [54] have more closely studied one of these examples and shown, by using enantiomeric donors, that unfavourable steric interactions prevented the acceptor from attack from the β-face. Here, the problem could very surprisingly be solved as described above by using the acceptor **29** instead, in which the release in steric strain compared to acceptor **27** is not obvious.

The disaccharide donor **31** was coupled to acceptor **27** to give tetrasaccharide **32**. Deprotection of **28, 30** and **32** then gave two *Haemophilus* core structure oligosaccharides and an analogue thereof [9, 52].

2.5 Synthesis of 3,4-Branched Heptose Structures

The synthesis of 3,7-branched core structures of *Salmonella* offered no problems (see Sect. 2.3). The synthesis of a 2,3-branched structure of the *Neisseria* core has also been described [5]. A synthetic 3,4-branched structure, β-D-Glc*p*-(1 $\rightarrow$ 4)-[L-α-D-Hep*p*-(1 $\rightarrow$ 3)]-L-α-D-Hep*p*-spacer, has been used in immunological experiments aiming towards vaccines against meningococcal diseases [55], but its synthesis has to our knowledge not been published. When we attempted the synthesis of similar 3,4-branched structure from the *Haemophilus* core, severe problems were encountered.

Removal of the isopropylidene group from **22** (Scheme 10, p. 182) gave a 2,3-diol, which was protected in the 2-position with an acetyl group (through the orthoester procedure) or a benzyl group (through regioselective reductive opening of a stereo-selectively formed *endo*-benzylidene acetal [56,57]). However, all glycosylations with these two acceptors and various heptosyl donors and promoters failed and gave only decomposition of the donors and no tetrasaccharide product. The reversed order of introduction of the two branching units was then tried, but once more no 3,4-branched structures was obtained from the 4-OH (1 $\rightarrow$ 3)-linked heptoside disaccharide acceptor. Also, several model experiments were performed on methyl α-D-mannopyranoside to try to construct a β-D-Glc*p*-(1 $\rightarrow$ 4)-[α-D-Man*p*-(1 $\rightarrow$ 3)]-α-D-Man*p* trisaccharide, but with little success [9].

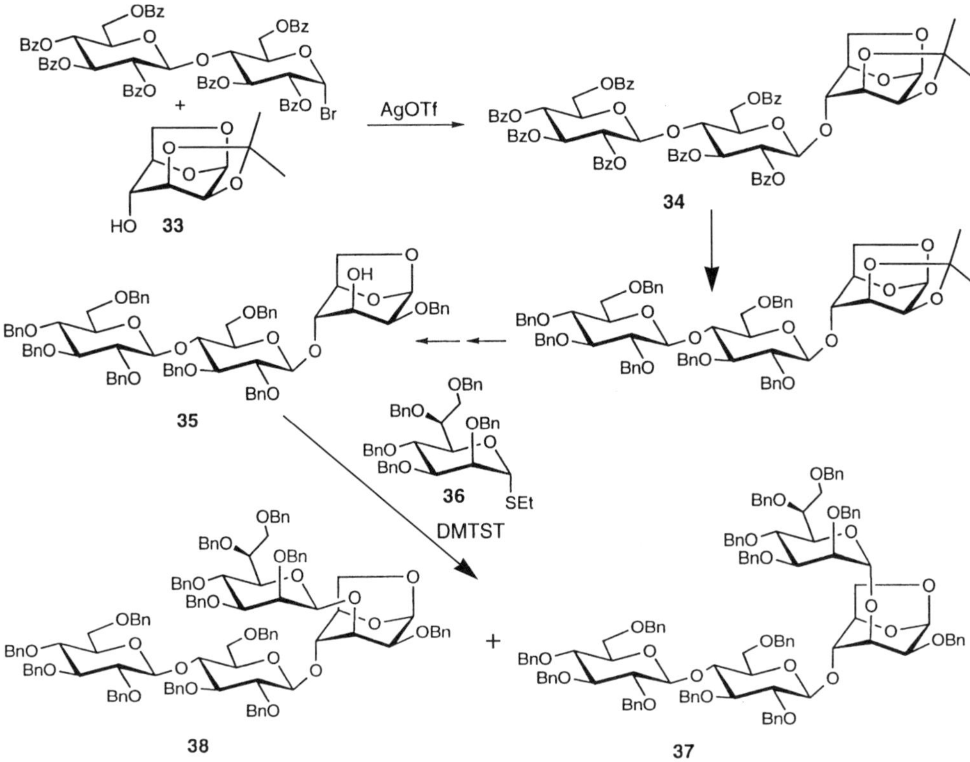

Scheme 12

Apparently, this branched structure is sterically very crowded, and it was considered that a conformational change of the acceptor could lead to a successful synthesis [54, 58]. The 1,6-anhydromannose derivative **33** [59] was therefore synthesised and used as a model acceptor in a silver triflate-promoted glycosylation reaction with benzobromocellobiose to give the trisaccharide **34** (Scheme 12). The benzoyl groups were changed to benzyl groups to enable a carbon-elongation through a Grignard reaction to be performed later, after which the isopropylidene acetal was removed and the 2,3-diol was tin-activated and regioselectively benzylated in the 2-position to give the acceptor **35**. Coupling of this acceptor with the perbenzylated ethyl thioheptoside donor **36** finally gave the desired 3,4-branched structure **37**, but also the β-(1$\rightarrow$3)-linked analogue **38** in about equal amount. Apparently, even this acceptor has problems of attack from the α-side of the donor, so that the normally unfavoured β-anomer is also formed in substantial amounts. Separation of these stereoisomers could, however, easily be performed by silica gel chromatography [60]. Future research will show whether a heptosyl donor with a participating 2-*O*-protecting group will improve the yield of α-linked product, and, more importantly, whether it is possible to perform the one-carbon elongation on the mannose residue of the tetrasaccharide after acetolysis of the 1,6-anhydro bridge and formation of a mannoside, or if it is more feasible to try to construct a 1,6-or a 1,7-anhydroheptopyranose derivative to be used as acceptor in the synthesis of 3,4-branched structures from the *Haemophilus* core.

2.6 Synthesis of Phosphorylated Heptose Structures

Lipopolysaccharide core structures are, as mentioned, phosphorylated. The phosphate groups complicate the separation and purification and consequently the structure elucidation of core structures, and these were therefore formerly often performed on dephosphorylated material. However, with the techniques of today it is generally possible also to establish the phosphorylation pattern of the native structures, giving an additional challenge to the synthetic chemist to synthesise phosphorylated oligosaccharide structures to be used as tools in the evaluation of the biological role of the phosphate groups.

Employing the same strategy and precursors as outlined above (Sects. 2.3 and 4), the phosphorylated heptoside disaccharides **38–40** [10] (Fig. 4), corresponding to *Salmonella* core structures (Fig. 1), were synthesised. A 7-*O*-phosphorylated structure (**41**) of the same disaccharide, corresponding to a *Neisseria* structure, has also been synthesised by van Boom [11]. This derivative was synthesised as the ethanolamine phosphate derivative and the oligosaccharide part was linked via a spacer to a peptide part, altogether re-

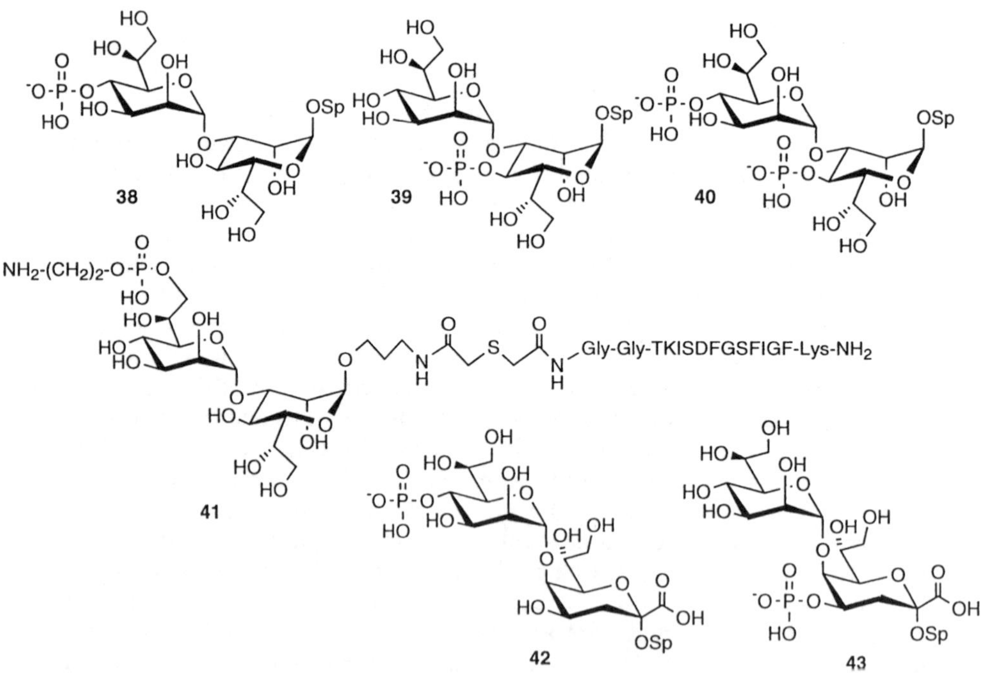

Fig. 4. Structures of synthesised phosphorylated heptose-containing disaccharides

presenting an excellent example of the complex structures which can be made in the laboratory today. Recently, the synthesis of phosphorylated heptosyl-Kdo-derivatives **42** and **43** (Fig. 4) [8] was reported.

3
Synthesis of Uronic Acid-Containing Oligosaccharides

3.1
Introduction

Uronic acids are common components of bacterial polysaccharides, especially in capsular polysaccharides [1, 2]. Glucuronic acid is also one of the components of carbohydrates from man, present in, inter alia, various proteoglycans [61] and metabolites [62]. Hydrolysis as well as glycosylation of uronic acids are strongly influenced by the presence of the electron-withdrawing carboxyl group. Although exceptions have been observed, the general effect is that both cleavage and formation of the glycosidic linkage are made more difficult, resulting, e.g., in the selective hydrolysis and easy isolation of disaccharides with the uronic acid residue at the non-reducing end from uronic acid-containing polysaccharides [63]. In glycosylations, the use of uronic acid donors is often hampered by the low reactivity resulting in low yields and orthoester formation. As a way to circumvent these problems, uronic acid-containing oligosaccharides are often synthesised using non-acidic donors, and the carboxyl group is introduced by a later oxidation step after the glycosylation (for examples see ref. [64]). This requires a donor regioselectively protected with a temporary group in the primary position or a regioselective oxidation method for primary alcohols. The strategy hence puts a lot of restraints on the protecting group pattern, and the subsequent oxidation on the oligosaccharide fragment is not always straightforward (see ref. [65], Sect. 3.3). The access to reactive stereospecific uronic acid donors would therefore be an asset in the synthesis of uronic acid-containing oligosaccharides.

3.2
Development of Reactive β-Selective Glucuronic Acid Donors

Cryptococcus neoformans, a fungus, is an opportunistic organism, which has emerged as one of the major causes of death in patients suffering from AIDS. The fungus is surrounded by a capsule, which is a necessary for the virulence of the organism. The major part of the capsule consists of a polysaccharide containing mannose, xylose, glucuronic acid and acetyl groups. A generalised structure of the polysaccharide is shown in Fig. 5 [66].

As a first attempt to synthesise uronic acid-containing structures from this polysaccharide and from the polysaccharide of *Streptococcus pneumoniae* type 3 (Fig. 6) [67], the common Koenigs-Knorr type donor methyl (2,3,4-tri-*O*-acetyl-α-D-glucopyranosyl bromide)uronate (easily obtained from the 3,6-glucuronolactone and recrystallised from ethanol! [68]) was tried in coupling reactions with the relevant acceptors **44** and **45** (for structures see Scheme 14, p. 189) and

$$
\begin{array}{ccc}
\beta\text{-D-Xyl}p & \beta\text{-D-Xyl}p & \beta\text{-D-Glc}p\text{A} \\
1 & 1 & 1 \\
\downarrow & \downarrow & \downarrow \\
2 & 2 & 2
\end{array}
$$
$$\rightarrow 3)\text{-}\alpha\text{-D-Man}p\text{-}(1 \rightarrow 3)\text{-}\alpha\text{-D-Man}p\text{-}(1 \rightarrow 3)\text{-}\alpha\text{-D-Man}p\text{-}(1 \rightarrow$$

Fig. 5. Generalised structure for the main polysaccharide in *C. neoformans* capsules

$$\rightarrow 3)\text{-}\beta\text{-D-Glc}p\text{A-}(1 \rightarrow 4)\text{-}\beta\text{-D-Glc}p\text{-}(1 \rightarrow$$

Fig. 6. Structure of the repeating unit of the capsular polysaccharide of *Streptococcus pneumoniae* type 3

various promoters with the formation of disaccharide products in very low yields. The need for development of a more reactive β-selective donor was therefore apparent.

A thioglycoside was chosen as the donor type, because of the stability of these donors during most protecting group manipulations, the many promoters known for their activation and their smooth conversion into other types of donors [44]. With uronic acids, one drawback of thio-glycosides might be the susceptibility to oxidants to give the sulfoxide or the sulfone derivative if an oxidation step is necessary to introduce the carboxyl group. The idea was to increase the reactivity of the donor by changing the protecting groups from the deactivating acetyl groups into activating benzyl groups. To avoid the oxidation step, the thioglycoside was synthesised from the above glucuronyl bromide in a two-step procedure involving the formation and rearrangement of an anomeric xanthate to give stereospecifically the β-linked ethyl thioglycoside as described in the literature [69]. After deacetylation, benzylation was attempted, but no conditions could be found for the smooth benzylation of this derivative [70]. Partly because of the possibility of β-elimination, alkylation of uronic esters is troublesome. The use of the acid as starting material facilitates the alkylation [71], but the yields are still not very high.

An alternative route by oxidation of a glucose derivative was therefore selected. The oxidation of ethyl 2,3,4-tri-*O*-benzyl-1-thio-β-D-glucopyranoside was accomplished by a two-step procedure, a DMSO-oxidation using the Pfitzner-Moffat procedure [72] followed by a PDC-oxidation of the aldehyde in the presence of methanol [73] to give the methyl ester **46** directly in good yield. The ^{13}C NMR data for the thioethyl group (δ 15.0, 25.1) in **46** unequivocally showed that no oxidation of the sulfur atom had taken place. This donor was tried in coupling reactions with the two acceptors **44** and **45** using DMTST as promoter and was found to give disaccharides in good yields but with low stereoselectivity (Scheme 14), which could not be improved to any great extent by changing the glycosylation conditions.

Benzyl-protected glucuronic acid derivatives, starting from the methyl glucoside, have been made earlier by Keglevic et al. [74]. After oxidation and hydrolysis, the compound was transformed into the bromide, which also gave low stereoselectivity in couplings. The corresponding α-linked trichloroacetimidate donor was synthesised by Schmidt et al. and used in coupling reactions to give mainly *β*-linked glycoside [75]. In our hands, this derivative gave lower yields compared to donor **46** and still low stereoselectivity in couplings with acceptors **44** and **45**. The synthesis of donor **46**, in a way analogous to ours but using sodium chlorite as oxidant, has recently been claimed by Misra and Roy [76]. However, inspection of the ^{13}C NMR shifts of the thioethyl group in this compound (*δ* 7.0, 42.2) proves that the sulfur atom has been oxidised to a sulfoxide. Anyway, the derivative functioned as a donor to give an *α*-linked disaccharide promoted by methyl triflate, a promoter, to our knowledge, never before used with sulfoxide donors.

To ensure *β*-selectivity, but still to have a reactive donor, a new protecting group pattern in the donor was constructed with a 2-*O*-participating group and benzyl groups in O-3 and -4, which hopefully would give enough reactivity to the donor. The benzylidene acetal in **47** was opened regioselectively to the 6-OH derivative [57], which was oxidized as described before to give **48**, with the desired design (Scheme 13). Deacetylation followed by acylation with various

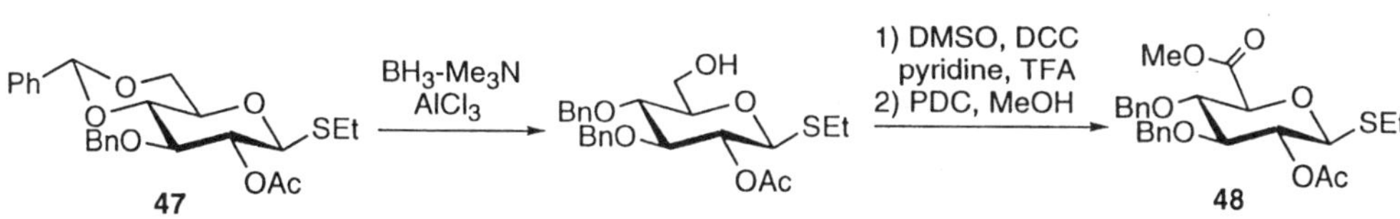

Scheme 13

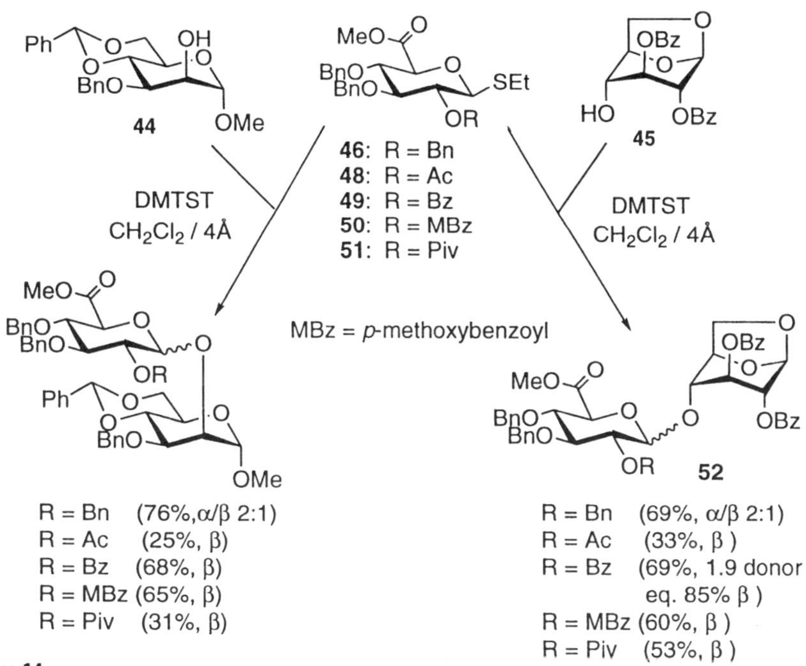

Scheme 14

acyl chlorides, benzoyl, pivaloyl, and *p*-methoxybenzoyl, gave other donors, **49–51**, with different 2-*O*-participating groups ready for test in glycosylations with acceptors **44** and **45**. The results, using DMTST as promoter, are summarized in Scheme 14 [77]. As found earlier, benzoyl participating groups gave higher yields than acetyl [46]. The pivaloyl group also gave inferior results to those obtained with the benzoyl group. The pivaloyl derivative might be of interest if competing orthoester formation is a problem during the glycosylation reactions, since pivaloyl groups have been introduced as participating groups to prevent this formation of orthoesters [78]. Although uronic acid donors were used, this was, however, not a severe problem in any of these couplings. The couplings were performed using 1.2–1.5 equivalents of the donor. If a larger excess of the donor was used, the yields could be further improved.

Thus, using the benzoylated donor **49**, high yields of disaccharides interesting for the synthesis of polysaccharide structures of *S. pneumoniae* type 3 and *C. neoformans* were obtained [77]. In addition to being a flexible reactive *β*-selective glucuronic acid donor or donor precursor, compound **49**, because of the protecting group pattern used, also contains the possibility for regioselective manipulations in the uronic acid part of the resulting oligosaccharide. For example, debenzylation of derivative **52** gives a 3′,4′-diol suitable as an acceptor or an acceptor precursor in the synthesis of oligomers of the *S. pneumoniae* type 3 repeating unit.

A more direct way to the same type of donors, which avoids the oxidation step, has also been developed [79]. Starting from the bromo sugar mentioned earlier, the ethyl thioorthoester was synthesised. Exchange of the acetyl groups for silyl groups followed by rearrangement of the orthoester gave in a very short synthesis the derivatives **53** and **54**, corresponding to donor **48** above (Scheme 15).

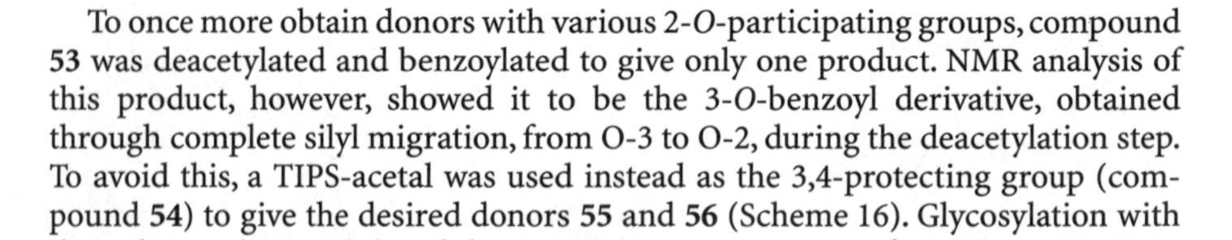

Scheme 15

To once more obtain donors with various 2-*O*-participating groups, compound **53** was deacetylated and benzoylated to give only one product. NMR analysis of this product, however, showed it to be the 3-*O*-benzoyl derivative, obtained through complete silyl migration, from O-3 to O-2, during the deacetylation step. To avoid this, a TIPS-acetal was used instead as the 3,4-protecting group (compound **54**) to give the desired donors **55** and **56** (Scheme 16). Glycosylation with these donors (1.5 equiv.) and the unreactive acceptors **44** and **57** using DMTST as promoter showed even more promising results than those obtained earlier with the benzylated donor. The results are summarised in Scheme 17.

Here, also the 2-*O*-acetylated donor gave high yields of disaccharide product, but the benzoyl analogue was still the most effective. These thioglycoside donors with their easy access and excellent glycosylation properties hence show very

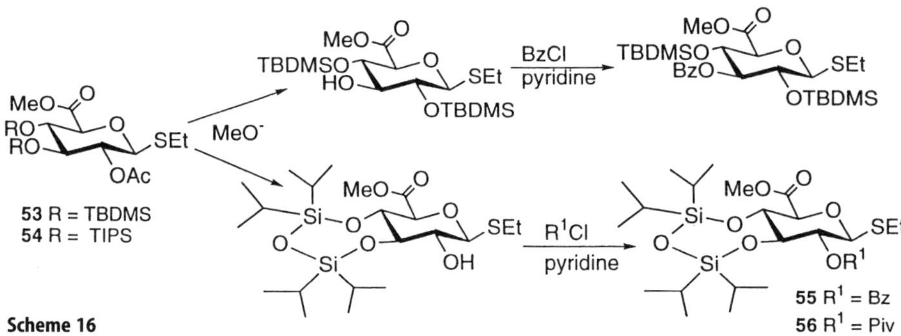

Scheme 16

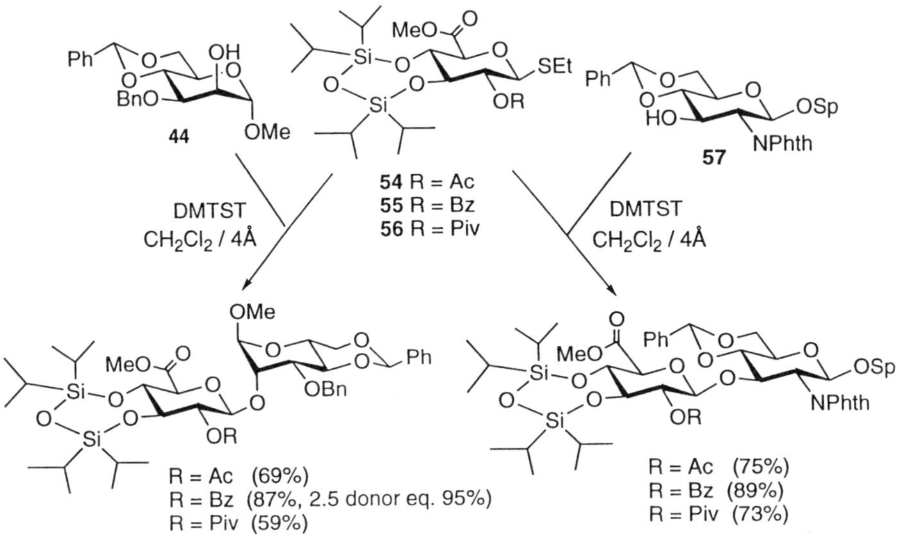

Scheme 17

promising results in model experiments, but their applicability as well as that of the benzylated analogue still have to be evaluated in the synthesis of glucuronic acid-containing oligosaccharides. This issue is currently being investigated in our laboratory in the syntheses directed towards *Cryptococcus*, *Streptococcus* and proteoglycan structures [80, 81]. Interestingly, Sinaÿ et al. recently presented an investigation on iduronic acid donors in which thioglycoside donors were found to be clearly inferior to pentenyl and trichloroacetimidate donors [82].

3.3 Synthesis of the Repeating Unit of *Streptococcus pneumoniae* Type 3 through a Regioselective Tempo-Oxidation

A parallel route to the synthesis of the *Streptococcus* type 3 structure (Fig. 6, p. 188) was designed [83], which would exploit the recent findings of regioselective oxidation of primary hydroxyl groups in the presence of secondary using 2,2,6,6-tetramethyl-1-piperidinyloxy (TEMPO) as the oxidant [84, 85]. Monomers and dimers of this repeating unit have earlier been synthesised by Chernyak et al. [86], who, among other methods, used a 3,6-lactone formation after oxidation to the glucuronic acid to construct a 3-OH acceptor in a very direct way.

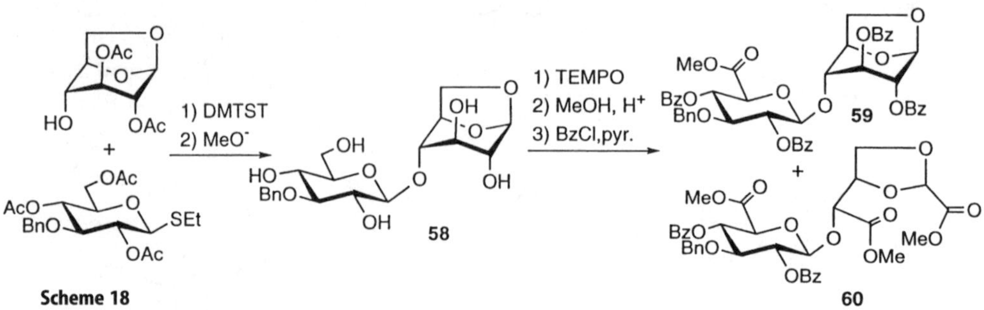

Scheme 18

From a 1,6-anhydro acceptor and a glycosyl donor, easily obtained from 1,2:5,6-di-*O*-isopropylidene-α-D-glucofuranose [87], a disaccharide was constructed, which after removal of the acetyl groups gave the pentaol **58** with one primary hydroxyl group and four secondary (Scheme 18). TEMPO-oxidation of **58** gave, after esterification and subsequent benzoylation, the expected uronic acid derivative **59**, but only in about 30% yield. Different solvents and different pHs were tried in the oxidation to improve the selectivity and yield, but without success. In all these experiments a by-product was formed in comparable yield to the desired product. NMR spectra of the isolated by-product proved it to be the over-oxidised derivative **60** containing three methyl esters. Similar products have been reported in the oxidation of octyl β-D-pyranosides using ruthenium complexes [88], but to our knowledge not in TEMPO oxidations. Whether the unexpected susceptibility for TEMPO-oxidation of the *trans*-diaxial glycol system in this anhydro-derivative also applies to similar glycol systems in other D-glycopyranosyl systems (i.e. *ido*- or *altro*-configuration) will be shown in the future.

The by-product could be avoided by choosing an alternative glycosyl donor, **61** [89], which, after coupling to 1,6-anhydro acceptor **45** and debenzylidenation, allowed the TEMPO-oxidation on a derivative with a protected 2,3-glycol

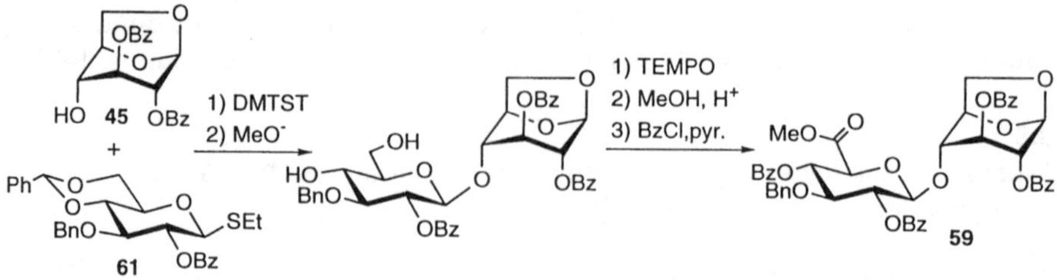

Scheme 19

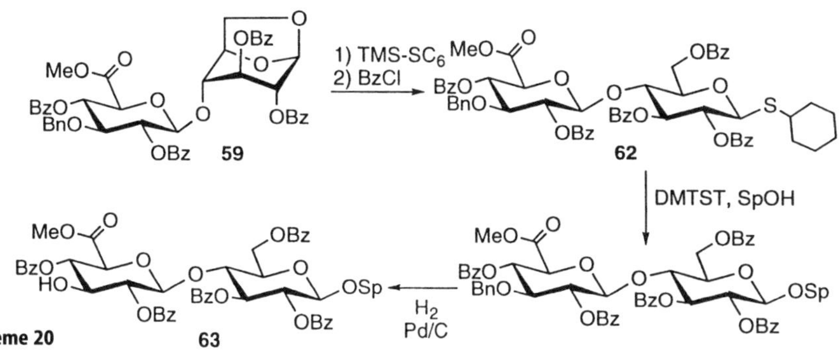

Scheme 20

system. Now the oxidation gave a high yield of the glucuronic acid compound **59** (Scheme 19). The cleavage of the anhydro bridge in **59** using zinc iodide in the presence of a cyclohexylthiotrimethylsilane (TMS–SC_6) [90] followed by benzoylation gave directly the thioglycoside donor derivative **62**, which was coupled to a spacer and debenzylated to give **63** (Scheme 20), suitable as a starting monomer acceptor together with donor **62** for synthesis of oligomers of the repeating unit of *S. pneumoniae* type 3.

4 Synthesis of Fructofuranosyl-Containing Oligosaccharides

4.1 Introduction

Fructofuranose is found in probably the most well-known and commercially available of all carbohydrates, sucrose. Various derivatives of sucrose are also known in nature, e.g. fatty acid derivatives [91] and agrocinopines [92]. Fructofuranosides are also present in various plant and bacteria polysaccharides, most often as fructans, but sometimes as a singular component in repeating units [2, 93]. Syntheses of sucrose or derivatives thereof are not very frequent [41, 94–96], and oligosaccharide synthesis with fructofuranosyl donors are even more scarce.

In a programme directed towards the synthesis and later antigenic evaluation of all the six different capsular types of *Haemophilus influenzae* (type a–f), the development of an effective synthesis of fructofuranosides became necessary, since this is a motif in the type e structure [97] (Fig. 7). The synthesis of the repeating unit without the fructofuranosyl residue has been reported [98].

→ 3)-β-D-Glc*p*NAc-(1 → 4)-β-D-Man*p*NAcA-(1 →
3
↑
2
β-D-Fru*f*

Fig. 7. The repeating unit of the *H. influenzae* type e capsular polysaccharide

Kochetkov et al. [99, 100] used the ethyl thioorthester, in a similar procedure to that with cyanoorthoesters [101], with tritylated acceptors and tritylium perchlorate as promoter to obtain high yields of α-linked fructofuranoside disaccharides. The thioorthoesters were also rearranged to thioglycosides, and various protecting group patterns were introduced. Using the same kind of acceptors and promoter, high yields of disaccharides were obtained from 3-*O*-benzylated derivatives as α/β-mixtures, which must be one of the earliest example of really successful glycosylations using thioglycoside donors. However, if the fully benzoylated thioglycoside donor was used, the promoter was not active enough and no disaccharide was produced.

More recently, Schmidt et al. [102] reported the use of fully benzoylated fructofuranose phosphites as effective glycosyl donors. Very high glycosylation yields were obtained. As for the stereochemical outcome, the authors claim that, with unhindered primary acceptors, the α-linked fructofuranosides were obtained exclusively, but with more hindered alcohols the β-linked saccharide was also formed, in some examples as the sole product. This phenomenon, formation of the *cis*-product in spite of the use of a participating group, was discussed in Sect. 2.4. However, since no carbon and only selected proton NMR values are reported, it is not clear from the publication of Schmidt how the anomeric configuration of the fructofuranosides has been determined. This is not trivial, since in a fructofuranoside there is no anomeric proton and thus neither proton-proton nor carbon-proton coupling constants can be used.

Angyal et al. [103] investigated ^{13}C NMR spectra of various unprotected *O*-fructofuranosides and found that the α-linked ones had anomeric C-2 signals at a lower field (~ 107–109 ppm) than did the β-linked (~ 103–105 ppm), and these different C-2-values are currently being used as criteria for the anomeric configuration of fructofuranosides in structural elucidation [97]. This finding is apparently also true for protected *O*-fructofuranosides independent of the protecting groups used [99, 100, 104], and the ^{13}C NMR C-2 values can therefore be used even on protected derivatives as a simple and reliable way to determine the anomeric configuration. This is not true for thiofructofuranosides, where the differences in C-2 values are small and the order sometimes exchanged. Too few *O*-linked fructofuranosides are known to be able to express an correlation about the connection between ^{13}C NMR data and anomeric configuration in these substances.

4.2 Synthesis of α-Linked Fructofuranosides (Fructofuranosides)

Considering the promising results already obtained by the Russian workers in the early eighties, a resurrection of thiofructofuranosides as glycosyl donors, this time combined with the knowledge and development of new and effective thiophilic promoters during the last decade [44], should be productive. Instead of using rearrangement of the anomeric thioorthoesters, the thioglycosides can be directly prepared from a peracylated fructofuranose by the common method using a Lewis acid and a mercaptan. The fully benzoylated ethyl thiofructofuranoside **64** was obtained in an almost quantitative yield as an inseparable

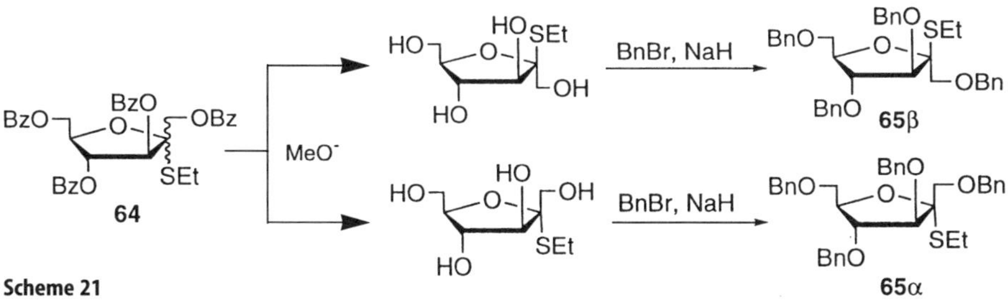

Scheme 21

α/β-mixture, which was further debenzoylated, separated and benzylated to give the two anomerically pure, benzylated thiofructofuranoside donors **65**α and **65**β (Scheme 21) [104].

The capacity of these thioglycosides as glycosyl donors was evaluated using various thiophilic promoters and three different acceptors, one primary (**66**) and two secondary (**67** and **68**), the latter relevant for the synthesis of *H. influenzae* type e structures. The yields obtained were extraordinary, both with the benzoylated and the benzylated donor (1.5 equiv.) and with all three acceptors; quantitative or nearly quantitative yields of disaccharide products were observ-

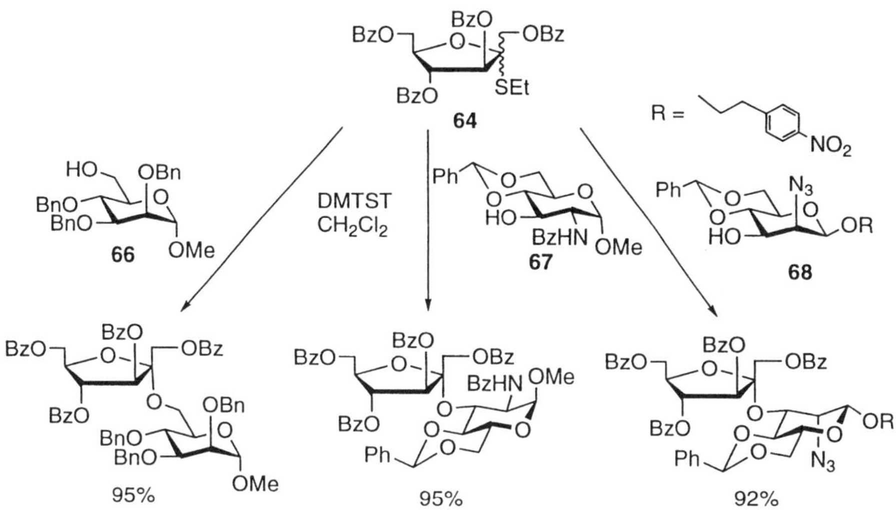

Scheme 22

ed when using DMTST as promoter in dichloromethane (Schemes 22 and 23) [104]. In contrast to peptide and nucleotide chemistry, such yields are rare in coupling steps in oligosaccharide chemistry. The development of solid phase synthesis of large oligosaccharides suffers severely from these comparably low yields in the coupling steps. In view of the yields obtained, the use of these thiofructofuranosides, especially those with a 3-*O*-benzoyl group, i.e. which ensure the stereospecificity in the coupling reaction, as model donors in a solid phase approach is tempting.

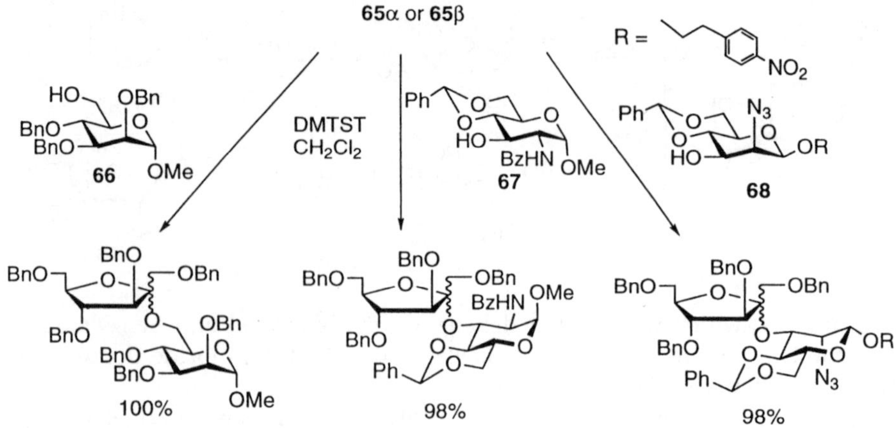

Scheme 23

Of other promoters tried, NIS also gave high yields of disaccharides, whereas methyl triflate, NBS, NCS and silver aluminium silicate gave inferior results. The benzoylated donor always gave stereospecifically the α-linked fructofuranoside, whereas the benzylated donors gave α/β-mixtures with the α-product generally in excess. The α/β-ratio and yield were independent of the anomeric configuration of the donor.

4.3 Synthesis of β-Linked Fructofuranosides

Thus, the synthesis of α-linked fructofuranosides apparently offers no real problem. However, most naturally occurring fructofuranosides, e.g. the one in *H. influenzae* type e, have the β-configuration, and therefore stereoselective or preferentially stereospecific glycosylation methods for the formation of β-linked fructofuranosides had to be found. Attempts to make the glycosylations with the benzylated donors **65**α and **65**β more β-selective by changing the conditions (solvent, temperature, promoter) gave only small and irregular effects. An additional problem was that the mixtures obtained were generally impossible to separate by chromatography [104]. If Schmidt's benzoylated phosphite donor was coupled to acceptor **68**, only the α-linked disaccharide was obtained [105].

The problem of synthesising the unfavourable β-*manno*-configuration has recently been given an elegant solution [106, 107]. Once more the 2-*O*-protecting group of the donor (generally a thioglycoside) is used as a stereo-auxiliary aid, this time not as a participating group, which would give the 1,2-*trans* α-configuration, but as a means to covalently link the incoming acceptor on the β-face through an acid-labile carbon or silyl acetal. Activation of the donor to form an electrophilic anomeric center is followed by a simultaneous cleavage of the acetal and internal delivery of the acceptor to the anomeric center from the β-face to give exclusively the 1,2-*cis* β-linkage. This or similar internal delivery approaches have also been reported in the synthesis of C-glycosides [108] and various other O-glycosides [109]. The problem with this approach to β-mannose linkages lies primarily in the formation of the intermediate labile acetal. However, with the introduction of *p*-methoxybenzylidene acetals, conveniently formed from a 2-*O*-*p*-methoxybenzyl group and an acceptor in the presence of DDQ, even oligosaccharide acceptors have been reported to give high yields of β-linked coupling products [110].

In order to try this approach on a furanosidic system, a 3-OH thiofructofuranoside was constructed. The benzoyl groups were exchanged for benzyl groups in the known orthoester derivative **69** [111], after which rearrangement of the orthoester in the presence of a large excess of ethyl mercaptan followed by debenzoylation and separation gave the desired compounds **70**α and **70**β, from which the different tethered acetals can be synthesised. The Ito and Ogawa method was chosen, and **70** was thus *p*-methoxybenzylated to give **71**α and **71**β (Scheme 24).

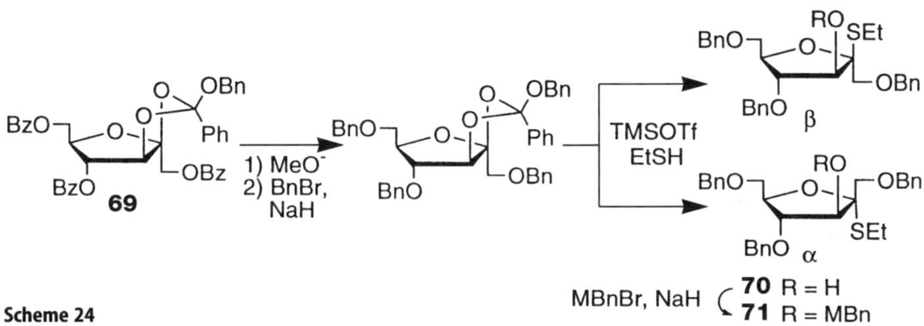

Scheme 24

With two of the earlier acceptors, **66** and **68**, together with **71**α in the presence of DDQ, the intermediate mixed acetals were obtained in high yields according to TLC. Work-up followed by immediate activation with different thiophilic promoters gave a reaction which was found to be as stereospecific for the fructofuranosidic system as earlier for pyranosides (Scheme 25) [112].

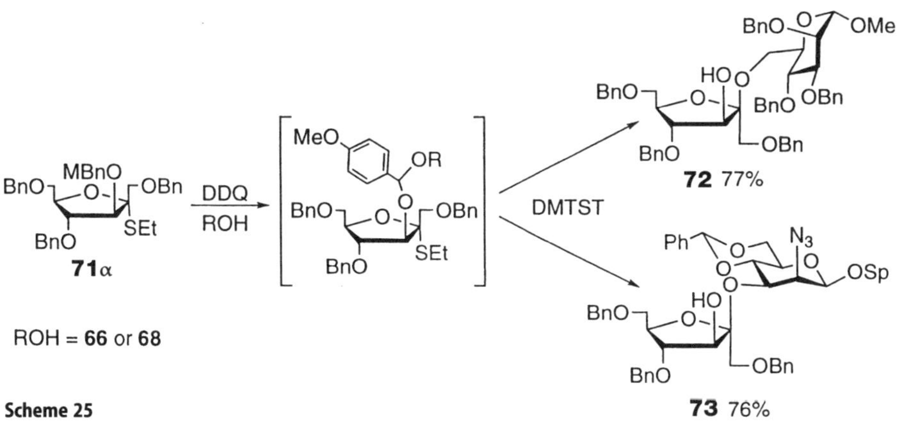

Scheme 25

Only β-linked fructofuranoside disaccharides **72** and **73** were observed in the coupling reactions. Once more, DMTST, was found to be a good promoter, high yields of both product disaccharides being obtained. If **71**β was used as donor, the reaction was still stereospecific but the yields were lower, especially with the primary acceptor (**72** 20%, **73** 50%) [105].

With NIS as promoter, comparable yields were observed but of different products. NMR showed these products still to contain signals from a *p*-methoxyphenyl group and in addition also *N*-succinimide signals, indicating the structure **74** of the products [105]. A mechanism for the formation of these quite stable *N*,*O*-acetals is suggested in Scheme 26. This usually unwanted effect of the nucleophilicity of the *N*-succinimide anion has also been found in other activation of glycosyl donors with NIS with or without acceptor [83, 104, 112, 113]. Hence, in couplings with reactive donors and unreactive acceptors, the *N*-linked succinimide glycoside of the donor can be formed as a by-product and sometimes as the main product. The use of iodonium type promoters with less nucleophilic counter ions, IDCP or IDCT, in the activation of the tethered acetals gave once more the 3-OH products **72** and **73** in comparable yields to DMTST.

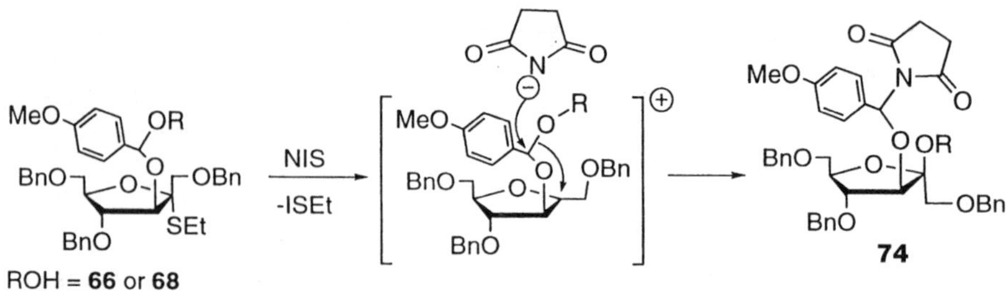

Scheme 26

An inverted procedure to construct the mixed acetals, i.e. with the *p*-methoxybenzyl group attached to the acceptor hydroxyl group and with a free 3-OH in the fructofuranoside donor derivative, would add flexibility to the protecting group manipulations in oligosaccharide synthesis. The *p*-methoxybenzyl group could be used as a temporary protecting group for the hydroxyl group to be glycosylated and in the glycosylation step directly transformed into the acetal and then rearranged into the glycoside. This approach has to our knowledge not been tried with pyranosides. When applied to fructofuranosides using a 6-*O*-*p*-methoxybenzylated analogue of **66** as acceptor, compound **70** as donor and DMTST as promoter, once more the β-linked product **72** was obtained, although in a lower yield (53%).

As a conclusion, the internal aglycon delivery approach works well in furanosidic systems and provides a convenient stereospecific route to β-fructofuranosidic disaccharides. The scope and limitations of this method will be further evaluated in the synthesis of larger oligosaccharides structures containing β-fructofuranosyl residues.

4.4
Synthesis of Sucrose Derivatives

Perhaps not from a biological, but most certainly from a chemical point of view, the synthesis of sucrose has always been of interest and a challenge. Since sucrose is a non-reducing sugar, two stereogenic centres are formed simultaneously in the coupling reaction, and the product configuration at both centres has to be controlled. Both classical and elegant solutions to this problem have been suggested [114–116]. Of the two linkages the β-fructofuranosidic is the more difficult one, being the less favoured configuration. Thus, with a good method to synthesise β-fructofuranosides from fructofuranosyl donors, the synthesis of sucrose should be feasible.

As a first attempt, the synthesis of sucrose was tried using the fully benzylated donor **65**α and the commercially available crystalline 2,3,4,6-tetra-*O*-benzyl-α-D-glucopyranose as acceptor. The conditions were optimised to give as much β-fructofuranoside as possible and at the same time not to cause anomerisation of the glucose acceptor. As found earlier with this approach, it was not possible to obtain a really good α/β-ratio for the furanosidic linkage. Only a 20% yield of perbenzylated sucrose was achieved, the main product (50%) being the α,α-isomer [104].

In view of the success of the internal glycosylation procedure in the earlier examples, this looked a most promising method for the synthesis of sucrose. Using the same glucose acceptor as above and donor **71**α, the mixed acetal was formed by treatment with DDQ. This time, with a more labile anomeric acetal as the intermediate product, the formation was not as high-yielding as before according to TLC. After activation of the acetal with DMTST, several products were formed. NMR investigation of some of these indicates that, in the activated complex, it is not the linkage between the benzylidene carbon and the oxygen that is being cleaved, but preferentially the anomeric linkage between the oxygen and the glucopyranosyl moiety, giving rise to, inter alia, a 2,3-*p*-methoxybenzylidene acetal and a 3-*O*-glucosylated fructose derivative (Scheme 27).

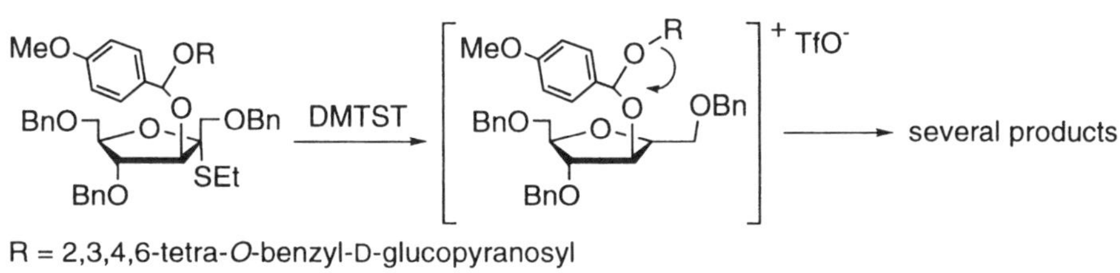

Scheme 27

Thus, sucrose, ubiquitous in everyday life, is still an elusive target substance in the chemical laboratory. Perhaps acyl-protecting groups in the glucosyl acceptor could stabilise the acetal and the anomeric linkage enough to enable sucrose to be finally produced in a simple procedure.

Acknowledgements. I would like to express my sincere gratitude to all the people who have been involved in the work presented in this article: Christian Bernlind, Kerstin Ekelöf, Per Garegg, Christian Krog-Jensen, Lars Olsson, Helena Ritzén, Maria Szönyi and Ulf Tedebark. The Swedish Research Council of Engineering Studies and the Swedish Natural Science Research Council are also gratefully acknowledged for financial support.

References

1. Lindberg B, Kenne L (1985) In: Aspinall GO (ed) The polysaccharides, vol 2. Academic Press, New York, p 287
2. Lindberg B (1990) Adv Carbohydr Chem Biochem 48:279
3. Szönyi M (1990) Doctoral Thesis, Stockholm University, Sweden
4. Brenken M (1991) Doctoral Thesis, Hamburg University, Germany
5. Boons GJPH (1991) Doctoral Thesis, Leiden University, The Netherlands
6. Ritzén H (1992) Doctoral Thesis, Stockholm University, Sweden
7. Höffgen EC (1992) Doctoral Thesis, Hamburg University, Germany
8. Ekelöf K (1996) Doctoral Thesis, Stockholm University, Sweden
9. Bernlind C (1996) Licentiate Thesis, Stockholm University, Sweden
10. Ekelöf K, Oscarson S (1995) J Carbohydr Chem 14:299
11. Boons GJPH, Hoogerhout P, Poolman JT, van Boom JH (1991) Biomed Chem Lett 1:303
12. Boons GJPH, Overhand M, van der Marel GA, van Boom JH (1989) Tetrahedron Lett 30:229
13. Garegg PJ, Oscarson S, Szönyi M (1990) Carbohydr Res 205:125
14. Paulsen H, Heitmann AC (1989) Liebigs Ann Chem 785
15. Paulsen H, Brenken M (1991) Liebigs Ann Chem 1113
16. Boons GJPH, Overhand M, van der Marel GA, van Boom JH (1989) Angew Chem Int Ed Engl 28:1504
17. Boons GJPH, Bank F, van der Marel GA, van Boom JH (1992) Tetrahedron 48:885
18. Boons GJPH, Overhand M, van der Marel GA, van Boom JH (1992) Recl Trav Chim Pays-Bas 111:144
19. Garegg PJ, Oscarson S, Ritzén H, Szönyi M (1992) Carbohydr Res 228:121
20. Dziewiszek K, Zamojski A (1986) Carbohydr Res 150:163
21. Oscarson S, Ritzén H (1994) Carbohydr Res 254:81
22. Roy R, Tropper FD (1988) Glycoconj J 5:203
23. Brimacombe JS, Kabir KMS (1986) Carbohydr Res 150:35
24. Cha JK, Christ WJ, Kishi Y (1984) Tetrahedron 40:2247
25. Brimacombe JS, Kabir KMS (1986) Carbohydr Res 152:329
26. Cram DJ, Wilson R (1963) J Am Chem Soc 85:1245
27. Dziewiszek K, Banaszek A, Zamojski A (1987) Tetrahedron Lett 28:1569
28. Grzeszczyk B, Zamojski A (1994) Carbohydr Res 262:49
29. Castro B (1986) Bull Soc Chim Fr 1533
30. Tamao K, Yshida N (1984) Tetrahedron Lett 25:4245
31. Fleming I, Philips EJ, Sanderson PEJ (1987) Tetrahedron Lett 28:4229
32. Boons GJPH, van der Klein PAM, van der Marel GA, van Boom JH (1988) Recl Trav Chim Pays-Bas 107:507
33. Boons GJPH, Elie CJJ, van der Marel GA, van Boom JH (1990) Tetrahedron Lett 31:2197
34. van Delft FL, van der Marel A, van Boom JH (1995) Synlett 1069
35. Raetz CRH (1990) Ann Rev Biochem 59:129
36. Dabrowski E, Hauck M (1988) Carbohydr Res 180:163
37. Lemieux RU, Driguez H (1975) J Am Chem Soc 97:4069
38. Oscarson S, Tedebark U (1996) J Carbohydr Chem 15:507
39. Oscarson S, Szönyi M (1989) J Carbohydr Chem 8:663
40. Oscarson S, Tidén A-K (1993) Carbohydr Res 247:323
41. Garegg PJ, Oscarson S, Ritzén H (1988) Carbohydr Res 181:89

42. Garegg PJ, Iversen T, Oscarson S (1976) Carbohydr Res 50:C12
43. Hanessian S, David S (1985) Tetrahedron 46:643
44. Norberg T (1996) In: Khan SH, O'Neill RA (eds) Modern methods in carbohydrate synthesis, vol 1. Harwood Academic, Amsterdam, p 82
45. Lindberg B (1948) Acta Chem Scand 2:426
46. Garegg PJ, Konradsson P, Kvarnström I, Norberg T, Svensson SCT, Wigilius B (1985) Acta Chem Scand B 39:569
47. Norberg T, Walding M, Westman E (1988) J Carbohydr Chem 7:283
48. (a) Phillips NJ, Apicella MA, Mcleod Griffiss J, Gibson BW (1992) Biochemistry 31:4515 (b) Phillips NJ, Apicella MA, Mcleod Griffiss J, Gibson BW (1993) Biochemistry 32: 2003
49. (a) Schweda EKH, Hegedus OE, Moxon ER, Lindberg AA (1993) Carbohydr Res 228:319 (b) Schweda EKH, Jonasson J, Jansson PE (1995) J Bacteriol 177:5316
50. (a) Jennings HJ, Beurret M, Gamian A, Michon F (1987) Antonie van Leeuwenhoek 53:519 (b) Michon F, Beurret M, Gamian A, Brisson JR, Jennings HJ (1990) J Biol Chem 256:7243 (c) Difabio JL, Michon F, Brisson JR, Jennings HJ (1990) Can J Chem 86:1029
51. Boons GJPH, van der Marel GA, Poolman JT, van Boom JH (1989) Recl Trav Chim Pays-Bas 108:339
52. Bernlind C, Oscarson S Carbohydr Res (in press)
53. (a) Ziegler T, Adams B, Kovác P, Glaudemans CPJ (1990) J Carbohydr Chem 9:135 (b) Nikolaev AV, Rutherford TJ, Ferguson MAJ, Brimacombe JS (1994) Bioorg Med Chem Lett 4:785
54. Spijker NM, van Boeckel CAA (1991) Angew Chem Int Ed Engl 30:180
55. Verhuel AFM, Boons GJPH, van der Marel GA, van Boom JH, Jennings HJ, Snippe H, Verhoef J, Hoogerhout P, Poolman JT (1991) Infection Immun 59:3566
56. Kerékgyártó J, Lipták A (1993) Carbohydr Res 248:361
57. Ek M, Garegg PJ, Hultberg H, Oscarson S (1983) J Carbohydr Chem 2:305
58. Shapiro D, Rabinsohn Y, Diver-Haber A (1969) Bichem Biophys Res Commun 37:28
59. Lafont D, Boullanger P, Banoub J, Descotes G (1990) Can J Chem 68:828
60. Bernlind C, Oscarson S. (unpublished results)
61. Kjellén L, Lindahl U (1991) Ann Rev Biochem 60:443
62. Caldwell J (1986) In: Paulson GD, Caldwell J, Hutson DH, Menn JJ (eds) Xenobiotic conjugation chemistry. American Chemical Society, Washington, p 2
63. Bochkov AF, Zaikov GE (1979) Chemistry of the *O*-glycosidic bond, Pergamon, Oxford p 192
64. (a) Slaghek TM, Hyppönen TK, Ogawa T, Kamerling JP, Vliegenhart JFG (1993) Tetrahedron Lett 34:7939 (b) Ichikawa Y, Ichikawa R, Kuzuhara H (1985) Carbohydr Res 141:273 (c) Zegelaar-Jaarsveld K (1995) Doctoral Thesis, Leiden University, The Netherlands
65. Garegg PJ, Nishida T, Oscarson S, Tidén A-K (1991) J Carbohydr Chem 10:1059
66. Cherniak R, Sundström JB (1994) Infection Immun 62:1507
67. Reeves RE, Goebel WF (1941) J Biol Chem 139:511
68. Bollenback GN, Long JJ (1955) J Am Chem Soc 77:3310
69. Sakata M, Haga M, Tejima S (1970) Carbohydr Res 13:379
70. Olsson L (1994) Licentiate Thesis, Stockholm University, Sweden
71. (a) Lönn H, Lönngren J (1984) Carbohydr Res 132:39 (b) Lindberg B, Lönngren J Thompson JL (1973) Carbohydr Res 28:351
72. Pfitzner KE, Moffat JG (1965) J Am Chem Soc 87:5661
73. O'Connor B, Just G (1987) Tetrahedron Lett 28:3235
74. Keglevic D, Ljevakovic D (1978) Carbohydr Res 64:319
75. Schmidt RR, Grundler G (1981) Synthesis 885
76. Misra AK, Roy N (1995) Carbohydr Res 278:103
77. Garegg PJ, Olsson L, Oscarson S (1995) J Org Chem 60:2200
78. Sato S, Nunomura S, Nakano T, Ito Y, Ogawa T (1988) Tetrahedron Lett 29:4097
79. Krog-Jensen C, Oscarson S (unpublished results)

80. Garegg PJ, Olsson L, Oscarson S (1996) Bioorg Med Chem (in press)
81. Olsson L (1996) Doctoral Thesis, Stockholm University, Sweden
82. Tabeur C, Machetto F, Mallet J-M, Duchaussoy P, Petitou M, Sinaÿ P (1995) Carbohydr Res 281:253
83. Tedebark U (1996) Doctoral Thesis, Stockholm University, Sweden
84. Davis NJ, Flitsch SL (1993) Tetrahedron Lett 34:1181
85. van Bekkum H, de Nooy AEJ, Besemer AC (1995) Carbohydr Res 269:89
86. Chernyak AY (1994) In: Kovác P (ed) Synthetic oligosaccharides. ACS Symposium series 560, p 133
87. Finan PA, Warren CD (1962) J Chem Soc: 3089
88. Boelrijk AEM, Jorna AMJ, Reedijk J (1995) J Molecular Catalysis A 103:73
89. Garegg PJ, Kvarnström I, Niklasson A, Niklasson G, Svensson SCT (1993) J Carbohydr Chem 12:933
90. Wang L-X, Sakairi N, Kuzuhara H (1990) J Chem Soc, Perkin Trans I 1677
91. (a) Cutler HG, Severson RF, Cole PD, Jackson DM, Johnson AW, (1986) ACS Symp Ser 296 178 (b) Wahlberg I, Walsh EB, Forsblom I, Oscarson S, Enzell CR, Ryhage R, Isaksson R (1986) Acta Chem Scand B 40:724
92. (a) Ellis JG, Murphy PJ (1981) Mol Gen Genet 181:36 (b) Ryder MH, Tate ME, Jones GP (1984) J Biol Chem 259:9704
93. McDonald EJ (1946) Adv Carbohydr Chem 2:253
94. Khan R (1976) Adv Carbohydr Chem Biochem 33:235
95. (a) Oscarson S, Ritzén H (1990) Carbohydr Res 205:61 (b) Oscarson S, Ritzén H (1996) Carbohydr Res 284:271
96. (a) Franzkowiak F, Thiem J (1987) Liebigs Ann Chem 1065 (b) Lindberg M, Norberg T (1988) J Carbohydr Chem 7:749 (c) Lindberg M, Norberg T, Oscarson S (1992) J Carbohydr Chem 11:243 (d) Lindberg M, Oscarson S (1993) J Carbohydr Chem 12:1139
97. Branefors-Helander P, Kenne L, Lindberg B, Petersson K, Unger P (1981) Carbohydr Res 88:77
98. Garegg PJ, Oscarson S, Tidén A-K (1992) Carbohydr Res 225:163
99. Balan N F, Backinowsky L V, Betaneli V I, Kochetkov N K (1981) Bioorg Khim 7:1566
100. Backinowsky L V, Balan N F, Betaneli V I, Kochetkov N K (1982) Carbohydr Res 99:189
101. Backinowsky LV (1994) In: Kovác P (ed) Synthetic oligosaccharides. ACS Symposium series 560, p 36
102. Müller T, Schneider R, Schmidt RR (1994) Tetrahedron Lett 27:4763
103. Angyal SJ, Bethell GS (1976) Austr J Chem 29:1249
104. Krog-Jensen C, Oscarson S (1996) J Org Chem 61:1234
105. Krog-Jensen C (1996) Licentiate Thesis, Stockholm University, Sweden
106. (a) Barresi F, Hindsgaul O (1991) J Am Chem Soc 113:9376 (b) Barresi F, Hindsgaul O (1992) Synlett 759 (c) Barresi F, Hindsgaul O (1994) Can J Chem 72:1447
107. Stork K, Kim G (1992) J Am Chem Soc 114:1087
108. (a) Vauzeilles B, Cravo D, Mallet JM, Sinaÿ P (1993) Synlett 522 (b) Stork G, Suh HS, Kim G (1991) J Am Chem Soc 113:7054
109. (a) Ziegler T (1995) Tetrahedron Lett 36:8973 (b) Valverde S, Goméz AM, Lopéz JC, Herradón B (1995) Tetrahedron Lett 37:1105 (c) Bols M (1992) J Chem Soc, Chem Commun 913
110. (a) Ito Y, Ogawa T (1994) Angew Chem Int Ed Eng 33:1765 (b) Dan A, Ito Y, Ogawa T (1995) J Org Chem 60:4680
111. Helferich B, Bottenbruch L (1953) Chem Ber 5:651
112. Krog-Jensen C, Oscarson S (1996) J Org Chem 61:4512
113. Lauplicher L (1992) Doctoral Thesis, Hamburg University, Germany
114. Lemieux RU, Huber GJ (1953) J Am Chem Soc 75:4118
115. Fraser-Reid B, Iley DE (1975) J Am Chem Soc 97:2563
116. Barett AGM, Melcher LM, Bezuidenhoudt BCB (1992) Carbohydr Res 232:259

Pyruvated Saccharides – Novel Strategies for Oligosaccharide Synthesis

Thomas Ziegler

Institut für Organische Chemie, Universität zu Köln, Greinstraße 4, D-50939 Köln, Germany

Pyruvic acid acetals are found as so-called non-carbohydrate groups among many bacterial polysaccharides and carbohydrate structures on the surface of higher cells. These groups are responsible for the distinctive physical and biological properties of the respective saccharides, and are thus attractive targets for chemical synthesis. In this review, methods for the efficient preparation of pyruvated building blocks for oligosaccharide synthesis are discussed in detail, and ways in which these building blocks are used in for the construction of important pyruvated bacterial oligosaccharides are described. Novel glycosylation strategies such as glycodesilylation or the approach via prearranged glycosides, as well as new protecting groups developed en route to pyruvated saccharides are also presented. The general applicability of these strategies and protecting groups is exemplified for the chemical synthesis of several complex oligosaccharide structures.

Table of Contents

1 Introduction

The last decades have provided a tremendous insight into the subtle biological functions of complex cell surface carbohydrates, and it is now well recognized that carbohydrate-protein interactions play a central role in many essential cell-cell interactions. Furthermore, the rapid development of modern analytical methods – namely mass spectrometry and NMR spectroscopy – has been a major driving force behind all the significant achievements in structure elucidation of complex carbohydrates. Thus, an immense number of oligo- and polysaccharides have been characterized in detail up today. Although the diversity of these sugars is already overwhelming, one has still to be astonished about new and unexpected structural features found among saccharides of biological origin. Taking into account the fact that even a small number of different monosaccharides can be combined to form a huge number of distinct oligosaccharides and that each oligosaccharide can also adopt several conformations, it seems certain that nature uses carbohydrates for encoding information for cell-cell interactions. Furthermore, the great structural diversity of cell surface saccharides is further increased dramatically by so-called non-carbohydrate groups (mainly acetyl groups) attached to various positions of the sugar chains [1]. However, it is not clear yet what biological functions these groups may perform in detail.

Among these non-carbohydrate groups, pyruvic acid acetals (*i.e.* 1-carboxyethylidene substituents) are frequently detected in capsular polysaccharides of bacteria and cell surface carbohydrate structures of higher organisms [2]. These groups are thought to "modulate" the biological properties of the respective saccharides and might be quite important constituents for carbohydrate-protein interactions. Thus, in order to study the biological functions of pyruvated saccharides, it appears to be highly desirable to have efficient methods for their selective chemical synthesis. Furthermore, acidic pyruvate acetals are responsible for interesting physical properties of some bacterial polysaccharides. For example, the acidic heteropolysaccharide produced by the bacterium *Xanthomonas campestis* (xanthan gum), containing a 4,6-*O*-(1-carboxyethylidene)-mannosyl residue, has gained significant importance as a food additive because of its pronounced gel-forming behaviour [3]. These gel-forming properties of xanthan gums are dependent on the amount of pyruvic acid acetals in the polysaccharide [4]. Similarly, the widely used polysaccharide from the red algae *Gelidium* (Agar) contains a 4,6-*O*-(1-carboxyethylidene)-galactosyl residue. Agar must be regarded as especially important in a historical sense because it was the first example of a polysaccharide containing pyruvic acid acetals [5].

Sect. 2 below summarizes some of the important structures containing pyruvic acid acetals. The structural diversity, methods for structure elucidation, biosynthesis and biological function of these oligo- and polysaccharides will be briefly discussed. Sect. 3 gives an overview of the synthetic methods for the stereoselective preparation of pyruvated building blocks for oligosaccharide synthesis. Here, special attention is paid to pyruvated glycosyl donors, and their application to the synthesis of selected structures is presented in Sect. 4.

Novel glycosylation strategies and novel protecting groups that were developed during the synthesis of pyruvated saccharides are presented in Sect. 5. The applicability of these methods is demonstrated, and some examples of pyruvated bacterial saccharides are given.

2 Structures

Over 70 pyruvated saccharide structures from natural sources are known today, but only a few representative examples are summarized here. The choice has been determined by the synthesis of pyruvated saccharides as discussed in the Sections below. Reference [2] gives a more concise overview of the great structural diversity found among naturally occurring pyruvated oligo- and polysaccharides.

2.1 Selected Structures of Pyruvated Polysaccharides

The majority of the pyruvated saccharides, the structures of which have been determined in detail, contain 4,6-*O*-(1-carboxyethylidene)-glycopyranosyl residues of D-glucose, D-galactose and to a lesser extent D-mannose (examples are given below). Most of these pyruvated glucosyl and mannosyl residues are β-glycosidically bound to the next sugar residue of the polysaccharide chain, whereas 4,6-pyruvated galactose is found α- and β-linked in almost equal amounts. Since the acetal carbon of the pyruvylidene residue is chiral, its configuration has to be determined as well. In all cases where this assignment was done, the carboxylate-group turned out to be axially oriented [i.e. (S)-configuration for glucose and mannose and (R)-configuration for galactose]. The preferred axial orientation of the carboxylic group is due to a strong anomeric effect in the 1,3-dioxolane ring of 4,6-pyruvated glycopyranoses. Thus, the axial carboxylate is about 3 kcal/mol more stable than the equatorial one [6], and nature obviously favours the thermodynamically more stable isomer.

3,4-Pyruvic acid acetals are found in β-D-galactopyranosyl residues of some bacterial polysaccharides and in a cerebroside of the Japanese snail

Aplysia kurodai. Here, the pyruvic acetal forms a 1,3-dioxolane ring which equalizes the differences in the stability of the two diastereomeric forms (i.e. endo- and exo-oriented carboxylate group at the dioxolane ring, respectively). Thus, both isomeric forms of 3,4-*O*-(1-carboxyethylidene)-*β*-D-galactopyranosyl residues are found among naturally occurring saccharides. In most pyruvated saccharides of this type, exo-oriented carboxylate groups have been identified [7]. However a 3,4-pyruvic acid acetal with an endo-oriented carboxylate group attached to a *β*-D-galactopyranosyl residue has undoubtedly been found in a phosphonoglycosphingolipid of *pomacea lineata* [8].

Only a few examples of 2,3-pyruvylidene residues of *β*-D-glucuronic acid and *α*-D-galactose and, more often, of 3,4-pyruvylidene residues of L-rhamnose have been reported so far. In these cases, the pyruvic acid acetal is formed by two trans-diequatorially-oriented oxygens of the respective pyranose, which makes them extremely unstable and prone to hydrolysis. It might be speculated that only a few examples of these unstable pyruvate acetals have been detected up to now simply because the pyruvate has been lost during isolation and purification of the polysaccharides. The only configurational assignment of the pyruvate acetal moiety has been made for the intracanenally bound 2,3-*O*-(1-carboxyethylidene)-*α*-D-galactopyranosyl residue of the capsular polysaccharide of *Streptococcus pneumoniae* type 4, which was revealed to have the (S)-configuration [9].

Rather uncommon are the two pyruvic acetal structures which have been identified in *Klebsiella* K12 capsular polysaccharides and in the teichoic acid of bacterium NCTC 9742 [10]. In the former case, a 4,5-*O*-(1-carboxyethylidene)-*β*-D-galactofuranosyl residue was found at the side chain terminus of the hexasaccharide repeating unit. In the latter case, the teichoic acid contains intracatenally bound 2,3-*O*-(1-carboxyethylidene)-D-mannitol phosphate. No data about the configuration of these pyruvic acetals are available.

Table 1 shows some examples of structures containing one or two pyruvated monosaccharide residues. For bacterial capsular polysaccharides, the pyruvic acid acetals are found either attached to intracatenally bound residues of the repeating unit or, more often, at the terminal residue of a side chain (*Klebsiella, Rhizobium, Escherichia coli, Streptococcus*). Other pyruvated bacterial cell-surface structures comprise glycolipid structures and glycopeptides (*Mycobacterium*). Similar glycoconjugates have been also found in higher organisms (*Aplysia, Microciona*).

2.2 Structure Determination

For the concise determination of the structure of pyruvated oligo- and polysaccharides, one has first to quantify the exact amount of pyruvic acid in relation to the monosaccharide distribution. Most conveniently, this is done by acidic hydrolysis of the polysaccharide followed by quantification of the liberated pyruvic acid either enzymatically or as its 2,4-dinitrophenyl hydrazone [19]. Next, the monosaccharide residue to which the pyruvate is bound and the posi-

Table 1. Some pyruvated oligo- and polysaccharide structures[a]

Rhizobium phaseoli [11]

→4)-β-D-GlcpA-(1→4)-β-D-GlcpA-(1→4)-β-D-Glcp-(1→4)-α-D-Glc-(1→
6
↑
1
β-D-Gal-(1→6)β-D-Gal-(1→6)-β-D-Glc-(1→6)-α-D-Gal-(1→4)-β-D-GlcA-(1→4)-β-D-Glc-(1→4)-β-D-Glc
3 4
X_{COOH}

Rhizobium trifolii [12]

→4)-α-D-GlcA-(1→4)-3-Ac-β-D-GlcA-(1→4)-α-D-Glc-(1→4)-β-D-Glc-(1→
6
↑
1
2,3-(D-3-hydroxybutanoyl)-β-D-Gal-(1→3)-β-D-Glc-(1→4)-β-D-Glc-(1→4)-β-D-Glc
6 4 6 4
(R) X_{COOH} (S) X_{COOH}

Mycobacterium smegmatis [13]

2,4-Me_2-2-eicosenoyl tetra- or hexadecanoyl
↓ ↓
4 6
3-*O*-Me-β-D-Glc-(1→3)-β-D-Glc-(1→4)-β-D-Glc-(1→6)-α-D-Glc-(1↔1)-α-D-Glc
6 4 6 4
(S) X_{COOH} (S) X_{COOH}

Mycobacterium avium [14]

3-*O*-Me-β-D-Glc-(1→3)-α-L-Rha-(1→2)-6-deoxy-α-L-Tal-(1→glycopeptidolipid) serovar. 8
6 4
(S) X_{COOH}
β-D-Glc-(1→3)-α-L-Rha-(1→2)-6-deoxy-α-L-Tal-(1→glycopeptidolipid) serovar. 21
6 4
(S) X_{COOH}

Klebsiella[15]

→3)-β-D-Glc-(1→4)-β-D-Man-(1→4)-α-D-Glc-(1→
3
↑
1
β-D-Gal-(1→4)-α-D-GlcUA
3 4
(S) X_{COOH}

Stretococcus pneumoniae (Typ 27) [16]

$Me_3N^+(CH_2)_2OPO_3^-$
↓
2
→3)-β-D-GlcNAc-(1→3)-α-D-Gal-(1→4)-β-L-Rha-(1→4)-β-D-Glc-(1→
6 4
(S) X_{COOH}

Escherichia coli (K47) [17]

→3)-β-D-GlcNAc-(1→2)-β-D-Gal-(1→4)-β-D-Man-(1→4)-α-D-Gal-(1→
3 4
(S) X_{COOH}

Microciona prolifera [18]

β-D-Gal-(1→4)-β-D-GlcNAc-(1→3)-L-Fuc-(1→protein)
6 4
(R) X_{COOH}

Agar-Agar (Gelidium) [5]

→3)β-D-Gal-(1→4)-3,6-anhydro-β-L-Gal-(1→
6 4
(R) X_{COOH}

[a] Fuc = fucopyranose, Glc = glucopyranose, GlcUA = glucopyranuronic acid, GlcNAc = 2-acetamino-2-deoxy-glucopyranose, Gal = galactopyranose, Gal*f* = galactofuranose, Man = mannopyranose, Rha = rhamnopyranose, Tal = talopyranose.

tion of the acetal must be determined. This is best done by NMR spectroscopy [7, 9a, 20] in combination with classical methods for sequence elucidation of oligosaccharides. Furthermore, NMR spectroscopy allows the assignment of the configuration of the pyruvic acid acetal. Confirmed by X-ray analysis [21], the chemical shifts of the methyl group of the pyruvate acetal are significant for its orientation in 4,6-*O*-(1-carboxyethylidene)-glycopyranoses. Typically, an equatorially oriented methyl group [i.e. (S)-configuration of the acetal carbon in D-glucose and D-mannose and (R)-configuration in D-galactose] shows an up-field shift in the proton NMR spectrum compared to an axially oriented (~1.5 ppm vs. ~1.7 ppm). In the carbon NMR spectrum, this effect is inverted, such that an equatorial methyl group shows a down-field shift compared to an axial one (~25 ppm vs. ~17 ppm).

The assignment of the configuration of 3,4-pyruvylated galactose derivatives is somewhat more complicated, since the differences in the chemical shifts are less significant. NOE effects have to be determined for the unambiguous assignment [17]. Diagnostic values for the methyl group of the acetal and for C-3 of the galactose residue for simple NMR spectra are only obtained if the carboxylic acid of the pyruvate is reduced to a hydroxymethyl group [20, 22]. However, significant differences for the signals of the methyl group and C-3 can be obtained for 2,6-benzoylated 3,4-pyruvated galactosides the assignment of which was confirmed by an X-ray study [23]. Similarly, the configuration of 2,3-pyruvic acetals of galactosyl and glucuronic acid residues can be determined by NOE experiments [9, 24].

An interesting alternative to NMR investigations for the assignment of the configuration of the cyclic acetal of pyruvated polysaccharides uses quantitative inhibition of specific immune precipitation techniques [25]. Thus, synthetic (R) and (S) 4,6-pyruvated methyl galactosides were applied as inhibitors for the precipitation reaction of *Klebsiella* polysaccharides with the respective antisera. The (R) isomer was revealed to be a potent inhibitor, and thus proved the *K.* serotype K11 and K21 polysaccharides to contain 4,6-*O*-[(R)-1-carboxyethylidene]-galactopyranosyl residues.

2.3 Biosynthesis and Biological Function

The biosynthetic pathway of pyruvated oligo- and polysaccharides is not clear yet. However, it has been demonstrated for rhizobial polysaccharides that pyruvic acetal formation occurs at a late stage of the biosynthesis of the sugar chain. Furthermore, the pyruvic acid acetal is most probably introduced by an initial transfer of a propenoate-2-yl residue from phosphoenolpyruvate to a glycosyl residue followed by spontaneous ring closure [26].

As for the biosynthesis of pyruvated saccharides, the exact biological function of these groups remains unclear. It has been speculated that pyruvylation might occur on lipid intermediates of the biosynthesis of polysaccharides in order to slow down this process [27]. However, further insight into the process is needed. The biological function of pyruvic acid acetals in oligo- and polysaccharides seems to be strongly linked to their immunological properties.

Heidelberger demonstrated that the pyruvate acetal substituents of rhizobial polysaccharides behave as immunodominant groups. Furthermore, it has been shown that the configuration of the acetal group also influences the immunological properties of such saccharides [25, 28], and the existence of cross reactivities between pyruvated saccharides of different origin but bearing the same puruvic acid acetal are common phenomena.

For rhizobial exopolysaccharides, a large number of papers deal with the possible involvement of these pyruvated saccharides in early processes of type-specific infection of the host plant (leguminoses). For this complex process, it seems to be most probable now that pyruvated saccharides play an essential role in the selective recognition of the plant by the *Rhizobium* bacteria [29]. Thus, pyruvated saccharides may be important for highly specific cell-cell recognitions. This was also recently demonstrated for the marine sponge *Microciona prolifera,* the pyruvated proteoglycan of which is responsible for the aggregation of dissociated sponge cells [18].

Synthetic pyruvated saccharides should therefore be excellent tools to shed more light on the complex biological functions and processes which these sugar derivatives participate in.

3 Pyruvated Building Blocks

For the efficient chemical synthesis of pyruvated oligosaccharides it is imperative to follow some prerequisites. First, the introduction of a pyruvic acetal to a suitable mono- or disaccharide building block has to proceed with high yield, since the acetalation is most often an early step of the complete saccharide synthesis. Furthermore, commonly used blocking groups for carbohydrate synthesis have to be tolerated in order to guarantee a broad spectrum of manipulations. Second, the acetalation should allow the selective preparation of both diastereomeric pyruvate acetal forms. In an ideal case, both isomers should be selectively available. Third, pyruvated building blocks have to be constructed in such a way that they can be used either as glycosyl donors or as glycosyl acceptors. Special attention has also to be paid to their reactivity in glycosylation reactions.

In the light of the above-mentioned prerequisites, the described procedures for the synthesis of pyruvylated saccharides will be briefly discussed.

3.1 Introduction of Pyruvic Acid Acetals

Since the formation of cyclic pyruvate acetals from alkyl pyruvates and diols under classical conditions (i.e. catalysis by acids) is expected to be unfavoured due to the necessity of the intermediate formation of a destabilized carbocation, early attempts to prepare 1-carboxyethylidene sugars used indirect procedures. Later, it was shown that under carefully controlled conditions the direct acetalation of a sugar diol with methyl pyruvate can also be used for preparative purposes.

3.1.1
Indirect Methods

A monosaccharide (**1**) is first treated with 2-acetoxy-acetone to give a diastereomeric mixture of the corresponding 1-hydroxy-2-propylidene derivatives **2** (Scheme 1). For the generation of the pyruvic acetal, the acetoxymethyl group is subsequently deacetylated and oxidized with platinum and oxygen, to give the pyruvylidene derivatives **3** [30]. Yields using this procedure are usually low, and the mixture of diastereomeric hydroxypropylidene derivatives has to be separated prior to the oxidation step. Therefore, this method was solely applied for the preparation of analytical samples of pyruvated monosaccharides for NMR investigations [7, 20c, 22, 31]. A similar but more efficient method uses 3,4-dimethoxyphenyl- [32] and 2-furyl-isopropylidene derivatives (**4**) [33], respectively, which are oxidized by ruthenium tetroxide to the corresponding pyruvated saccharides **5** (Scheme 1). These indirect procedures furnish the intermediate 1-aryl-propylidene-glycopyranose derivatives **4** in a diastereoselective fashion (i.e. equatorial methyl group in 4,6-acetalized glucosides) in good yield. However, the oxidation step does not allow the presence of other oxidable blocking groups (for example, allyl and thio groups) and the method has not yet been used for the synthesis of 3,4-acetals of galactose.

Based on the Noyori-acetalation, 4,6-*O*- and 3,4-*O*-bis-silylethers of simple glycosides can be acetalized with alkyl pyruvates [34]. In general, mixtures of the corresponding diastereomeric 1-(1-alkoxycarbonylethylidene)-glycosides are obtained. This procedure for the preparation of pyruvated saccharides provides an easy entry to variously blocked derivatives. In addition, protecting groups which are rather sensitive can be used. A major drawback of the Noyori-acetalation is, however, that the acetalation is usually performed under kinetic control, and therefore, the diastereomers formed, have to be separated by tedious chromatography. This problem can be circumvented if thermodynamic conditions are applied [35]. For example, methyl and benzyl 2,3-di-*O*-benzoyl-

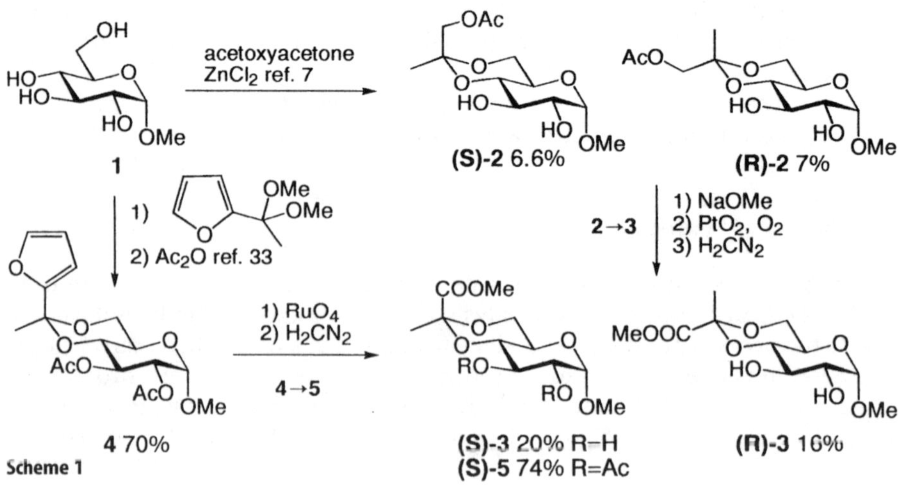

Scheme 1

4,6-*O*-bis(trimethylsilyl)-α-D-glucopyranoside (**6**), respectively, afford upon treatment with 2 molar equivalents of methyl pyruvate and trimethylsilyl trifluoromethanesulfonate (TMSOTf) the corresponding 4,6-*O*-pyruvate acetals **7** in 68.5% and 76% yield (Scheme 2). When a catalytic amount of TMSOTf (10 mol-%) is used, a 2.3:1 mixture of the corresponding (S)- and (R)-acetals is obtained [34a]. In contrast, when a larger amount of TMSOTf (33 mol-%) is used, only the more stable (S)-isomer is formed [35a].

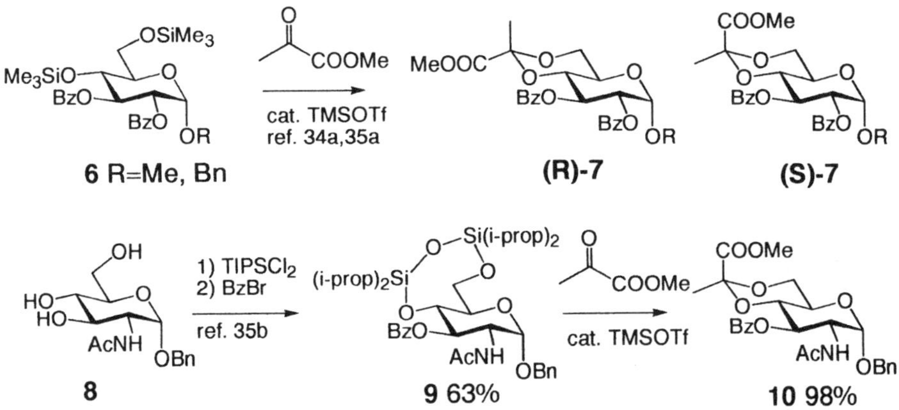

Scheme 2

Further improvements in the synthesis of pyruvated saccharide building blocks are achieved when the 1,1,3,3-tetraisopropyl-1,3-disiloxane-1,3-diyl group (TIPS) is used for the Noyori-acetalation. 4,6-TIPS-protected glycosides such as **9** are conveniently prepared from the corresponding alkyl glycosides **8** in two steps and can be converted into the respective pyruvated glycosides **10** under essentially thermodynamic conditions [35b] (Scheme 2).

Another alternative to the Noyori-acetalation uses pyruvyl thioacetals as precursors, which are activated with thiophilic reagents such as methyl triflate, nitroso tetrafluoroborate or SO_2Cl_2-trifluoromethanesulfonic acid [36]. Under these conditions, 2,3-blocked alkyl or aryl glycosides in the D-gluco and D-galacto series afford the corresponding diastereomeric pyruvate acetals in good yield and sometimes high selectivity. However, the method is not applicable to 1-thioglycosides, because of the thiophilic activation.

3.1.2
Direct Acetalation of Sugar Diols

The most convenient procedure for the efficient preparation of pyruvated building blocks for oligosaccharide synthesis is the direct acid-catalysed acetalation of a glycoside with methyl pyruvate. As mentioned above, this procedure affords carefully optimized conditions. Thus, early attempts in this direction showed that acceptable yields of pyruvated glycosides can only be obtained if small amounts are prepared [16, 28b, 37]. The conditions are not suitable for the pre-

Table 2. Condensation of D-glycopyranosides **11** with methyl pyruvate under BF_3-catalysis

starting material 11	product 12	yield	starting material 11	product 12	yield
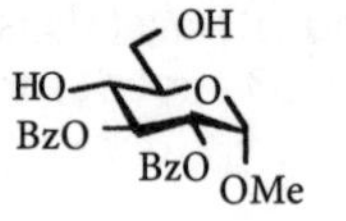 **11a**	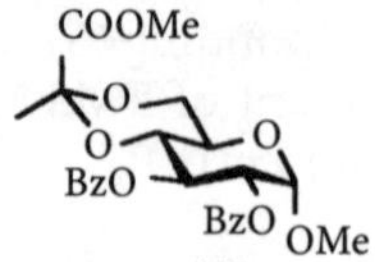 **12a**	85%	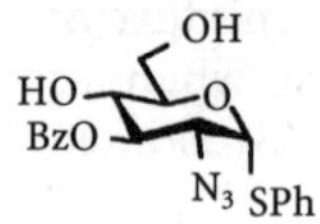 **11g**	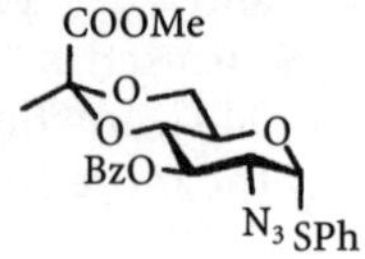 **12g**	67%
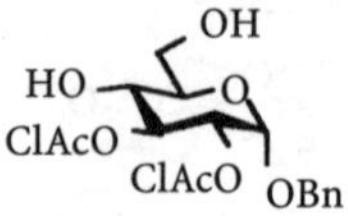 **11b**	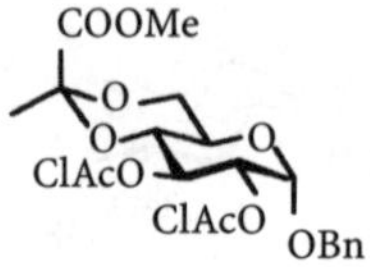 **12b**	78%	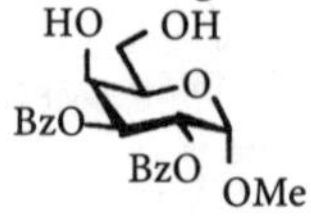 **11h**	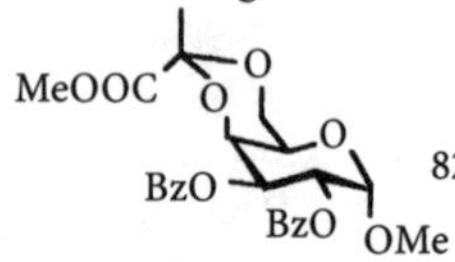 **12h**	82%
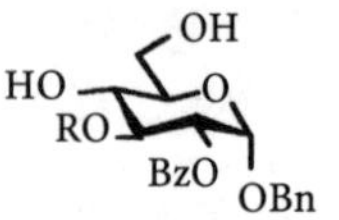 **11c,c′** R = Bz, Me	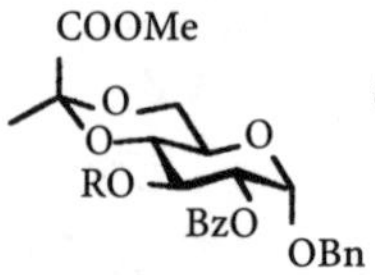 **12c,c′** R = Bz, Me	86%	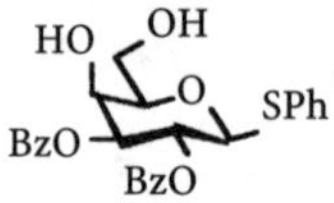 **11i**	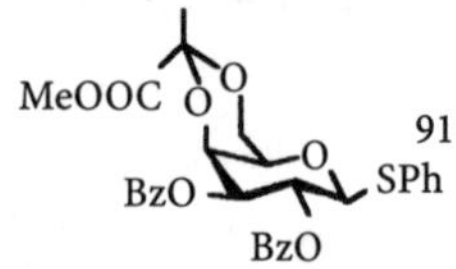 **12i**	91%
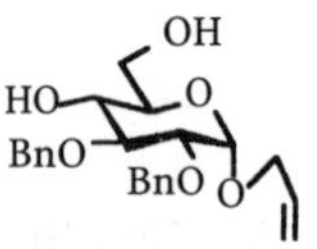 **11d**	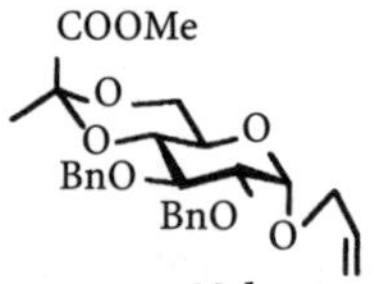 **12d**	65%	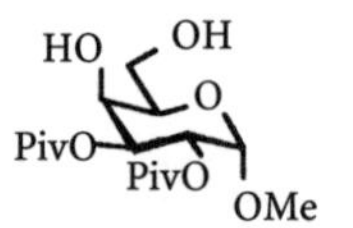 **11j**	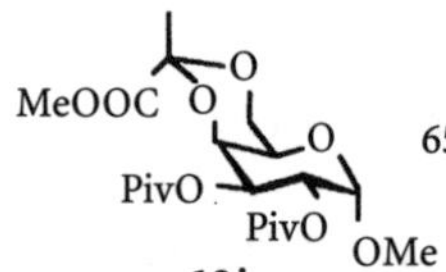 **12j**	65%
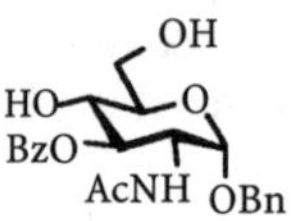 **11e**	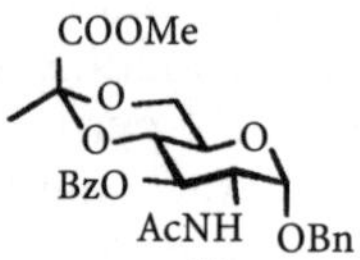 **12e**	83%	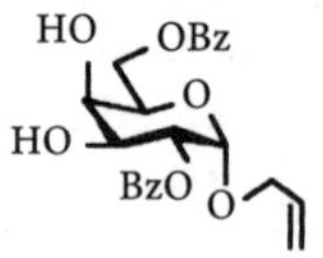 **11k**	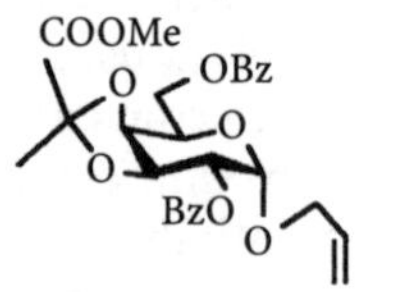 **12k** 59% (R), 15% (S)	74%
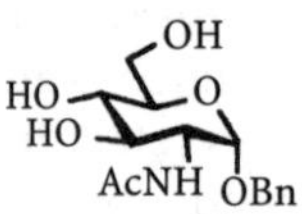 **11f**	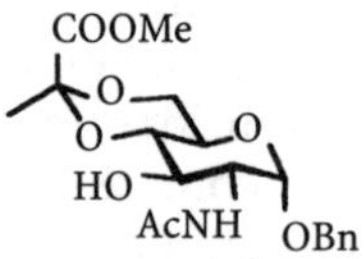 **12f**	65%	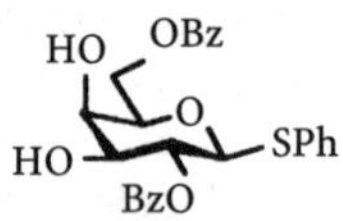 **11l**	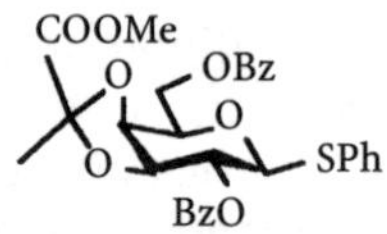 **12l** 54% (R), 17% (S)	71%
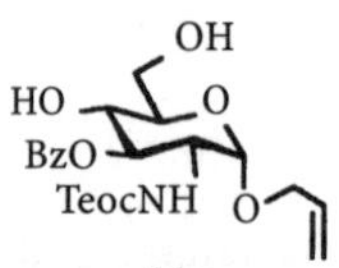 **11g**	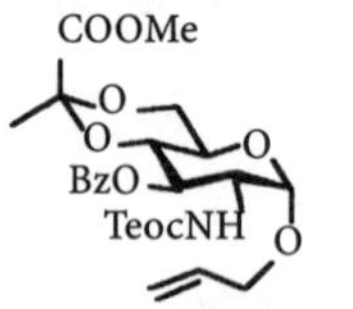 **12g**	68%	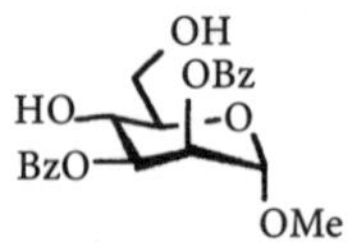 **11m**	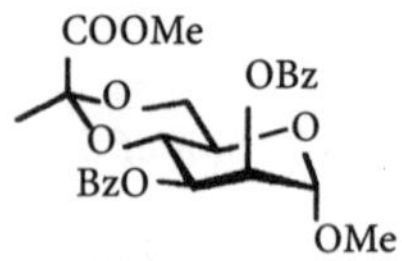 **12m**	81%

paration on a larger scale [37], because by-products which are difficult to separate are formed here [38]. Reproducible good results are obtained, however, with partially protected sugar diols (**11**) in acetonitrile or methyl pyruvate as the solvent and with BF_3-ether as the catalyst [39, 40]. A broad spectrum of protecting groups is tolerated by this procedure, and yields of pyruvated glycosides **12** are usually good. Table 2 summarizes some representative examples. For 4,6-acetals, the thermodynamically more stable diastereomers having an equatorial methyl group are formed almost exclusively. For 3,4-acetals of galactosides, a mixture of the diastereomers is usually obtained.

The direct pyruvylation of sugar diols also allows the selective synthesis of both diastereomers of 4,6-acetals [39c]. Thus, in acetonitrile as the solvent, the thermodynamically favoured diastereomers are formed, whereas in methyl pyruvate and with shorter reaction times, the kinetically favoured acetals are obtained.

3.2
Pyruvated Glycosyl Donors

Pyruvated glycosyl donors (halides and trichloroacetimidates) are conveniently prepared from the corresponding alkyl or aryl *O*- and *S*-glycosides or from 1-*O* unprotected glycoses using standard methods [39d, 41]. In addition, 1-thioglycosides can be used directly as pyruvated glycosyl donors when adequate thiophilic activations are applied [39b, d].

In general, 4,6-pyruvated D-glucopyranosyl donors are significantly less reactive than their counterparts in the D-galacto series. For example, 2,3-di-*O*-benzoyl-4,6-*O*-(1-methoxycarbonyl-ethylidene)-α-D-glucopyranosyl chloride requires, under promotion with silver trifluoromethane-sulfonate, 1 h at room temperature for complete reaction with methanol [41], whereas the corresponding pyruvated galactosyl chloride reacts almost instantaneously under identical conditions [6a]. This is attributed to the conformationally restricted skeleton of pyruvated glucose derivatives, which resembles a *trans*-decalin system. In contrast, 4,6-pyruvated galactose derivatives represent a more flexible *cis*-decalin-like system, and are thus significantly more reactive in glycosylation reactions. Corresponding observations have been made for other conformationally restricted glycosyl donors [42]. Nevertheless, under carefully optimized conditions, pyruvated glycosyl donors can be applied for highly effective coupling reactions even with less reactive acceptors [36, 39, 41, 43]. Some representative examples are presented in the following Section.

4
Synthesis of Pyruvated Oligosaccharides

As mentioned above, special attention has to be paid to the distinct properties of pyruvated glycosyl donors in order to achieve the efficient synthesis of related oligosaccharides. This Section will give a glimpse of the numerous problems which may be encountered during "classical" couplings of pyruvated donors.

4.1 Rhizobium Saccharides

Rhizobial exopolysaccharides show a highly diverse pattern of differently pyruvated structures (see Table 1), and are therefore attractive targets for chemical syntheses. Some examples using pyruvated glycosyl donors are presented here in order to further demonstrate the special features of these donors for the synthesis of complex rhizobial oligosaccharides.

4.1.1 *R. leguminosarum Biovar trifolii*

The repeating unit of the polysaccharide of *R. leguminosarum* biovar *trifolii* displays a doubly pyruvated side chain, which is probably important for the species-specific infection of leguminoses by *Rhizobium* bacteria (see Sect. 2). For the synthesis of fragments related to this pyruvated side chains, a disaccharide building block is needed. Therefore, several 4,6-pyruvated galactosyl donors have to be tested for glycosylation of a suitably protected 4,6-pyruvated glucosyl acceptor in order to find "good" conditions for its preparation. A series of galactosyl donors **13** (Scheme 3) are first coupled with benzyl 2-*O*-benzoyl-4,6-[(S)-1-methoxycarbonylethylidene]-α-D-glucopyranoside (**14**). Table 3 summarizes the results [44].

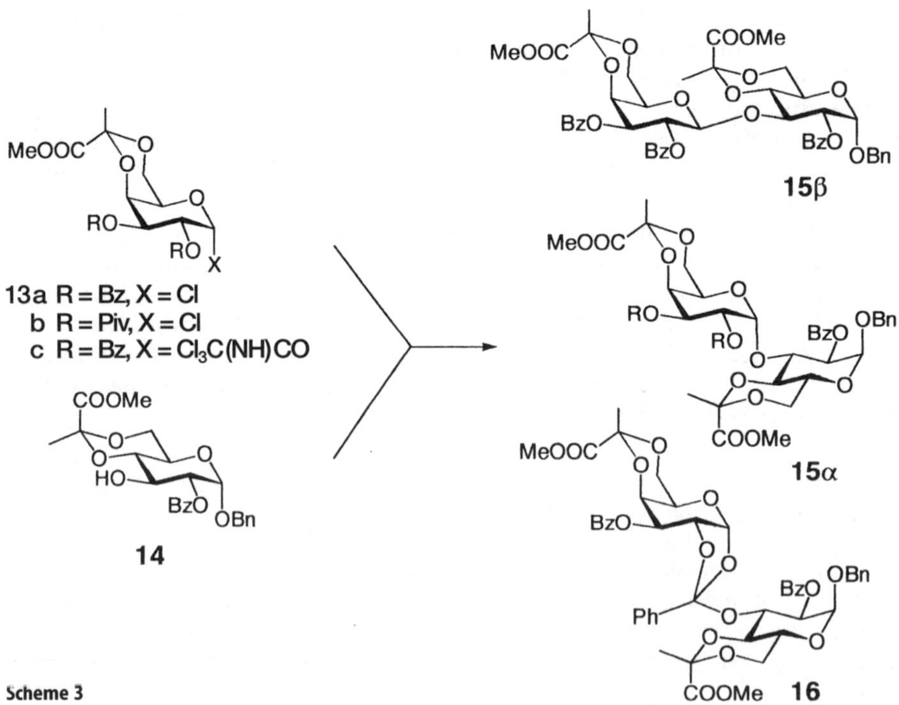

Scheme 3

Table 3. Reaction of donors **13** with acceptor **14**

Entry	Donor	Reaction conditions	Products (yield)	
1	**13a**	AgOTf, *sym*-collidine, CH_2Cl_2, 20 °C	**15**β (trace)	**15**α (68%)
2	**13b**	AgOTf, *sym*-collidine, CH_2Cl_2, 20 °C	**15**β (-)	**15**α (82%)
3	**13a**	Ag-silicate, CH_2Cl_2, 20 °C	**15**β (28%)	**16** (12%)
4	**13c**	cat. TMSOTf, MeCN, –20 °C	**15**β (68%)	**15**α (12%)
5	**13c**	cat. TMSOTf, MeCN, 0 °C	**15**β (80%)	**15**α (20%)

Rather unexpectedly, the chloride **13a** gives almost exclusively the corresponding α-linked product **15a** (entry 1, Table 3). This may be attributed to the formation of a mismatched pair [45] between the donor **13** and the acceptor **14** during glycosylation. The matched-mismatched pair formation is supported by the finding that a sterically more demanding pivaloyl group in the donor, such as **13b**, increases the yield of the α-linked disaccharide (entry 2). Alternatively, a similar effect to that which was found for 4,6-*O*-isopropylidene-protected galactosyl donors, where the nucleophile attacks the intermediate glycosyl cation from the convex side [46], may be operative in the case of the 4,6-pyruvated galactosyl donor **13a**. This unpleasant situation is, however, not improved by the use of Paulsen's insoluble silver silicate (entry 3) where a significant amount of the corresponding orthoester **16** is formed. Best results for the coupling with respect to β-selectivity is only obtained with the reactive trichloroacetimidate **13c** (entries 4 and 5). However, a strong temperature dependence of the galactosylation is observed. This example highlights impressively the fact that a careful optimization of glycosylations with pyruvated donors is always necessary.

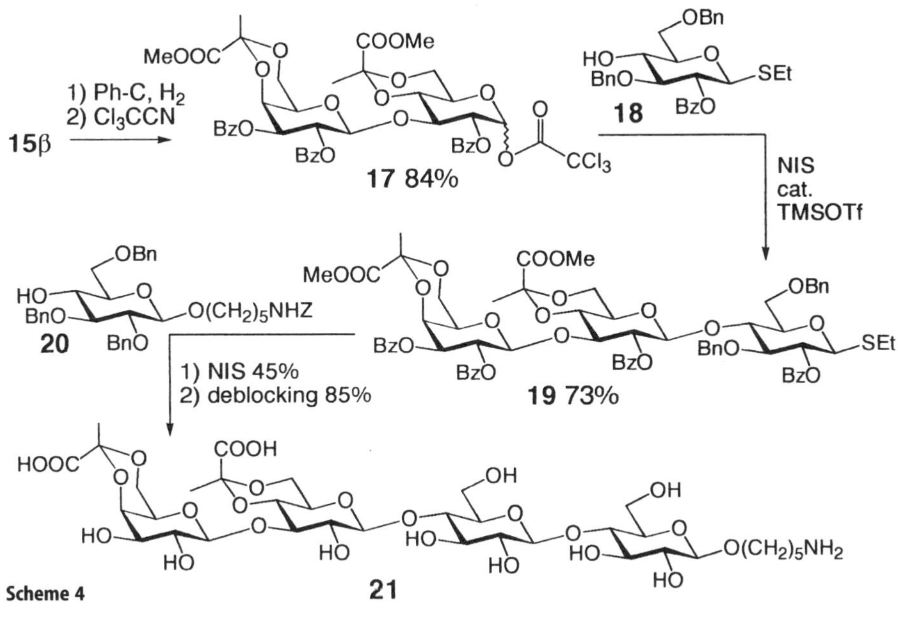

Scheme 4

Further elongations of the sugar chain should therefore be performed with trichloroacetimidates when pyruvated glycosyl donors are involved. Thus, disaccharide **15***β* is converted into imidate **17** and coupled to the thio-glycoside **18** (Scheme 4) [44]. Now, for the formed trisaccharide **19**, thiophilic activation can be applied without any problems, since here a "normal" donor is involved. Coupling to acceptor **20** followed by deblocking furnishes the tetrasaccharide **21**, which can be coupled to a protein via its aminopentyl aglycon.

4.1.2
R. leguminosarum Biovar Phaseoli

The capsular polysaccharide of *R. leguminosarum* biovar *phaseoli* contains a linear side chain with a single 3,4-pyruvated galactosyl residue in its repeating unit (see Table 1). Unfortunately, the conformation of the 3,4-pyruvic acetal has not been determined [47]. Therefore, it appears to be desirable to prepare both diastereomers of the corresponding oligosaccharides, since this will give the opportunity to determine the influence of the configuration of the acetal moiety in biological studies.

As the 3,4-pyruvated donors, both diastereomers of phenyl 1-thio-*β*-D-galactopyranoside **12l** and allyl *α*-D-galactopyranoside **12k** (Table 2) [23] can be used advantageously. As shown in Scheme 5, (S)-**12l** is first converted into the corresponding pyruvated galactosyl bromide **22** and coupled without isolation to benzyl 2,3,4-tri-*O*-benzoyl-*β*-D-galactopyranoside **23a** and thiogalactoside **23b**, respectively, to give the disaccharide blocks (S)-**24a** and (S)-**24b**. Similarly, (R)-**12k** is converted into imidate **25**, coupling of which to **23a** furnishes (R)-**24a** in 78% yield. Thus, both coupling procedures are equally well suited for 3,4-pyruvated galactosyl donors.

Next, both diastereomers are converted into the corresponding trichloroacetimidates **25a** and into the bromide **25b**, and are condensed with position 6′ of the disaccharide allyl glycoside **26** (see Sect. 5). Final desilylation, debenzoylation and saponification of the methyl pyruvate moiety affords the two diastereomeric tetrasaccharides **27a** that can be coupled to a protein via the allyl aglycon. However, when the pyruvated disaccharide bromide (S)-**25b** is used instead of imidates **25a**, only the corresponding *α*-linked tetrasaccharide **27b** is obtained. It is difficult to interpret this finding – however, the complete change in anomeric selectivity must be due to the presence of a pyruvated residue next to the reacting one. No *α*-linked products can be detected if donors corresponding to **25** but devoid of pyruvic acid acetals are applied.

4.2
Microciona Prolifera Saccharides

Special attention has to be paid towards the pyruvated building blocks if the pyruvated glycosyl residue is linked to a position which is expected to be highly unreactive in glycosylations. For example, the proteoglycan of the marine sponge *Microciona prolifera* (Table 1) exhibits a 4,6-pyruvated lactosamine moiety. Since position 4 of glucosamine acceptors is such an unreactive position,

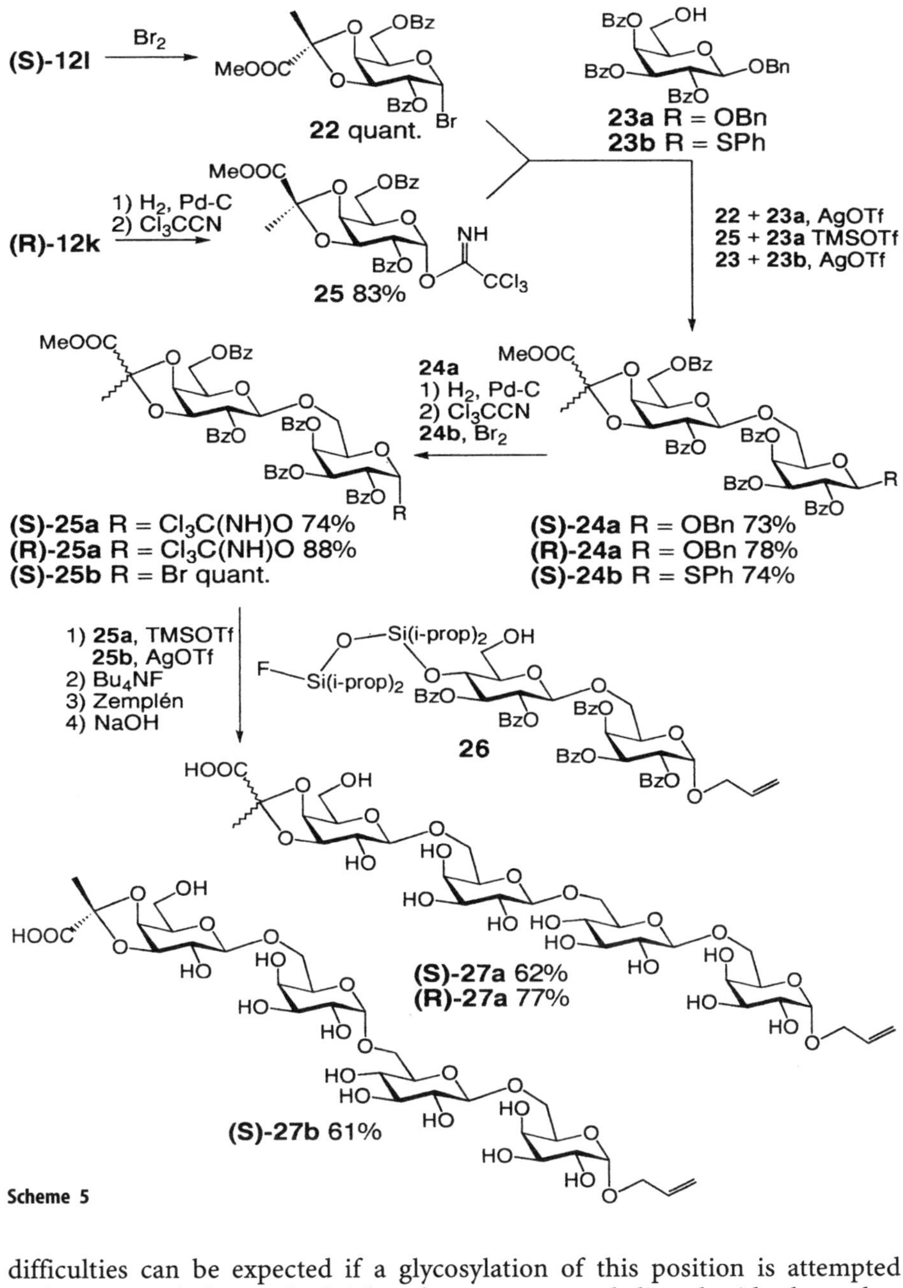

Scheme 5

difficulties can be expected if a glycosylation of this position is attempted with pyruvated donors [48]. Therefore, a pyruvated disaccharide donor has to be constructed which allows the efficient extension of the oligosaccharide chain. For that purpose, the Teoc(2-trimethylsilylethyl carbamate)-protected lactosamine trichloroacetimidate (**30**) (Scheme 6) is prepared as follows. Pyruvation of the disaccharide block **28** proceeds with high yield and high

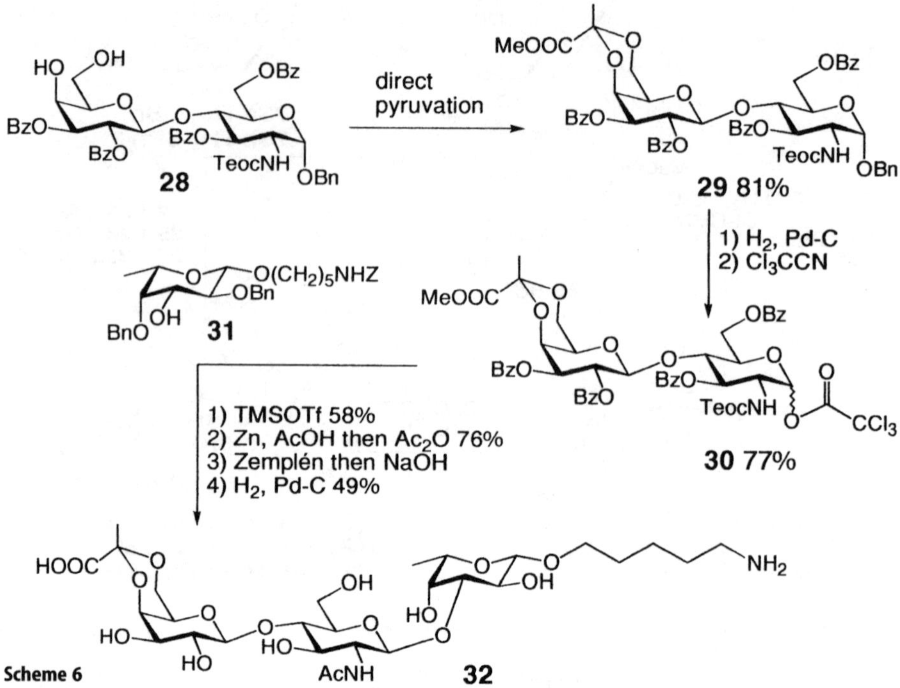

Scheme 6

diastereoselectivity if the direct acetalation procedure (see above) is used. Conversion of the thus formed 4,6-pyruvated disaccharide **29** into the corresponding imidate **30** can then be achieved by classical manipulations. The latter imidate is a suitable donor and is coupled to the L-fucose derivative **31** to give the corresponding trisaccharide. Final conversion of the Teoc group into an acetyl group and sequential deblocking furnishes the desired *M. prolifera* fragment **32** [48].

5 Novel Strategies and Protecting Groups

Since pyruvated glucosyl donors appeared to be rather unreactive, problems were also expected for the efficient construction of contiguously pyruvated oligosaccharide structures as found in the mycobacterial glycolipids (Table 1). Therefore, a novel glycosylation procedure based on the glycosylation of silylated alcohols with glycosyl fluorides has been developed for the solution of this problem and is discussed here.

5.1 Glycodesilylation

Glycosyl fluorides can easily be coupled to silylated alcohols when activation with Lewis acids is applied [49]. In combination with the 1,1,3,3-tetraisopropyl-

1,3-disiloxane-1,3-diyl (TIPS) group previously used for the synchronous protecting of two hydroxyl groups in carbohydrates and for generating pyruvated glycosides (see Sect. 3), this method can be extended towards a regioselective glycosylation protocol. The possibility of selectively introducing the TIPS group to positions 4 and 6 of unprotected glycosides and the ease of rearranging these intermediates to the corresponding 3,4-TIPS derivatives [50] opens up a highly flexible strategy for the construction of pyruvated oligosaccharides.

5.1.1
Regioselectivity

When the 4,6-TIPS-protected methyl glucoside **33** is treated with acetofluoroglucose **34** (Scheme 7) and a catalytic amount of BF_3-ether, a regioselective β-(1→6)-glycosylation is observed, affording the gentiobioside **35** [51]. The fluoride is transferred to the TIPS residue, and thus results in a 1-fluoro-1,1,3,3-tetraisopropyl-1,3-disiloxane-3-yl (FTIPS) group at position 4 of the disaccharide. The FTIPS group can be selectively cleaved for further elongation of the sugar chain at this position. The 3,4- and 2,3-TIPS derivatives **36** and **38** are both glycodesilylated at position 3 to give the laminaribiosides **37** and **39**. Both disaccharides can be further manipulated at positions 4 and 2 respectively, and a wide variety of similar couplings can be realized in this fashion [50b].

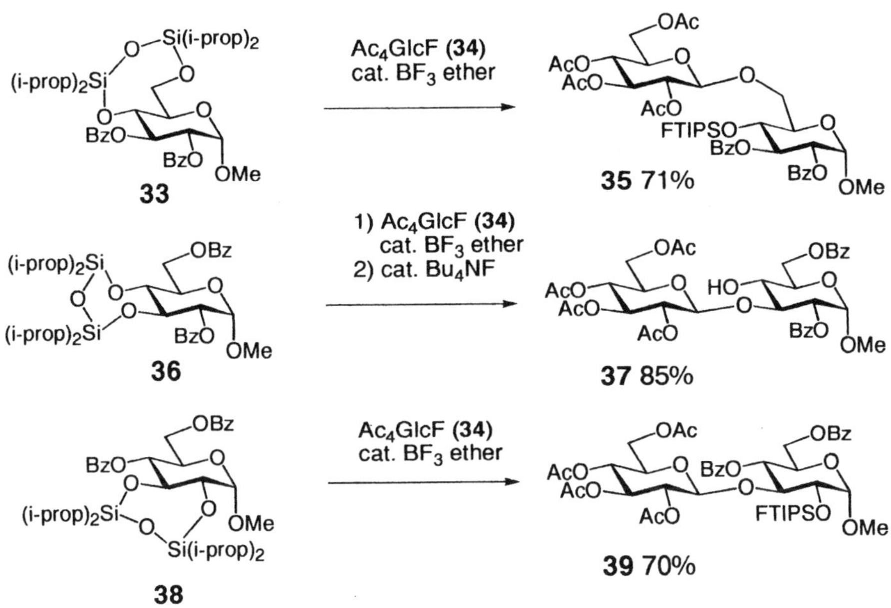

Scheme 7

The flexibility of the glycodesilylation protocol is further increased when the 4,6-TIPS group, as shown for the mannoside **40** (Scheme 8), is opened regioselectively with HF-pyridine complex to give the mannosyl acceptor **41**.

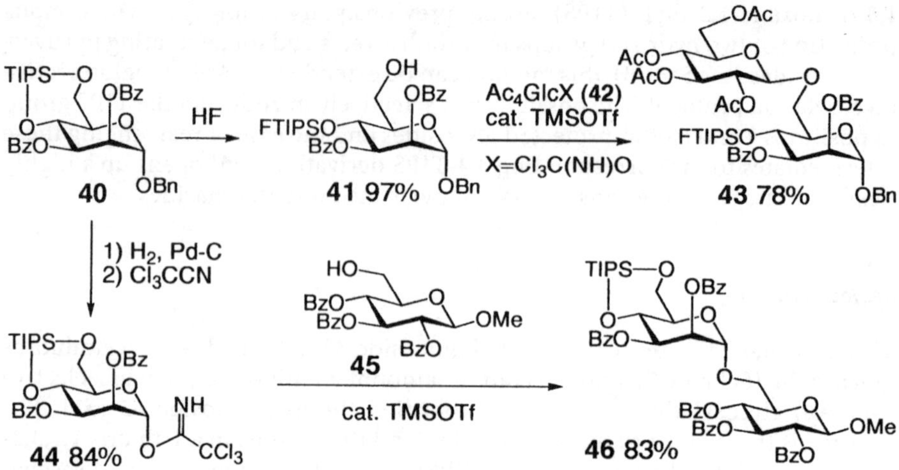

Scheme 8

Now other glycosyl donors, e.g. imidate (**42**) are efficiently coupled to the free hydroxyl group, giving **43**. Furthermore, compound **40** is easily converted to the corresponding imidate **43**, which, upon coupling to a suitable glycosyl acceptor such as **45** furnishes disaccharide **46** [50b]. The latter is once again a possible substrate for glycodesilylations as outlined above.

5.1.2
Application to the Synthesis of Mycobacterial Saccharides

The immunodominant surface glycolipid of *Mycobacterium smegmatis* comprises a tetrasaccharide containing a doubly 4,6-pyruvated laminaribiosyl terminus [13] (Table 1). Thus, it should be an attractive synthetic target for the glycodesilylation protocol. As mentioned above, 4,6-pyruvated glucosyl donors and acceptors react rather sluggishly in glycosylation reactions. This restriction, however, can be elegantly overcome when pyruvated glucosyl fluorides are coupled to pyruvated 2,3-TIPS-protected glucosyl acceptors as outlined in Scheme 9 [52].

The conveniently prepared pyruvated benzyl glycosides **12c** and **12c′** (Table 2) are first converted by a two-step sequences into the fluoride **47** and the TIPS-derivative **48**, respectively. The latter are subsequently condensed, using the glycodesilylation-protocol, to give the desired laminaribiosides **49** in excellent yield.

This approach is highly convergent and allows the efficient preparation of building block **49**. The latter is further transformed into the corresponding trichloroacetimidate **50** and used for the construction of di- to pentasaccharide fragments of *M. smegmatis* glycolipids [41, 52]. Furthermore, 5-aminopentyl di- and trisaccharides, useful precursors for neoglycoconjugates, are available as well using this method. For example, **50** is first condensed with the ethyl 1-thio-glucoside **51** to give **52**, followed by coupling to N-protected 5-aminopentanol and sequential deblocking to give the trisaccharide 5-aminopentyl glycoside **53**.

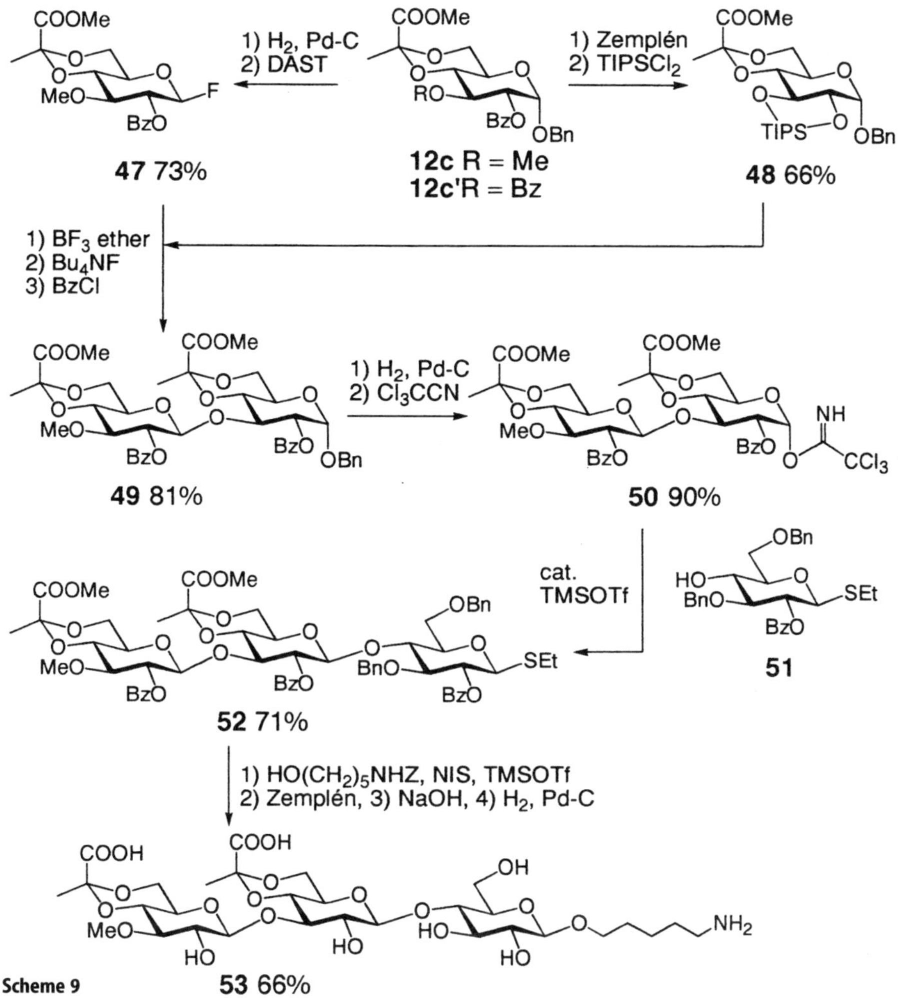

Scheme 9

5.2 Novel Protecting Groups

The design of orthogonal protecting group strategies must be regarded as one of the central prerequisites of efficient oligosaccharide syntheses. One has to chose an adequate protective group pattern for both glycosyl donor and glycosyl acceptor in order to achieve sufficient reactivity for the formation of the glycosidic bond and to allow later on the selective deblocking of any desired position in the formed oligosaccharide. Furthermore, the protecting group at position 2 of the glycosyl donor strongly affects the diastereoselectivity of the glycosylation reaction. For example, a neighbouring group active residue (acyl groups) forces the condensation towards a 1,2-*trans*-selective glycosylation. However, especially if less reactive acceptors are used in combination with 2-*O*-acetyl-halogenoses, a transesterification of the acetyl group to the nucleophile can occur and thereby significantly decrease the yield of coupling product [53]. Therefore, novel protecting groups which circumvent these problems are desirable. Furthermore, groups that are fully compatible with other commonly used acyl groups such as acetyl and benzoyl are needed in order to allow the synthesis of β-(1→2)-linked oligosaccharides. These prerequisites are fulfilled by the novel 2-chloroacetoxymethyl (CAMB) and 2-chloroacetoxyethylbenzoyl (CAEB) groups.

5.2.1
The CAMB Group

There are several protecting groups available which fulfil the above-mentioned prerequisites [53]. Of these, the chloroacetyl group has found broad application in oligosaccharide synthesis since it can be cleaved under essentially neutral conditions which leave other acyl groups intact [54]. However, the tendency of chloroacetates to undergo transesterification from position 2 of a glycosyl donor under common glycosylation conditions is very strong [53]. The 2-chloroacetoxymethyl-benzoyl group (CAMB, Scheme 10) combines the stability of the benzoyl group with the selective fissionability of the chloroacetyl group [55].

The CAMB group can easily be introduced into position 2 of 1,3,4,6-tetra-*O*-acetyl-α-D-glycopyranoses **54** via its acid chloride **55**. The latter is conveniently prepared on a 10 g scale from commercially available phthalide [55]. The intermediate 2-*O*-CAMB-protected glycoses **56** can be converted to the

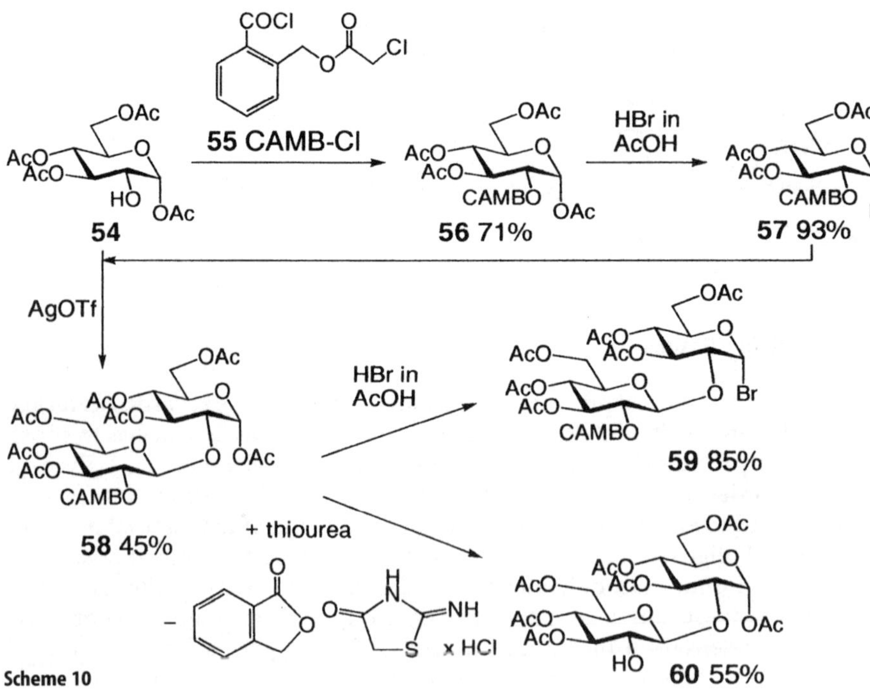

Scheme 10

corresponding halogenoses **57** by standard procedures and can be coupled with **54** to give *β*-(1→2)-linked saccharides **58**. Iterative formation of a disaccharide glycosyl bromide **59** or assisted cleavage of the CAMB moiety then affords either a disaccharide donor or the disaccharide acceptor **60** for further elongation of the sugar chain [55]. Effective synthetic strategies towards *β*-(1→2)-linked oligosaccharides can thus be elaborated with the aid of the CAMB group.

5.2.2 *The CAEB Group*

A major drawback of using the CAMB group is its incompatibility with benzyl groups, since CAMB is hydrogenated under the conditions usually used in the cleavage of benzyl groups [55a, 56]. Thus, the 2-chloroacetoxy-ethyl-benzoyl group (CAEB) is an alternative. CAEB is introduced as a temporary protecting group, essentially like CAMB, via the corresponding acid chloride and is cleaved under similar conditions. CAEB-chloride is prepared from isochromane, which is oxidized and saponified to 2-(2-hydroxyethyl)benzoic acid [55a, 57]. This is followed by chloroacetylation and treatment with $SOCl_2$ [56]. CAEB is essentially stable under conditions which are prone to cleave benzyl and benzylidene groups respectively. Thus, CAEB can be advantageously applied as a *β*-directing blocking group for glycosyl donors in those oligosaccharide syntheses where benzyl protection cannot be avoided.

5.2.3 *Application to the Synthesis of Saccharides Related to Escherichia coli*

The exopolysaccharide of *Escherichia coli* K47 contains an intracatenally bound 3,4-pyruvated *β*-D-galactopyranosyl residue to position 2, to which a 2-acetamino-2-deoxy-*β*-D-glucopyranosyl moiety is attached [17]. Thus, the synthesis of a disaccharide 5-aminopentyl glycoside containing the structure *β*-D-GlcNAcp-(1→2)-3,4-(R)-pyruvate-*β*-D-Galp requires a pyruvated galactosyl donor that can be used for *β*-selective couplings and, in addition, bears a temporary protecting group at position 2. The latter protecting group must be cleavable besides containing a benzoyl group, which is necessary for the diastereoselective introduction of the pyruvic acid acetal. The CAMB group is therefore well suited for this purpose [23] (Scheme 11).

Starting from phenyl 2,6-di-*O*-benzoyl-3,4-[(R)-1-methoxycarbonylethylidene]-1-thio-*β*-D-galactopyranoside **121** (Table 2), debenzoylation and selective rebenzoylation of position 6 furnishes compound **61**, which is further acylated with CAMB-Cl **55** to give the donor **62**. The latter is smoothly coupled to the unreactive N-protected 5-aminopentanol to give glycoside **63** without any transesterification. Next, the CAMB group is removed with thiourea, and the intermediate **64** is condensed with the glucosamine donor **65** [58] to give the disaccharide **66**. Final conversion of the Teoc group to an acetyl group and deblocking furnishes the free pyruvated disaccharide 5-aminopentyl glycoside **67** [23].

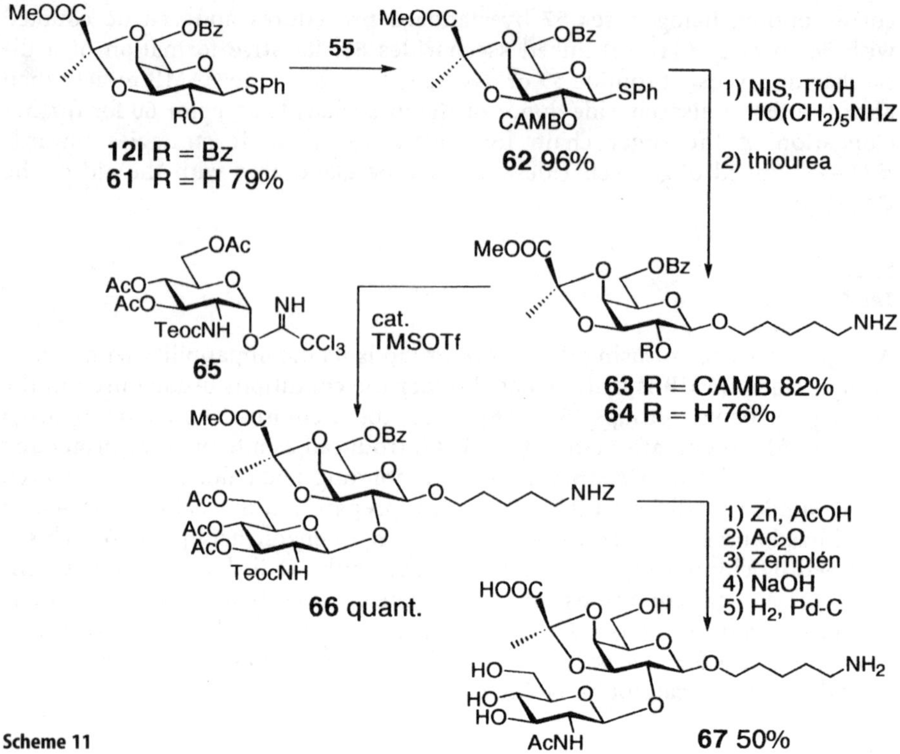

Scheme 11

5.3
Prearranged Glycosides

The selective preparation of 1,2-*cis*-linkages in the D-manno and L-rhamno series is still one of the major challenges in oligosaccharide synthesis. Several new and promising approaches deal with this problem [59]. Since a number of pyruvated oligo- and polysaccharides also contain β-D-mannosyl and β-L-rhamnosyl residues (Table 1), efficient strategies for their preparation are also needed here.

5.3.1
Synthesis of β-L-Rhamnosides

β-L-Rhamnosidic linkages can be established with good selectivity if a suitably protected rhamnosyl donor is linked by a spacer to the glycosyl acceptor (prearranged glycosides) [60]. If the spacer group is finely tuned with respect to its length, flexibility and position of attachment at the rhamnosyl donor and the glycosyl acceptor, the subsequent intramolecular glycosylation step is governed by double diastereoselection. Thus, the anomeric outcome of the rhamnosylation reaction is influenced to a large extent by the spacer group, and other factors usually influencing the selectivity (i.e. neighbouring group participation)

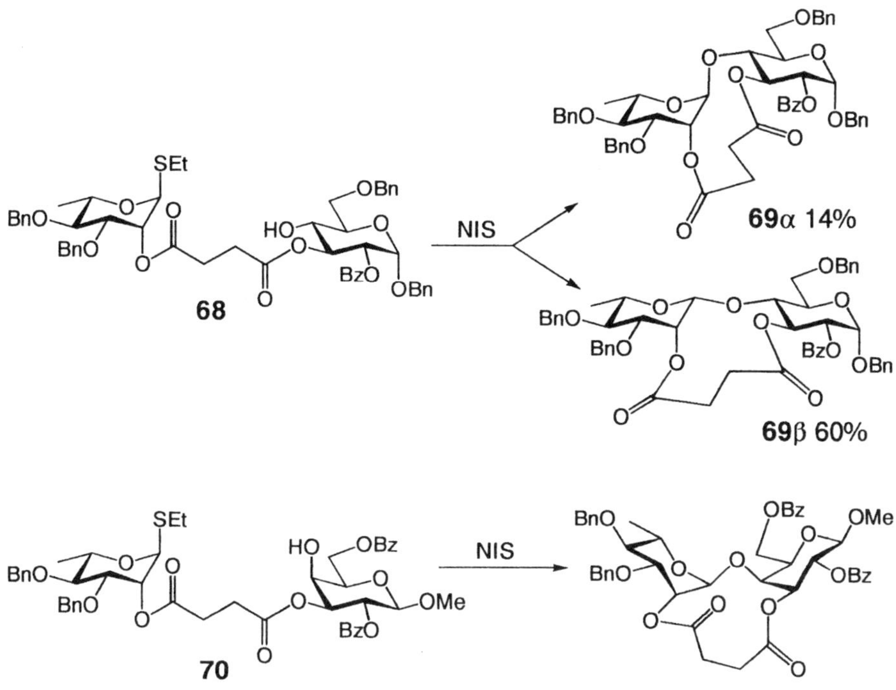

Scheme 12

play a minor role. Furthermore, since the glycosylation proceeds intramolecularly, less reactive hydroxyl groups in the acceptor moiety are also rhamnosylated very well. For example, the succinyl-bridged glycoside **68** results in 14% yield of the corresponding α-linked disaccharide **69α** and 60% of **69β** (Scheme 12) [61]. In contrast, if the relative stereochemistry of the acceptor moiety is changed to a galactosyl residue, as in compound **70**, only the corresponding α-linked disaccharide **71** is formed in high yield.

Although the approach via prearranged glycosides is in principle applicable to glucosylations and β-mannosylations [62], the latter case still requires further optimisations, since the best α/β-selectivity of 1:2 yet obtained is not sufficient for preparative purposes.

5.3.2
Application to the Synthesis of Saccharides Related to Pneumococcus

Prearranged glycosides offer the opportunity of efficiently preparing fragments related to the capsular polysaccharide of *Streptococcus pneumoniae* type 27 (Table 1). For example, the synthesis of the tetrasaccharide 5-aminopentyl glycoside **78** [63], useful for the preparation of glycoconjugates and artificial vaccines, is outlined in Scheme 13. For the convergent block synthesis of the latter, the pyruvated disaccharide donor **74** is first prepared from the 4,6-pyruvated imidate **72**. Coupling of the latter with galactoside **73** proceeds in almost

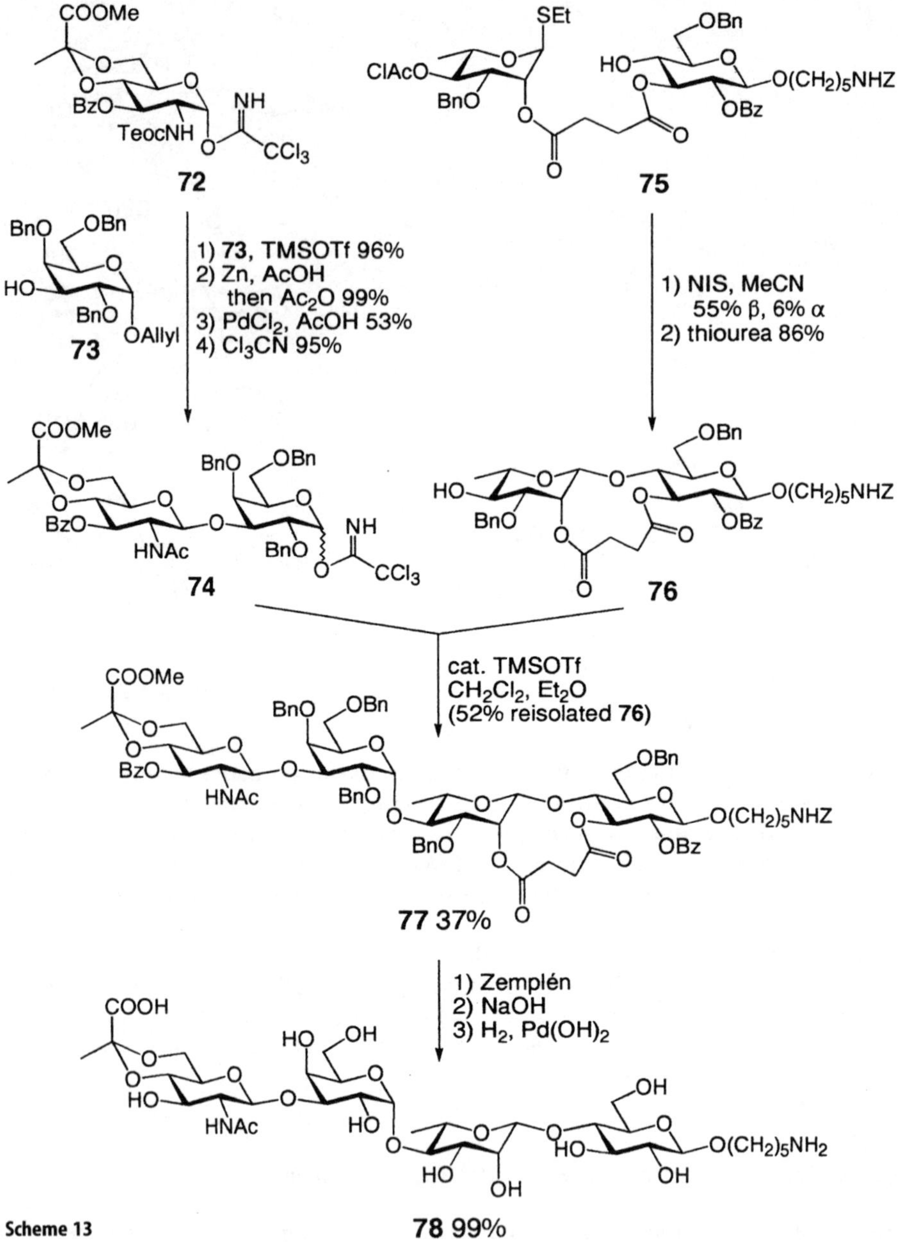
COOMe
BzO
TeocNH
NH
CCl3
72
SEt
ClAcO
BnO
OBn
HO
O(CH2)5NHZ
OBz
75
BnO
OBn
HO
BnO
OAllyl
73
1) 73, TMSOTf 96%
2) Zn, AcOH
then Ac2O 99%
3) PdCl2, AcOH 53%
4) Cl3CN 95%
1) NIS, MeCN
55% β, 6% α
2) thiourea 86%
HNAc
CCl3
74
76
cat. TMSOTf
CH2Cl2, Et2O
(52% reisolated 76)
77 37%
1) Zemplén
2) NaOH
3) H2, Pd(OH)2
COOH
O(CH2)5NH2
78 99%

Scheme 13

quantitative yield. Next, the prearranged glycoside **75** is used for the construction of the β-L-rhamnosidic bond. Intramolecular glycosylation of **75** affords 55% of the desired β-linked disaccharide along with 6% of the α-counterpart. Cleavage of the chloroacetyl group then furnishes the required acceptor **76**. The condensation of the two building blocks **74** and **76** is for some reason sluggish and gives the tetrasaccharide in only 37% yield. However, 52% of unreacted acceptor **76** can be reisolated. The final sequential deblocking then affords the tetrasaccharide 5-aminopentyl glycoside **78**.

Other variations of the tetrasaccharide repeating unit (i.e. with the rhamnosyl residue at the reducing end) are best prepared by classical approaches [64].

References

1. (a) Jann K, Westphal O (1975) In: Sela M (ed) The antigens. Academic Press, New York, p 1 (b) Jann K, Jann B (1977) In: Sutherland I (ed) Surface carbohydrates of the procariotic cell. Academic Press, New York, p 247
2. (a) Sviridov AF, Arifchodzaev ChA, Cizov OS, Kochetkov NK (1980) Bioorganiceskaja Chimija (in Russian) 6:165; (b) Kojima N, Kaya S, Araki Y, Ito E (1988) Eur J Biochem 174:255 (and references cited therein)
3. (a) Glicksman M (1969) Gum technology in the food industry. Academic Press, New York (b) Jeanes A, Pittsley JE, Senti FR (1961) J Appl Polymer Sci 5:519
4. (a) Shatwell KP, Sutherland IW, Dea ICM, Ross-Murphy SB (1990) Carbohydr Res 206:87 (b) Gamini A, de Bleijser J, Leyte JC (1991) Carbohydr Res 220:33
5. Hirase S (1957) Bull Soc Chim Jpn 30:68
6. (a) Ziegler T, Eckhardt E, Herold H (1992) Liebigs Ann Chem 441 (b) Bailey WF, Eliel EL (1974) J Am Chem Soc 96:1798
7. Garegg PJ, Lindberg B, Kvarnström I (1979) Carbohydr Res 77:71 (and references cited therein)
8. Araki S, Abe S, Ando S, Kou K, Fujiwara N, Satake M (1989) J Biol Chem 264:19922
9. (a) Jones C (1990) Carbohydr Res 198:353 (b) Jones C, Currie F, Foster M (1991) Carbohydr Res 221: 95
10. (a) Beurret M, Joseleau JP, Vignon M, Dutton GGS (1989) Carbohydr Res 189:247 (b) Anderton WJ, Wilkinson SG (1985) Biochem J 226:587
11. (a) Dudman WF, Franzén L-E, McNeil M, Darvill AG, Albersheim P (1983) Carbohydr Res 117:169 (b) McNeil M, Darvill J, Darvill AG, Albersheim P (1986) Carbohydr Res 146:307
12. Hollingsworth RI, Dazzo FB, Hallenga K, Musselman B (1988) Carbohydr Res 172:97
13. (a) Saadat S, Ballou CE (1983) J Biol Chem 258:1813 (b) Kamisango K-i, Saadat S, Dell A, Ballou CE (1985) J Biol Chem 260:4117
14. Brennan P, Aspinall GO, Shin JEN (1981) J Biol Chem 256:6817
15. Niemann H, Frank N, Stirm S (1977) Carbohydr Res 59:165
16. Bennett LG, Bishop CT (1977) Can J Chem 55:8
17. de Bruin AH, Parolis H, Parolis LAS (1992) Carbohydr Res 233:195
18. Spillmann D, Hard K, Thomas-Oates J, Vliegenthart JFG, Miservic G, Burger MM, Finne J (1993) J Biol Chem 268:13378
19. (a) Hadjivassiliou AG, Rieder SV (1968) Clin Chim Acta 19:357 (b) Duckworth M, Madden JK (1993) Methods Carbohydr Chem 9:123
20. (a) Choy YM, Dutton GGS, Stephen AM, Yang MT (1972) Anal Lett 5:675 (b) Garegg PJ, Jansson PE, Lindberg B, Lindh F, Lönngren J, Kvarnström I, Nimmich W (1980) Carbohydr Res 78:127 (c) Gorin PAJ, Mazurek M, Duarte HS, Iacomini M, Duarte JH (1982) Carbohydr Res 100:1 (d) Jones DNM, Sanders JKM (1989) J Chem Soc, Chem Commun 167 (c) Jansson PE, Lindberg J, Widmalm G (1993) Acta Chem Scand 47:711
21. Eckhardt E (1991) Diploma Thesis, University of Stuttgart

22. Gorin PAJ, Mazurek M, Duarte HS, Duarte JH (1981) Carbohydr Res 92:C1
23. Ziegler T, Herold G (1994) Liebigs Ann Chem 859
24. Jansson PE, Lindberg B, Lindquist U (1981) Carbohydr Res 95:73
25. (a) Heidelberger M, Kvarnström I, Eriksen J, Nimmich W, Dudman WF (1980) Proc Natl Acad Sci USA 77:4244 (b) Rao AS, Liao J, Kabat EA, Osserman EF, Harboe M, Nimmich W (1984) J Biol Chem 259:1018
26. (a) Kvarnström I (1980) Chem Commun Univ Stockholm 7:3 (b) Ielpi L, Couso RO, Dankert M (1981) Biochem Biophys Res Commun 102:1400 (c) Tolmasky ME, Staneloni RJ, Leloir LF (1982) J Biol Chem 257:6751
27. Sutherland IW (1969) Biochem J 115:935
28. (a) Dudman WF, Heidelbereger M (1969) Science 164:954 (b) Bennett LG, Bishop CT (1977) Immunochem 14:693 (c) Kabat EA, Liao J, Bretting H, Franklin EC, Geltner D, Frangione B, Koshland ME, Shyong J, Osserman EF (1980) J Exp Med 152:979 (d) Rao AS, Kabat EA, Nimmich W, Osserman EF (1982) Mol Immun 19:609
29. Gray JX, Rolfe BG (1990) Mol Microbiol 4: 1425 (and references cited therein)
30. (a) Gorin PAJ, Ishikawa T (1969) Can J Chem 45:521 (b) Fontana JD, Duarte JH, Iacomini M, Gorin PAJ (1982) Carbohydr Res 108:221
31. Rolf D, Gray GR (1982) J Am Chem Soc 104:3539 (b) Zeller SG, Gray GR (1990) Carbohydr Res 198:285
32. Aspinall GO, Ibrahim IH, Khane NK (1990) Carbohydr Res 200:247
33. Collins PM, McKinnon AC, Manro A (1989) Tetrahedron Lett 30:1399
34. (a) Hashimoto H, Hiruma K, Tamura JI (1988) Carbohydr Res 177:C9 (b) Lipták A, Szabo L (1989) J Carbohydr Chem 8:629
35. (a) Ziegler T, Eckhardt E, Herold G (1992) Liebigs Ann Chem 441 (b) Ziegler T, Eckhardt E, Neumann K, Birault V (1992) Synthesis 1013
36. (a) Lipták A, Szabo L (1988) Carbohydr Res 184:C5 (b) Bajza I, Kerekgyarto J, Hajko J, Szilagyi L, Lipták A (1994) Carbohydr Res 253:111
37. Hind CRK, Collins PM, Renn D, Cook RB, Caspi D, Baltz ML, Pepys MB (1984) J Exp Med 159:1058
38. Ziegler T, Vollmer M, Oberhoffner S, Eckhardt E (1993) Liebigs Ann Chem 255
39. (a) Ziegler T, Eckhardt E, Herold G (1992) Tetrahedron Lett 33:4413 (b) Ziegler T, Herold G (1994) Liebigs Ann Chem 859 (c) Ziegler T (1994) Tetrahedron Lett 35:6857 (d) Ziegler T, Eckhardt E, Strayle J, Herzog H (1994) Carbohydr Res 253:167
40. Schüle G (1996) Dissertation, University of Stuttgart
41. Ziegler T, Eckhardt E, Birault V (1993) J Org Chem 58:090
42. Fraser-Reid B, Wu Z, Andrews CW, Skowronski E, Brown JP (1991) J Am Chem Soc 113:1434
43. (a) Ziegler T (1992) Angew Chem Int Ed Engl 31:1358 (b) Ziegler T (1994) Carbohydr Res 253:151
44. Eckhardt E, Ziegler T (1994) Carbohydr Res 264:253
45. Spijker NM, van Boeckel CAA (1991) Angew Chem Int Ed Engl 30:180
46. Nakahara Y, Ogawa T (1987) Tetrahedron Lett 28:2731
47. Albersheim P (personal communication)
48. Ziegler T (1995) Liebigs Ann Chem 949
49. (a) Hashimoto S, Hayashi M, Noyori R (1984) Tetrahedron Lett 25:1379 (b) Kunz H, Sager W (1985) Helv Chim Acta 68: 283 (c) Nicolaou KC, Chucholowski A, Dolle RE, Randall JL (1984) J Chem Soc, Chem Commun 1155 (d) Ernst B, Wagner B (1989) Helv Chim Acta 72:165 (e) Kreuzer M, Thiem J (1986) Carbohydr Res 149:247 (f) Ziegler T, Seidl U (1991) J Carbohydr Chem 10:813
50. (a) Thiem J, Duckstein V, Prahst A, Matzke M (1987) Liebigs Ann Chem 289 (b) Ziegler T, Eckhardt E, Pantkowski G (1994) J Carbohydr Chem 13: 81 (and references cited therein)
51. Ziegler T, Neumann K, Eckhardt E, Herold G, Pantkowski G (1991) Synlett 699
52. Ziegler T, Eckhardt E (1992) Tetrahedron Lett 33:6615
53. Ziegler T, Kovac P, Glaudemans CPJ (1990) Liebigs Ann Chem 613 (and references cited therein)

54. (a) Ziegler T, Kovac P, Glaudemans CPJ (1992) In: Ogura H, Hasegawa A, Suami T (eds) Carbohydrates, synthetic methods and applications in medicinal chemistry. VCH, Weinheim, p 357 (b) Ziegler T (1990) Liebigs Ann Chem 1125
55. (a) Pantkowski G (1996) Dissertation, University of Stuttgart (b) Ziegler T, Pantkowski G (1994) Liebigs Ann Chem 659
56. Ziegler T, Pantkowski G (1995) Tetrahedron Lett 36:5727
57. Watanabe Y, Ishimura M, Ozaki S (1994) Chem Lett 11:2163
58. Paulsen H, Helpap B (1991) Carbohydr Res 216:289
59. (a) Paulsen H (1990) Angew Chem Int Ed Engl 29:823 (b) Kaji E, Lichtenthaler FW (1993) Trends Glycosci Glycotech 5:121 (c) Barresi F, Hindsgaul O (1996) In: Khan SH, O'Neill RA (eds) Modern methods in carbohydrate synthesis. Harwood, Amsterdam p 251 (and references cited therein)
60. (a) Ziegler T, Lau R (1995) Tetrahedron Lett 36: 1417 (b) Lau R, Schüle G, Schwaneberg U, Ziegler T (1995) Liebigs Ann Chem 1745
61. This is the highest β-content for a rhamnosylation published so far in the literature
62. (a) Valverde S, Gomez AM, Hernandez A, Herradon B, Lopez JC (1995) J Chem Soc, Chem Commun 2005 (b) Valverde S, Gomez AM, Lopez JC, Herradon B (1996) Tetrahedron Lett 37:1105 (c) Ziegler T, Lemanski G, Rakoczy A (1995) Tetrahedron Lett 36:8973
63. (a) Schüle G (1996) Thesis, University of Stuttgart (b) Schüle G, Ziegler T (1996) Liebigs Ann Chem 1599
64. Schüle G, Ziegler T (1996) Tetrahedron 52:2925

Author Index Volumes 151–186

Author Index Vols. 26–50 see Vol. 50
Author Index Vols. 51–100 see Vol. 100
Author Index Vols. 101–150 see Vol. 150

The volume numbers are printed in italics

Springer and the environment

At Springer we firmly believe that an international science publisher has a special obligation to the environment, and our corporate policies consistently reflect this conviction.
We also expect our business partners – paper mills, printers, packaging manufacturers, etc. – to commit themselves to using materials and production processes that do not harm the environment. The paper in this book is made from low- or no-chlorine pulp and is acid free, in conformance with international standards for paper permanency.

Springer

Printing: Saladruck, Berlin
Binding: Buchbinderei Lüderitz & Bauer, Berlin